THE FASCINATION
OF STATISTICS

POPULAR STATISTICS

a series edited by

D. B. Owen
Department of Statistics
Southern Methodist University
Dallas, Texas

Nancy R. Mann
Biomathematics Department
UCLA
Los Angeles, California

Other Volumes in Preparation

THE FASCINATION
OF STATISTICS

edited by

Richard J. Brook and Gregory C. Arnold

Massey University
Palmerston North, New Zealand

Thomas H. Hassard

University of Manitoba
Winnipeg, Ontario, Canada

Robert M. Pringle

Ministry of Agriculture and
Fisheries
Palmerston North, New Zealand

Marcel Dekker, Inc.   New York and Basel

Library of Congress Cataloging-in-Publication Data
Main entry under title:

The Fascination of statistics.

Includes index.
1. Statistics. I. Brook, Richard J.
QA276.12.F37   1986   519.5   85-29171
ISBN 0-8247-7329-2

MARCEL DEKKER, INC.
270 Madison Avenue, New York, New York 10016

Current printing (last digit):
10 9 8 7 6 5 4 3 2 1

PRINTED IN THE UNITED STATES OF AMERICA

Preface

People have always been fascinated by numbers. From earliest times they have learned to count and then to keep records of the number of animals in their flocks and herds or of their financial transactions.

Throughout history, numbers have often taken on a mystical quality, such as in biblical times when 1 was associated with God, 6 with evil, 7 with completeness or perfection, and 12 with the tribes of Israel or number of disciples. Before we judge too harshly, we should note the almost idolatrous awe in which the results of political opinion polls are held today, particularly by politicians.

One source of a large amount of data is the population census. This was popular with the Romans but known well before their time. The census has often aroused strong feelings, not usually of fascination but of fear and suspicion, because it formed a basis for taxation or subscription into the army. After the Romans, there was a lull in census-taking until the middle of the eighteenth century, when the practice of regular census-taking was resumed in most countries.

The upsurge of science in the nineteenth century resulted in the collection of large amounts of data. In their different fields of endeavor, Charles Darwin and Florence Nightingale were avid collectors of numbers, organizing them to bolster scientific theories or galvanize public opinion behind programs for social change.

The theory and practice of statistics, however, is a twentieth-century phenomenon with ever-increasing areas of application to which the chapters in this book bear testimony. It is this wide range of possible applications that motivated us to compile this book. In doing so we hope to provide

Students of statistics with a view of the many uses of statistics

Teachers of statistics with a range of examples on which to draw

People who use specialized fields of statistics with illustrations of techniques applied in other fields

Numerate readers with ideas of the scope and power of statistical methodology

The authors are experts in their fields, which differ widely. Many of them would not claim to be statisticians, but they share in common an interest in statistics or other quantitative methods, which they employ in their professional activities. They were asked to write in a readable fashion for nonexperts on an aspect of statistics that they find fascinating. We believe that they have accomplished the task admirably, and our thanks go out to them for their efforts and for the cheerful manner in which they accepted the occasional suggestion from the editors. The illustrations by Tim Brook speak for themselves, for they are not only delightful but add further dimension to the chapters.

The chapters are arranged in seven groups around common statistical themes, and each group has an overall introduction. The reader may find it helpful to read the articles carefully group by group, to use the index, or to be led by the whim of the moment. Whatever way is chosen, we are sure that the reader will also find these topics fascinating, stimulating, and enjoyable.

Richard J. Brook
Gregory C. Arnold
Thomas H. Hassard
Robert M. Pringle

Contents

Contents

Contributors

Allan M. Anderson   Development Manager, Food Division, Salmond Industries, Ltd., Palmerston North, New Zealand

Robert D. Anderson   Professor, Department of Animal Science, Massey University, Palmerston North, New Zealand

Gregory C. Arnold   Senior Lecturer, Department of Mathematics and Statistics, Massey University, Palmerston North, New Zealand

Judith A. Brook   Senior Lecturer, Department of Psychology, Massey University, Palmerston North, New Zealand

Richard J. Brook   Reader, Department of Mathematics and Statistics, Massey University, Palmerston North, New Zealand

Kerry Chamberlain   Senior Lecturer, Department of Psychology, Massey University, Palmerston North, New Zealand

David A. Dickey   Associate Professor, Department of Statistics, North Carolina State University, Raleigh, North Carolina

E. Jacquelin Dietz   Associate Professor, Department of Statistics, North Carolina State University, Raleigh, North Carolina

Howard P. Edwards   Senior Lecturer, Department of Mathematics and Statistics, Massey University, Palmerston North, New Zealand

Brian S. Everitt   Reader in Statistics in Behavioral Science, Biometrics Unit, Institute of Psychiatry, University of London, London, England

Hans J. Eysenck   Professor, Institute of Psychiatry, University of London, London, England

Philip J. Gendall   Senior Research Officer, Market Research Center, Massey University, Palmerston North, New Zealand

David J. Hand   Senior Lecturer, Biometrics Unit, Institute of Psychiatry, University of London, London, England

Thomas H. Hassard   Associate Professor, Department of Social and Preventive Medicine, University of Manitoba, Winnipeg, Ontario, Canada

Richard G. Heerdegen   Senior Lecturer, Department of Geography, Massey University, Palmerston North, New Zealand

Michael D. Hendy   Senior Lecturer, Department of Mathematics and Statistics, Massey University, Palmerston North, New Zealand

Mary Casey Jacob   Research Psychologist, Department of Surgery, King's College Hospital Medical School, University of London, London, England

Alan J. Lee   Senior Lecturer, Department of Mathematics and Statistics, University of Auckland, Auckland, New Zealand

Richard B. Le Heron   Senior Lecturer, Department of Geography, Massey University, Palmerston North, New Zealand

Bryan F. J. Manly   Associate Professor, Department of Mathematics and Statistics, University of Otago, Dunedin, New Zealand

R. Hugh Morton   Senior Lecturer, Department of Mathematics and Statistics, Massey University, Palmerston North, New Zealand

David Penny   Reader, Department of Botany and Zoology, Massey University, Palmerston North, New Zealand

Keith Petrie   Clinical Psychologist, Department of Psychological Medicine, Waikato Hospital, Hamilton, New Zealand

Kenneth H. Pollock   Associate Professor, Department of Statistics, North Carolina State University, Raleigh, North Carolina

Samuel M. Putnam   Associate Professor, Department of Medicine, St. Mary's Hospital, Rochester, New York

Arthur A. Rayner   Emeritus Professor, Department of Statistics and Biometry, University of Natal, Pietermaritzburg, South Africa

David A. Rhoades   Scientist, Applied Mathematics Division, Department of Scientific and Industrial Research, Wellington, New Zealand

Alastair J. Scott   Professor, Department of Mathematics and Statistics, University of Auckland, Auckland, New Zealand

Barry Singer   Graduate Teaching Fellow, Department of Computer and Information Science, University of Oregon, Eugene, Oregon

William B. Stiles   Professor, Department of Psychology, Miami University, Oxford, Ohio

Peter J. Thomson   Senior Lecturer, Department of Mathematics, Victoria University, Wellington, New Zealand

Alan L. Tyree   Senior Lecturer, School of Law, University of Sydney, Sydney, Australia

Paul van Moeseke   Professor, Department of Economics, Massey University, Palmerston North, New Zealand

Bruce S. Weir   Professor, Department of Statistics, North Carolina State University, Raleigh, North Carolina

P. Brent Wheeler   Deputy City Planner, City Planning Department, Palmerston North City Corporation, Palmerston North, New Zealand

THE FASCINATION
OF STATISTICS

I
Probability

We all have to make decisions in our lives and, while some may enjoy this activity, most of us approach decisions with some trepidation. The difficulty is that we live in a world in which events are rarely certain, and we have no precise way of measuring this uncertainty. Primitive people would have been able to compare lengths and weights long before rulers or scales were invented, but you and I have no way of deciding whether we agree precisely on the likelihood of rain tomorrow. Nevertheless we may be able to reach some consensus on a rough figure that in some way describes the way we feel about the occurence of some future event, and this figure could be called a probability. Probability theory shows how the probabilities of simple events can be combined to give the probabilities of complex events. Not many of us would imagine that probability theory could be applied to very personal areas of our lives. The consequences of doing so are sometimes dramatic, as Barry Singer shows in his light-hearted discussion of one of the most important decisions we ever make.

Our feelings of uncertainty about a particular event in the future may be vague, but when similar events can be observed repeatedly a pattern can be built up permitting probabilities to be calculated. When probability arguments are used in official regulations, however, the legal jargon seems very stilted in trying to cope with the intricacies of this topic.

Dick Brook considers regulations on the packaging of goods and explores some of the implications of these laws for consumers and manufacturers.

Have you ever been at a party and recognized a person's face but completely forgotten his or her name? It is quite common, of course, to have these lapses of memory which can be embarrassing in a social setting. Tom Hassard explains how it is even more important to match a name to a person in a medical situation, particularly if that person has had a number of tests in different clinics at different times. Although it is not obvious, determining the most economical way of storing information is a development of probability.

1

Probabilities, Meeting, and Mating

Barry Singer
University of Oregon
Eugene, Oregon

THE NEGATIVITY OF SEARCHING

Most unmarried people dislike the process of searching for dating or marriage partners. We are all aware of people who hang on in awful relationships because they fear being alone, or because they dread the strains of having to meet and date again. Maybe that's even you, from time to time. It's not easy to put yourself on the line and face rejection. It's not even easy to admit that you're looking. Most of all, singles dislike the fact that no matter how hard they look, they can't seem to find *The One.*

Surveys have shown that almost all single people are frustrated by their inability to find the dating or marriage partners they want. A few singles, overly impressed by a few chance successes, believe that they have found an answer:

> Look, Susie, it's easy. You just go down to the laundromat. Wear jeans, but make them tight, you know. Do your hair. Take something to read that people can talk about, like maybe *Newsweek* . . . but don't look too interested in it. Works every time.

It doesn't work every time. It may work well for particular people, or may work for others occasionally. But there is no magic answer in laundromats or dating services or any other *technique.* Do not be totally

discouraged, though. I believe there is some useful wisdom on this matter.

THE ARITHMETIC OF SEARCHING

The basic issue is: Why is everyone having such a frustrating time finding a relationship? I know why. For you to understand why, you'll need to participate in some brief mental gymnastics. First, jot down some brief phrases that describe essentials—ingredients you feel need to be there— for someone you would contemplate a relationship with. Here's my list:

Age: Between 25 and 45
Intelligence: Very bright
Values: Liberal
Religious beliefs: Few and low-key
Occupation: Must be self-supporting
Kids: No kids
Personality: Funny, warm, outgoing, considerate, sense of humor
Sexuality: Must like to cuddle, be sexually assertive

Attractiveness: Relatively attractive, medium height and weight
Bad habits I can't stand: Smoking, excessive drinking

We must add to the list above, as to any list, that the person needs to be unattached, and that she likes me as well as I like her. My list is probably a typical sort of list; it looks relatively reasonable.

Other categories that singles often add to their lists are: must have excellent health; must share most of my important interests, must be sensitive and passionate; must keep in good physical shape. If any of these categories or other categories that I've left out apply to you, add them to your list.

Next, we're going to consider some probability theory. Below are schematics (diagrams) of two hypothetical electric generators.

Generator I:	A	B		Produces 70 kW
	90%	90%		
Generator II:	A	B	C	D	Produces 100 kW
	90%	80%	75%	90%	

Generator I has two components, A and B, each of which functions with 90% reliability. It can produce 70 kW. Generator II has four components, whose reliability varies between 70 and 90%, and it can produce 100 kW. Which is the better generator?

The answer is generator I. If any one component in either machine is "missing"—isn't working—the generator as a whole won't work. The probability that the machines will be working at any given time is thus equivalent to the probability that all components will be working at once. This probabiltiy, called a *joint probability*, is obtained by multiplying the probabilities of all the components together.

For generator I, the probability of its working at any given time is 90% × 90% = 81%. It will produce an average of 81% of 70 kW, or 56 kW. For generator II, the probability of its working is 90% × 80% × 75% × 90% = 42%. It will produce 42% of 100 kW on the average, or only 42 kW. Generator I is a much better machine.

Here's why we multiply to obtain the joint probability. We'll take generator I as an example. When we turn the machine on, we should expect the following. Nine times out of 10 component A will work; 1 time out of 10 it will not. The same is true for component B. For each of the nine times component A will work, therefore, there are nine times that component B will work. Therefore, there will be a total of 9 × 9 = 81 times that both components will be working, on the average, out of 10 × 10 = 100 times that we turn the machine on. When a series of less-than-perfect (less than 100%) probability events must happen together for some outcome to be true, for the fraction of the time that any given event comes true, there will be only a fraction of the time that any other given event comes true, and so on down the line.

An analogy may help. Suppose that you are shopping for a new purse, and make the following list: It must be soft pink; have a brass hasp; have a strap at least 16 in. long; be no less than 20 in. in circumference but no more than 40 in.; have an inner pocket to hold a mirror; and be lined with vinyl or similar wear-resistant fabric. Each of these criteria, considered separately, is a reasonable and common attribute of purses. But when we require them to all occur together in a single purse, we will probably never find that purse. You may want to prove this principle for yourself by constructing an actual purse criteria list and going on a shopping expedition.

It really is important to understand the above. Play with it for a while. It means that as we add on parts and complexity, we drastically reduce the possibility of achieving the whole.

Now let's return to my list of relationship essentials. For each essential, I'm going to write down a reasonable probability that a given woman I'll meet will satisfy that essential. For instance, considering all the women I might meet day to day, how many are going to be between 25 and 45? About half, I'd say. So I assign the first item on my essentials list a probability of 1/2. Here's my list again, this time with the probabilities tacked on:

Essential	Probability
Between 25 and 45	1/2
Very bright	1/25
Liberal	1/3
Relatively nonreligious	1/3
Self-supporting	1/2
No kids	1/3
Funny, warm, considerate, sense of humor	1/3
Cuddles; sexually assertive	1/2
Attractive, medium height and weight	1/2
Doesn't drink or smoke	1/2
Is not presently attached	1/2
Likes me	1/5

To find the probability that all these qualities will be combined in one person that I meet, multiply all the probabilities together. Do the same for your list. For my list, the probability is

$$\frac{1}{2} \times \frac{1}{25} \times \frac{1}{3} \times \cdots = \frac{1}{648,000}$$

What does this mean? It means, Good Luck, Barry. It means that the chances are less than 1 in 500,000 that a new woman I meet will have all my required characteristics. Even though I thought my list was reasonable, I'm clearly expecting too much. If I meet three new women every day, I will meet *the Person of the List* only once every 600 years, on the average. Yet I didn't put too many picky things on my list, and the probabilities I assigned were reasonable.

GETTING BETTER ODDS

Aha, you say, the big problem is my requirement that the person be very bright, to which I assigned a probability of 1/25. That is much lower than any other probability on the list. What if I were to compromise on brightness? That is a good idea, and I have thought of it. If I am willing to settle for someone in the upper 20% in intelligence instead of the top 4%, my probability now becomes 1/5. But that doesn't help much. Now, on average I will meet a person with all the essentials every 120 years instead of every 600. Are there other ways to get better odds, search strategies that will increase the probability of finding a suitable partner?

It has probably occurred to you that you can better the odds by focusing your search within particular social environments. For instance, even though about half of the people you happen to meet on a given day will be single, if you go to a pub during the evening, perhaps two-thirds of the people there will be single. Or, if you're looking for people who share your religious views, church would be a good place to look. Searching within such specific social environments has the effect of removing one of the components in the chain of joint probabilities. Thus, if we require that our potential mate be a practicing Episcopalian, a trait that has a probability of one-third in the general population, searching at the Episcopal church socials on Saturday will remove the 1/3 probability component. This has the effect of multiplying our odds of finding a mate by 3. However, remembering that the odds of finding our dream person were so low to begin with, this would reduce the chances from 1 in 600,000 to 1 in 200,000—which is still not very encouraging.

You may have spotted another useful idea. Some traits tend to be grouped together. Just for the sake of discussion, let's assume that being liberal correlates positively with being bright. If being liberal and being bright were traits that had nothing to do with each other—as, for instance, hair color and shoe size have nothing to do with each other—we say that such traits are *independent*. However, in my experience, most bright people are in fact liberal. If we assume that two-thirds of the people who are bright are liberal, we have established what is called a *conditional* probability. That is, even though only one out of three people are liberal, if we know beforehand that a person is bright, then two-thirds of such persons are going to be liberal. Thus, as we go down my list, and we take as "given" that I've met a person who is bright, we can change the probability of "liberal" from 1/3 to 2/3. This has the effect of doubling the chances of finding the mate of the list. Conditional probabilities are also known as *dependent* probabilities, since one probability depends on the other. The more dependent probabilities (correlated traits) there are on my list, the better off I am.

But wait! There's a problem. Sometimes traits can be negatively correlated as well, in which case the presence of one lowers the probability of the other. This is especially true in the case of "sex roles," the tendency for men and women to be raised differently and to have differing personalities. Men are supposed to be assertive, self-confident, practical, experienced, and unemotional; women are supposed to be pretty, emotionally expressive, soft, and shy. We are taught to expect, to admire, and to like these traits in the opposite sex. The problem comes because we sometimes also want the opposite sex to be like us as well—to share our personalities—for traditional *male traits* would be negatively correlated with *female traits,* thus reducing the probability that a person would have a mixture of traits, some *male* and some *female.*

As an illustration, I gave a seminar based on this chapter to a singles group. Each person in the group made up their own "list" and then calculated the probabilities of meeting their *list person.* One man found that his probability was 1 in 1 billion. Most found that their chances were only one in several hundred thousand. One attractive young brunette, Carol, saw that her probability was only 1 in 160 million. Her list went something like this: "He *must* be over 6 feet 2 inches tall. I want someone who has accomplishments to his credit, who is very ambitious. He must be sensitive, warm, and passionate . . . .

Carol's list seemed reasonable to her, but let's examine it. As soon as you specify even one low-probability item, such as being over 6 foot 2, you are cutting down your chances almost to the vanishing point. Only one man in 20 is over 6 foot 2, yet you are asking that within that very small group of men who are that tall, all of these other qualities are also going to be found in one person. Not likely. Second, and more important, asking that the dream man have status and ambition, and that he also be warm and sensitive, is asking for two things that do not often go together.

Men do not come in these kinds of packages. They tend to be either ambitious and achieving, or warm and sensitive, but not both.

What Carol wants is what many single women want and is a basic cause of women's frustration with men, whether the woman is single or married. Women tend to be attracted to ambitious men with status; but such men often did not get there by being warm and sensitive, by being responsive to others' needs and feelings, by taking time to cuddle and talk with their dating partners if their partners were upset about something. They got there by drive and singlemindedness and the willingness to compete, even if the competing hurt others.

There is no reason men cannot develop both sides of themselves. But it takes extra work, and men are not socialized to do so. The man Carol wants, that so many single women want, who is good-looking and tall, ambitious and sensitive, is as scarce as peaches in winter.

Men are guilty of the same kinds of incompatibilities. For instance, in my list, I asked for someone who was warm and lively and liked to cuddle, but who had her own career and was self-supporting. I'm not sure that these qualities go together very well. Men also typically want a woman who is sexually assertive, who enjoys sex, is adventurous in bed, and who is physically warm. But they don't want a woman who has grown in these directions through sexual relationships with other men. They would rather find a near-virgin who, when they touch her, magically bursts into sexual aliveness for the first time in her life. The world tends not to work that way.

Look at your list again and see whether you, too, have items that are *incompatibles*.

RETOOLING OUR ATTITUDES ABOUT SEARCHING

We have now established that finding a mate who meets all criteria is a low-probability endeavor, and, mind you, we have defind the task as finding someone who fulfills all the *essentials*. If we were to go beyond essentials and make a list describing our "ideal" mate, we would have to leave the planet and scour the galaxy at lightning speed to have a reasonable chance of success. Also, don't forget that last, important probability on everyone's list: You may find your person of the list, but then that person must also like you. In my experience, that last essential can be quite problematical. We have also seen that there are search strategies that can better the odds, although the odds will still remain formidably high.

It may be, however, that you are not convinced. You still have the feeling that the search will not be in vain, that you've come close to finding your Person of the List several times already, so that you're about due to hit the jackpot. Or, you may feel that you'll do better than the average person in the searching process because you're trying harder.

Let's examine where these feelings come from. Many people have the experience of meeting one or two people a year who seem at first to meet all the essentials, but upon better acquaintance, are lacking two or three criteria. I call such people *Almosts*.

After sorting through three or four Almosts, one can get the feeling that the Person of the List really exists, since you've come so close several times: or since you've had a run of Almosts, you might think you are going to meet a winner soon. Neither perception is accurate.

Think about a population of 100,000 people of the opposite sex, and suppose that your list contains eight *essentials*. From the population, how many people are going to have one or two of your essentials? Almost everyone will. How many are going to have four or five of your essentials? That number will be much less, but it will still be large—perhaps 2000 or 3000. These are the Almosts. And how many Persons of the List will there be? We have seen that there will be perhaps one, or maybe none. If you were to draw a picture or graph representing the information above, which we call a *distribution*, it would be clear that random sampling will frequently turn up Almosts, but almost never the dream person.

Nor does a string of Almosts make it any more likely that the next person you meet will be The One. Having flipped a coin five times in a row and gotten heads does not make it more likely that the next time will result in tails. The probability of tails is always 1/2 on any given flip. The contrary belief is termed the *gambler's fallacy*. Life is not *fair* in the ways people would like or expect it to be. Rather, life is fair in the same sense that a roulette wheel is fair. If you have been searching for two years and have turned up only a dozen Almosts, it is no more likely that the next candidate will be The One than it was when you started your search. The next candidate will almost certainly also turn out to be an Almost.

Research in America has shown that people often fail to look at themselves objectively and to apply statistics and probability to themselves, because they feel they are special and different. This is a reasoning fallacy that overlooks the fact that each of us feels that we are special and different. For instance, statistics show that marriage in America is a faltering institution. A couple who marry around the age of 20 may expect to get divorced, typically in about seven years, and that there will be frequent periods of marital discord even if they do not get divorced. They may also expect that the man will commit adultery, with a high probability, and that the woman will be unfaithful with a probability of 50%.

As a university professor, I meet many young couples about to be married. I am also a minister and have performed perhaps 30 weddings for students. I have never yet met a young couple who say about their impending marriage, "Well, I guess this marriage won't last that long. We'll probably get divorced in a few years. There are sure to be lots of fights. One or both of us will probably commit adultery."

Research shows that young people do know about the discouraging data on marriage, but they are sure that the data do not apply to them, because they are different. Their marriage will be better, they believe, because they will try harder than most and because they have good communication with their mate. The fallacy is that *everyone* thinks that; and most of those everyones will fail.

Similarly, many people search unremittingly for their Person of the List, because they believe that they are an especially worthy person, that they deserve only the best, and that they will succeed in their search because they are trying very hard. However, the actuality is that virtually every single person believes that he or she is especially worthy and deserving and is searching hard for a mate who meets all criteria. You are no different in those respects. Although it is deflating to the ego, it is also practical and realistic sometimes to regard yourself as a piece of data, a single statistic who will probably fall squarely among the averages. That may not be a pleasant or ennobling thought, but it may be a useful one.

What to do? What to do? How should one alter one's search strategies or attitudes about acceptable mates? What are the final recommendations?

Here we leave the realm of statististics and probabilities and enter the domain of personal values and psychology. Each of us must reach a private decision. Most of us, if we take this chapter to heart, will eventually decide that we are, after all, only ordinary, or at least less than perfect, and that we will only become frustrated by looking for a collection of traits in another person that is very demanding and unlikely even to exist. We will settle for someone who is an Almost, who is missing a few of what we had thought to be *essentials*. For instance, I may have to settle for someone who is overweight, smokes, and is religious. As a matter of fact, that sounds a lot like Tracy, with whom I had a marvelous relationship last year (before another fellow won her affections).

For others of us, the romantics, the search is the thing. It is the quest that impels us. After all, to seek the precious and rare, to persevere, is an old and exciting human adventure. It can lend a sense of meaning and purpose. Further, we can always switch from being romantic to being practical when we decide the time is right. As I finish writing this paragraph, I see an attractive woman waiting at the elevator. She is not smoking or drinking, and she is carrying a worn volume of political philosophy. This looks intriguing. Perhaps she is also funny, warm likes to cuddle, is nonreligious, sexually assertive, and without children. Life, are you ready? I'm about to roll the dice.

2

How Much Does A Kilogram of Milk Powder Weigh?

Richard J. Brook
Massey University
Palmerston North, New Zealand

THE PROBLEM

Clearly, a kilogram of milk powder weighs 1 kg! However, when we buy a package of milk powder which bears a stamp claiming that the net weight is 1 kg, would we expect to obtain exactly 1 kg of milk powder? On reflection, we would probably expect a weight close but not exactly equal to 1 kg. In fact, if we used a very sensitive weighing machine the chances of the package weighing exactly 1 kg would be very small indeed. If the package weighed more than a kilogram, we would not become agitated, but if it was considerably less than the stated amount we might become annoyed and write to the manufacturer or even begin legal proceedings. But if the package contains less than 1 kg by a small amount, what should, or could, be done about it? Besides, what would be considered a "small amount" in this context?

The manufacturer, or packing firm, could argue that it would be impossible to check every item, so that some will be underweight even if the majority are slightly overweight. As consumers do not complain if the package is too heavy, they should not be too quick to complain if one happens to be slightly underweight. The manufacturer or packager may claim that, on average, packages weigh at least 1 kg each and they are following the spirit of the statement on the container that the net weight is 1 kg.

In this area of possible conflict, most countries have set up laws to protect the consumer and the manufacturer. In New Zealand, the current law is defined in the Weights and Measures Bill, but this is very brief and does not stipulate when an article is deemed to be underweight; it merely states that it is illegal to sell such goods. If this law were strictly applied, manufacturers would have a legitimate grievance. It appears, however, that inspectors more or less follow the pattern set down in the packaging acts of the Australian states, and these allow the manufacturer some leeway.

In this chapter we consider some aspects of these packaging laws. In particular, we consider the chances that a manufacturer may run foul of these laws and be liable to prosecution.

DISTRIBUTION OF WEIGHTS

Fluctuations in the weights of packaged goods will always occur even if the fluctuations are small and require sophisticated scales to detect. These days, most packaging is performed by machines. These are set to deliver a certain weight of product, such as milk powder, which we will call the target weight, and the weights of packages will vary around this

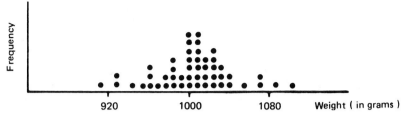

FIGURE 1   Possible frequency graph of the weights of 50 packages. Target weight = claimed weight = 1 kg.

mean value. The amount of variation will depend on the efficiency of the machine itself as well as certain properties of the powder, such as the degree of granulation, for the machine may deliver a certain volume rather than weight of powder. The manufacturer, or packager, may be able to reduce this variation, but no amount of expertise or effort could lead to its complete removal.

If the scales were set to a target weight of 1 kg, which is also the claimed weight on the package, a sample of 50 packages may give a distribution of weights such as that shown in Figure 1. If we now consider a much larger sample of packages and if the packages were weighed much more accurately, a smoother curve would result. On the vertical axis we now plot probability, which can be thought of as the relative frequency of occurrence. The resulting distribution would undoubtedly look very like the bell-shaped curve of a so-called normal curve (Figure 2).

Many naturally occurring phenomena approximately follow such a normal curve. The areas under the curve between any two points indicate the probability, or proportion of times, that the weight of the package will be between those two points. For example, the shaded area in Figure 2 indicates the probability, or proportion of times, that a package weighs less than 920 g. The theoretical normal curve becomes ever closer to the horizontal axis as it extends in both directions but never quite reaches it. This suggests that there is even a small probability of very large or very small weights occurring. In practice, the graph may be realistic for a certain range of weights, as shown in Figure 2. With this assumption, we can use statistical tables to find the areas under the theoretical curve, that is, the probability of packages being underweight.

More accurate machines giving less variation in weights would result in a steeper graph of similar shape, for the area under the graph must be 1 unit (or 100%), as it indicates the total frequency of occurrence. Actually, the graph in Figure 2 would represent a rather sloppy weighing

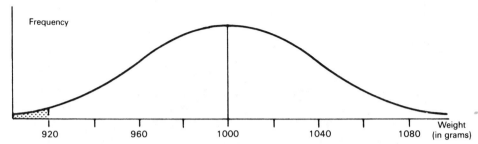

FIGURE 2 Distribution of weights when the target weight and the claimed weights is 1000 g.

machine, but it will allow us to compare the chances of packages being underweight.

There are two main points to notice about the graph: the middle called the mean, which equals 1000 here, and the variation about the mean, which is measured by a quantity called the standard deviation and in this case equals 40. Nearly all of the probability of occurence occurs in the interval of three standard deviations from the mean (880 to 1120), where these numbers refer to weights in grams.

PROBABILITY THAT A PACKAGE WILL BE LESS THAN THE CLAIMED WEIGHT

If the target weight of the filling machine is set equal to the claimed weight of 1 kg, half of the packages would be underweight and half would be overweight. This may seem perfectly reasonable to the manufacturer but consumers may feel differently, particularly if they happen to buy the underweight goods.

To make the customer happy, the manufacturer may decide to overfill the packages slightly so that the target weight of the machine is more than the claimed weight. Of course, it would still be possible for a customer to obtain an underweight package if the weights follow a normal distribution, but the probability of this occurring will be small if the target weight is much more than the claimed weight. In Figure 3, the target weight has been set at 1020 g whereas the claimed weight is 100 g. The probability that a package is underweight is indicated by the shaded areas.

It is easy to calculate the probability. To do so, we convert the weight to a standard normal score, Z, by the formula

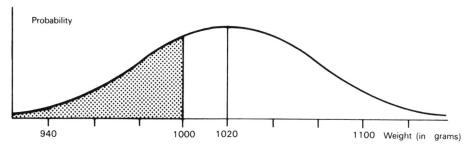

FIGURE 3 Distribution of weights when the target weight (1020g) is more than the claimed weight (1000 g).

$$Z = \frac{\text{weight} - \text{mean}}{\text{standard deviation}}$$

We have now set the mean, or target weight, to be 1020 g and we are interested in the package being less than the claimed weight of 1000 g.

Pr [a package is less than the claimed weight]

$$= \Pr\left[Z < \frac{1000 - 1020}{40} \right] \text{ or } [Z < -0.05]$$

$$= 0.309 \text{ or nearly 31 chances in 100}$$

Note that the symbol "<" stands for "less than" and we have looked up the probability in standard normal tables, which give, for a given value of Z, the area under the standard normal curve to the left of Z.

PACKAGING LAWS IN AUSTRALASIA

It appears that there will always be a possible conflict of interest between the customer and the seller. If the weights do approximately follow a normal curve, there will always be a possibility that a package will be underweight regardless of how high the target weight has been set. For the manufacturers of large quantities of milk powder, even a small increase in the target weight represents a loss of many thousands of dollars.

Packaging laws must take into account the conflicting interests of both parties and the variations in weight of the packages. The U.S. Code of Federal Regulations (see note 1 of the appendix to this chapter) states

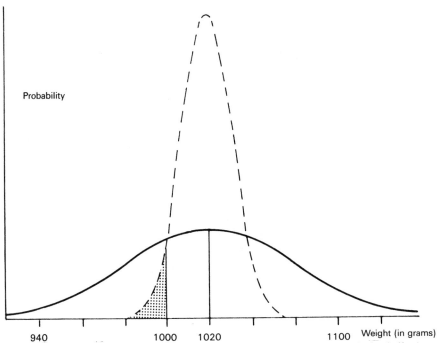

FIGURE 4 Distribution of the average weight of 12 items (dashed line) compared with the distribution of a single item (solid line). Target weight = 1020 and claimed weight = 1000.

that "Variations from stated quantity of contents shall not be unreasonably large." In Australia, the packaging laws vary slightly from state to state (see note 2 of the appendix) but on the whole are very similar. The South Australian Packages Act, for example, allows one article to be underweight provided that the average of 12 similar articles is not below the claimed weight. In fact, there are a few more "ifs" and "buts," but this is a convenient starting point for this discussion. Clearly, this gives some protection to the manufacturer, for it will be more unlikely for the mean of 12 packages to be underweight than for a single package to be underweight.

Probability That the Average of 12 Articles Is Underweight

It is well known that the variability of the average of a number of packages will be less than the variability of individual packages. Indeed, the average of four packages is only half as variable as the weight of individual packages, Instead of the standard deviation being 40, as above for individual packages, the standard deviation of the average of 12 packages is $40\sqrt{12} = 11.6$. The mean value will be unchanged at the target weight of 1020, giving the steeper graph in Figure 4. By applying the formula to find the standard normal score Z, and looking up the appropriate tables, we find that

Pr[average of 12 packages underweight]

= 0.042 or about 4 chances in 100

This is shown by the shaded area under the curve in Figure 4. This is considerably less than the probability of one package being underweight (31 chances in 100).

For an infringement of the law, two events have to take place—an initial package must be underweight and the average of 12 other similar packages must be underweight. Provided that this second sample is drawn independently of the initial package, the probability of an infringement of the law will be the probability of both these events occurring, which can be found by multiplying the probabilities.

Pr[one article underweight and the average of 12 others underweight]

= 0.309×0.042

= 0.013 or about 1 chance in 100

The Packages Act points out that if 12 articles are not available, the average of the articles present, provided that there are at least six, may be used. With six packages, the standard deviation of the average is $40\sqrt{6} = 16.3$ and the probability that this average is less than the claimed weight of 1000 is 0.106, or about 11 chances in 100. The probability of both events occurring (one article and the average of another six being underweight) becomes higher, 3 chances in 100. The probability, then of an infringement of the law is higher if there are fewer articles present in the shop or warehouse when they are inspected.

Probability of an Article Being at Least 5% Underweight

The Packages Act acutally leans more toward the manufacturer by requiring only that the individual article be not less than 95% of the claimed weight (see notes 3 and 4 of the appendix). For a kilogram of milk powder, the package must not weigh less than 950 g. From Figures 3 and 4 it can be seen that the probability of this occurring is not large. This is a reasonable provision of the law, as it would prevent silly charges against manufacturers when the deficiency of the container is only very small.

$$\text{Pr [1 article less than 950 grams]}$$

$$= \text{Pr}\left[Z < \frac{950 - 1020}{40} \right]$$

$$= 0.040 \text{ or 4 chances in } 100$$

This article would only be deemed to be underweight if the average value of at least six others was less than the claimed value, and from the preceding section we see that this would occur with a probability of nearly 11%. Thus the probability of both occurring is about 4 chances in 1000. Even with this rather sloppy weighing machine set at 1020 g the manufacturer is fairly safe from breaking the law.

Selection of the Sample of Additional Packages

A question that could be raised at this point is how the additional articles are to be selected to obtain their average weight. Some regulations (e.g., those of the Northern Territory and Australian Capital Territory) require that they be selected "at random."

This has a special meaning in statistical usage which is not equivalent to "haphazardly" or "in a disorganized manner." A statistician selects items at random with the aid of a randomizing device, such as tossing a coin or consulting a random number table. It is hard to imagine an inspector of weights and measures following such a procedure. Instead, an inspector is likely to select a sample of articles more purposefully, and this could bias the weight. This could happen if he or she deliberately selected packages with a particular batch number.

The wording of the regulations often does not indicate clearly to the layperson whether the original offending underweight article should be included in the sample. This could make quite a difference to the average weight, of course, if the first article was very deficient. In note 5 of the appendix we show that including the first deficient article in the average of six considerably increases the probability (from 4 to 12 chances in 1000) that the manufacturer will be found to have breached the law.

Selection of the First Deficient Package

In all packaging acts there is little, or no, indication of how the first package is selected for testing. Quite possibly the most common situation is that on receipt of a complaint, an inspector samples one or two individual packages and weighs the contents to determine if they are more than 5% deficient. Alternatively, regular checks may be made by inspectors. The important point here is that only a small number of packages of any one brand and size are likely to be checked. In the unlikely event that an inspector became fixated on a particular grocery item and tested a large number of these, the probability of finding at least one underweight would increase considerably.

Manufacturers could well fear overzealous inspectors or consumers who weighed each package. The wording of the packaging laws offers no protection to the manufacturers, but their lawyers could, no doubt argue that this practice was not in the spirit of the law. There may be a problem, though, in proving that such dastardly behavior was used by the inspectors or consumers.

Other Problems in Packaging

One problem which is very obvious to the consumer is that of powdery substances settling in transit. This should not affect the weight of the contents, but it can make the buyer very suspicious. By the composition of the solid, it may either take up water or lose water over time even if it is packed in a polythene bag. Milk powder, fortunately for the packer, is hygroscopic, so that it takes up water and increases slightly in weight over time.

The packaging of liquids can pose more problems than that of solids, as more variation often results. A discerning youngster, for example, can spend many minutes studying soft-drink bottles to select the one filled to the highest level. To allow for this variation, most packaging acts in Australia allow a deficiency of $7\frac{1}{2}\%$ rather than only 5%. Volatile liquids would have other packaging problems, not to mention aerosol cans of fly spray and the like.

One could imagine a somewhat ludicrous situation of inspectors squeezing toothpaste from a tube to verify its weight. Ideally, one would like to weight the tube plus the contents, but evidently this has its problems, as one manufacturer who was interviewed insisted that the weights of these tubes vary considerably.

Quality Control

Perhaps the greatest benefit of packaging laws is to make the manufacturer conscious of the need for quality control. The manufacturer may not be able to control the variability of weighing machines but could continually check on the setting of the target weight. To do this, the manufacturer may sample a product by choosing packets at random and weighing them. Alternatively, if the packets are placed in larger cartons, some cartons could be weighed. If a carton was very deficient in weight, it may be rejected. This may affect the distribution of weights so that it may not follow the normal curve as closely as we have suggested. The area of quality control is a large one and is the other side of the coin to the packaging laws.

POSTSCRIPT

In New Zealand, for the year ending March 31, 1982, court actions resulted from four breaches of the regulations for short-weight goods, seven breaches for short-weight bread, two breaches for using imperial instead of metric scales, and two cases of obstruction of an inspector. Clearly, inspectors play mainly a deterrent role in controlling under-weight goods.

The other main point is that the regulations only apply to retail establishments. Perhaps in the future the law will be changed to be more in line with the 1979 U.K. Weights and Measures Bill (see note 6 of the appendix), which changes the emphasis from testing packages at retail outlets to testing and surveillance at manufacturers' premises. In this bill the emphasis is not on the weight of individual packages but on the mean and variance in weight of large samples of packages.

APPENDIX: NOTES AND REFERENCES

1. U.S. Code of Federal Regulations (January 1982), Title 16, Commercial Practices (No. 500,22)

2. Australian Packaging Acts include:

Weights and Measures Act, 1915 (NSW)
Weights and Measures Act, 1958 (Vic)
Weights and Measures Act, 1934 (Tas)
Packages Act, 1967–1972 (SA)
Weights and Measures Act, 1951–1972 (Qld)
Weights and Measures Act, 1915–1976 (WA)
Weights and Measures (Packaged Goods) Ordinance, 1974 (ACT)
Weights and Measures (Packaged Goods) Ordinance, 1970–1973 (NT)

3. In a circular to its inspectors, the New Zealand Labour Depart-ment states: "The permissible error in deficiency in any single package may not be in excess of 5% provided that if this short weight package is weighed together with five similar packages, there is no deficiency on the six."

4. Section 20, subsection 3, of the South Australian Packages Act, 1967–1972, states that:

For the purposes of this section an article will be deemed to be of the weight or measure stated on the pack containing the article if

(a) any deficiency of weight or measure does not exceed five parts per centum of the stated weight or measure of where the article is contained in a bottle, the stated contents of which do not exceed eight fluid ounces, eight ounces, 250 milliliters or 250 grams seven and one-half parts per centum of the stated contents.

and

(b) there is no average deficiency in the contents of twelve packs containing the article selected by an Inspector from amongst the packs containing that article on the premises of the packer of where there are less than twelve such packs all the packs on those premises being not fewer than six.

5. The effect of including the offending article in the second sample of size 6 can be shown as follows. Let the six weights be designated $x_1, x_2, x_3, x_4, x_5, x_6$, with x_1 being the weight of the first deficient article. If $x_1 = 950$, we are interested in the average of x_1 through x_6 being less than the claimed value of 1000. That is, we are interested in the average of six weights, which is

$$\frac{x_1 + x_2 + x_3 + x_4 + x_5 + x_6}{6} < 1000$$

This can be rewritten as

$$\frac{x_2 + x_3 + x_4 + x_5 + x_6}{5} < \frac{6000 - x_1}{5}$$

$$\text{average of weights} < \frac{6000 - 950}{5}$$

$$= 1010$$

This gives

Pr[one article less than 950 g and the average of this article with five others less than the claimed value of 1000 g]

= Pr[x_1 < 950] × Pr[average of five < 1010]

= 0.040 × 0.288

= 0.012 (12 chances in 1000)

Including the first deficient article in the average of six has considerably increased the probability (from 4 to 12 chances in 1000) that the manufacturer will be found to have breached the law.

6. The U.K. bill is Statutory Instrument (1979) 1979:1613, the Weights and Measures (Packaged Goods) Regulations 1979 (London: Her Majesty's Stationery Office).

3

Writing the Book of Life: Medical Record Linkage

Thomas H. Hassard
University of Manitoba
Winnipeg, Ontario
Canada

THE BOOK OF LIFE

In 1946, H. L. Dunn, Chief of the U.S. National Bureau of Vital Statistics, summed up the health experiences of all of us in the following graphic phrases.

> Each person in the world creates a Book of Life. The book starts with birth and ends with death. Its pages are made up of the records of the principal events in life. Record linkage is the name given to the process of assembling the pages of this book into a volume.
>
> The person retains the same identity throughout the book. Except for advancing age, he is the same person. Thinking backwards he can remember pages of the book even though he may have forgotten some of the words. To other people, however, his identity must be proven.

The need to prove an individual's identity lies at the heart of record linkage and represents a much more complex challenge that we might imagine. I know, for instance, that the Thomas Henry Hassard mentioned on my birth certificate and the T. Hazard who, according to the records of an overworked general practitioner, had a recent health checkup are, in fact, one and the same person. Would that, however, be obvious to anyone else? In this article we explore the way we reach such

decisions and how we can automate human decision making, thus enabling us to piece together the pages of the book of life for thousands or even millions of people.

PROMISE AND PROBLEMS

Each individual, uncompleted book of life is of course of vital interest to its author. Within its pages, however, are many clues to the ways our health patterns will subsequently develop. Exposure to certain drugs during pregnancy may lead to the birth of a malformed child; obesity in middle age may lead to coronary heart disease. Taken collectively these individual clues can become dramatic pointers to health problems and their possible solution. The collected life health histories of an entire population is therefore potentially one of the most powerful tools for improved care available to medical research.

There are, however, major practical problems in compiling a complete life medical history for even a single person—never mind the population of an'entire province or country. Each of our contacts with the health profession generates a record of our health status at that point in time (the pages in our book of life); an obstetric record when we are born, records of routine medical examination during our schooldays, hospital records relating to, say, an operation for appendicitis, and ultimately, of course, a record detailing the cause of our death.

Each such record consists of two parts; one detailing the relevant medical experiences and one that details who the record relates to. These details—our name, date of birth, address, and so on—should, we would like to think, identify us uniquely for posterity. Sadly for the sake of our egos, this is rarely the case. Some records are not as carefully kept as we might wish, names get misspelled, dates of birth get omitted. Other factors intervene to obscure our true identities; other people may have the same name as ours; our address is extremely likely to change as our life evolves. The apparently simple task of telling the world exactly who we are is therefore far from trivial.

The human clerk charged with the task of compiling a book of life for even one person therefore has a series of difficult decisions to make. Does this series of medical records, created perhaps over a period of 50 or 60 years, in fact relate to the one person and should therefore be linked together to form a medical life history; or is it possible that some of the records relate to different persons, possibly with some similar names, and should therefore be left unlinked? This is the essential problem of medical record linkage.

ASSESSING THE EVIDENCE

If all the identifying items—name, date of birth, sex and so on—are recorded on all the records and are all in complete agreement, then the decision is clear cut. As we have pointed out, however, this is rarely if ever the case. In real life the human clerk will have to assess the evidence available and use his or her experience and intuition to reach a decision as to whether two records have in fact been created by the same person and therefore should be linked together. In reaching this decision he will attach different weight or importance to different pieces of evidence. If, for instance, two records both relate to a person called "Smith," it is far from conclusive evidence that they both refer to the same person. After all, Smith is a fairly common name and it is possible that the two records might refer to two distinct persons who happen to share this surname. On the other hand, a human clerk would attach great weight to the fact that two records both related to someone called "Verblunsky." This is a very uncommon surname and the chances of two distinct persons sharing this unusual surname is remote. Agreement between the records on such an unusual item is powerful evidence that the two medical records belong together and should be linked.

Similar reasoning can be applied to other identifiers. We would, for example, attach little weight to the fact that the persons referred to on two records were both born in London. In such a large population center the odds of two quite separate persons having the same place of birth are extremely high. If, however, the place of birth recorded on both records was remote and exotic, such as Ulan Bator, we would regard this a very strong indication that the records do indeed relate to the same person, the odds of such an unusual and distinctive place of birth being shared by two people being extremely small. Disagreements, as well as agreements, offer us varying degrees of evidence as to whether or not we should link two records together. If two records disagree on the persons' current address it is not particularly damning evidence against linking them. Most of us change our address several times during our lifetime and the odds on two records relating to the same person having differing addresses must be reasonably high. On the other hand, we would attach a great deal of weight to the fact that two records had different sexes recorded on them. Agreement between the sex recorded on records relating to different people will occur quite often (we would expect it to happen, say, 50% of the time) and is therefore of little positive evidence in favor of linking two records. Disagreement between the sex recorded on two records relating to the same person will, however, occur very, very infrequently (after all, we can generally always remember what sex we are and hopefully, it is always apparent to our doctors). A disagreement of this type is therefore a very powerful indication that the two records do in fact relate to different people and should therefore not be linked together. We would, of course, assess this evidence together with the other evidence of the agreement or disagreement of the surname, and so on, on the two records, but we would certainly attach a great deal of weight to this discrepancy.

THE BALANCE OF THE ODDS

It is instructive at this stage to reflect on exactly what we do when we decide on the weight or importance we will attach to the outcome (agreement or disagreement) that results when we compare two items of identification. Indeed, as we have seen, the weight we attach to an agreement will often vary with the nature of the agreement (e.g., Smith versus Verblunsky). We ask ourselves "What is the chance of this outcome occurring if the two records genuinely belonged to the same person (matched records)?" and "What is the chance of this outcome occurring if the two records in fact belonged to different people (unmatched records)?" and we then strike a balance between the two possibilities. If the odds make it more likely that such an outcome would occur when the records are matched, we regard it as positive evidence in favor of linking the two records together. If the odds favor the possibility that such an outcome would occur when the records were unmatched, we regard this as negative evidence (evidence against linking the two records). The amount of importance or weight we attach to a particular outcome depends on how much the odds favor one conclusion or the other. We must, of course, work out this "balance of the odds" subjectively by calling on our own experience and common sense. Luckily, these are quantities with which most people are remarkably well supplied. To reach a final conclusion as to whether or not two medical records do indeed belong together, we then draw together all the individual items of evidence, decide which conclusion the overall weight of evidence points to, and act accordingly.

The following pairs of records both show a fairly comparable degree of agreement in the identifying items recorded on them.

Record pair A:

Surname	Forename/initial	Place of birth	Sex
Verblunsky	K.G.	Ulan Bator	M
Verblonsky	G.	Ulan Bator	M

Record pair B:

Surname	Forename/initial	Place of birth	Sex
Smith	C.R.	London	M
Smith	Carol	London	F

Yet, using the intuitive assessment process we have discussed, we would almost certainly come to quite different conclusions about them. Despite the discrepancies in the initials and the slightly different spelling of the surname, we would feel very confident that the two records in pair A refer to the same person. This conclusion would be based largely on the fact that the birthplace and general structure of the surname are so unusual that it is highly unlikely that they could refer to different people.

On the other hand, there must be very grave doubts that the records in pair B really belong to two distinct persons. The surnames and place of birth do indeed agree, but they are both so common that this agreement could quite easily have happened accidentally. The fact that the recorded sex on the two records differs is, however, a very powerful suggestion that two separate persons are involved.

(a)

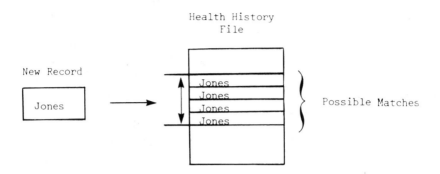

(b)

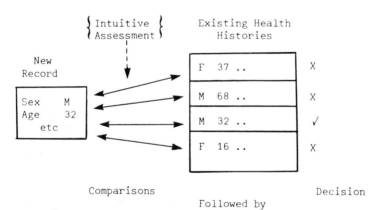

FIGURE 1   Updating a health history: Manual linkage.
(a) Searching step. (b) Matching step.

SEARCHING AND MATCHING

In reality, of course, it is very unlikely that a clerk would be employed
solely to compare two particular health records. A much more likely
situation is one in which the clerk is charged with keeping up to date the
health histories of a whole population. When a new health record arrives,
the clerk must therefore decide which person (if any) in the population
file the new record relates to and then add the new medical data to that

person's file. (If no corresponding person exists in the files, the clerk will create a completely new health history for this new person.)

This adds a whole new dimension to the problem. Health files of this type usually contain tens of thousands of individual histories, quite frequently hundreds of thousands. It is therefore an impossibility to check the identification information on the new record against the identification information on each of the records in the file. This would, of course, be highly inefficient anyway, as the vast majority of records in the file will not have the slightest resemblance to the new record. The solution normally employed in practice is to restrict our interest to a small number of records in the main health file which seem to have special relevance to the new record. Normally, these would be records with the same surname as the new record. Each of these records is compared in detail with the new record, and the one that shows an intuitively acceptable pattern of agreement is accepted as the person to whom the new record relates and the new health data are added to this file. If none of the existing records shows an acceptable pattern of agreement, we conclude that the new record must relate to someone not already included in the file and we create a new record in the main health file.

The selection of a small group of records for further comparison is usually referred to as the searching step, and detailed comparison of the new record with each of the existing records in this group or block is usually referred to as the matching step. These two steps in the record linkage process are illustrated in Figure 1.

It is absolutely vital, of course, that the criterion on which the searching step is based should be recorded reliably. If this is not the case, the wrong section of the file may be searched and, since none of the other records in the file are now examined, the person to whom the new record refers will never be found. It would, in addition, be a nice bonus if the criterion used split the main file up into fairly small blocks of persons. After all, the whole point of the searching step is to cut down as much as possible the number of detailed comparisons we have to make. It would be of little real comfort, for example, to reduce the number of comparisons needed from 100,000 to 10,000. Luckily, surname fulfills both these requirements quite well. It is recorded on nearly everyone's health records. It is usually recorded accurately, so we can be fairly confident that the new record must refer to somebody with the same surname (with one important exception, which we discuss below). It is fairly specific in that not very many people share the same surname (although, of course, this will vary quite a lot from surname to surname), so comparisons with all persons of the same surname is reasonably practical. This strategy of a search based on surname followed by detailed matching will inevitably result in some mistakes being made and people missed. This is, however, a fairly small price to pay for converting an insoluble problem to a very solvable one.

In the case of a married woman who changes her surname, an additional step is required. If no matching record is found under her current surname, a similar series of comparisons is carried out with those health records held under her previous surname. If a match is found here, the new health data are added to her file and her surname is updated to her current surname. If no match is found under either name, a new record is created under her current surname.

THE COMPUTERIZED CLERK

Most health history files are at present maintained by human "updaters" in very much the manner we have just described. However, these files are usually restricted to one particular aspect of a person's health experience (e.g., the medical records held by a particular hospital or the medical histories of the patients registered with a certain health care clinic). Such records are, of course, of interest and use to the agencies who hold them, but by themselves are of limited value in telling us how patterns of health and disease evolve over the course of a lifetime. To do this we need to be capable of linking together all the pages in a person's book of life, whether these relate to hospital operations, treatment by a general practitioner, or any other health source. The number of health records involved in such a scheme are, however, likely to be extremely large even for a relatively small community. Keeping patient health histories up to date in a busy hospital is a full-time job requiring a considerable number of clerical staff. The task of keeping a population-wide file of general health histories up to date with new health information flooding in daily from hospitals, health clinics, school medical services, and so on, is quite simply beyond the scope of routine human operation and requires some form of computer-based information processing system to make it feasible.

The primary task that such a system would be concerned with is deciding which person in the master health history file a new health record actually refers to. Once this decision is reached, the actual incorporation of the new health information into the appropriate master history is a trivial operation. As we have seen, the way in which human beings tackle the problem of deciding whether or not two records relate to the same person involves some very sophisticated assessments of the available information. It certainly differs dramatically from the crude "do

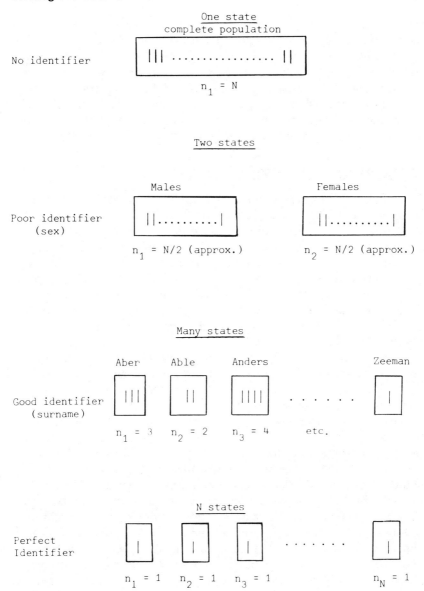

FIGURE 2 Identification potential of identifiers.

they agree or don't they" approach that we might expect a machine to take. Yet if a computer-based health record system is to do an acceptable job, it must somehow employ exactly the same skills that come instinctively to a human being. This may seem an awful lot to ask. However, by taking each of the basic steps and through processes that we have outlined and using basic statistical ideas to mimic them, we can, in fact, create an automatic system for linking together medical records which behaves in almost exactly the same way as a human clerk.

MEASURING IDENTIFYING POWER

The identification information that is available to a computerized record linkage system (and indeed to a human-based system) is determined by the identifiers which are commonly recorded on most medical records wherever they have been created. The list of potentially available identifiers is fairly obvious: surname, forenames (or initials), address, age and/or date of birth, place of birth, current marital status, and birth surname. Before we decide on a strategy for utilizing this information in a decision-making process, it is profitable to consider exactly what we mean by identification and how we might measure how well a particular identifier does this (see Figure 2).

In the absence of any identification information, a population of N persons is effectively a single faceless mass from which we are unable to pick out any specific person. An identifier such as sex splits the total population up into two approximately equally sized groups.
Knowledge of a person's sex and nothing else does therefore make it a little (but not very much) easier to identify this person in a population. The group of people who share this identity is still so large as to be almost meaningless. On the other hand, an identifier such as surname will split the total population up into a very large number of groups. The number of persons in each group will, of course, vary; rare or unusual surnames will be represented by only one or two individuals; common surnames, a few hundred. In general, therefore, surname is a powerful identifier that reduces the number of people we cannot distinguish between from a very large number, N (the total population), to a relatively small number, n (the number of people who share a particular surname). A perfect identifier would take the process to the extreme and split the total population up into N separate groups each containing only one uniquely identifiable person. Identifiers like this do indeed exist, the identity numbers that government agencies delight in bestowing on us being cases in point. Since few of us can ever remember what on earth our identity numbers are (a very reassuring situation to those of us who

believe identity numbers to be faceless, soulless, and just a little bit frightening), these identifiers are much less useful in practice than they are in theory.

The key measure of an identifier's power is clearly the extent to which it reduces the number of persons we cannot distinguish between. Using N to denote the total population and n_x to denote the number of people sharing a particular identity, x (possibly a certain surname), this reduction is measured by the ratio

$$\frac{n_x}{N} = P_x \qquad \text{proportion of persons sharing identity } x$$

Since the number of persons in each identity state (e.g., sharing a surname) will vary from state to state (e.g., surname to surname), we clearly need to construct an overall measure of identification that includes information on all the identity states. If a particular identifier splits the population up into k groups, our initial instinct would be to total the ratios for all these groups together:

$$\sum_{x=1}^{k} P_x$$

However, this index fails to distinguish at all between perfect identification and no identification.

Perfect identification:

$$k = N \qquad n_x = 1$$

$$\sum P_x = \frac{1}{N} + \frac{1}{N} + \cdots + \frac{1}{N} = \frac{N}{N} = 1$$

No identification:

$$k = 1 \qquad n_x = N$$

$$\sum P_x = \frac{N}{N} = 1$$

If, instead, we total the squares of the ratios, we, however, find that we do indeed get a logical distinction between these two extremes.

Perfect identification:

$$\sum P_x^2 = \frac{1}{N^2} + \frac{1}{N^2} + \cdots + \frac{1}{N^2}$$

$$= \frac{N}{N^2} = \frac{1}{N}$$

No identification:

$$\sum P_x^2 = \frac{N^2}{N^2} = 1$$

The only problem with this measure is that an increase in the identifying power of an identifier is reflected by a decrease in the size of this index. A more intuitively meaningful measure is therefore given by

$$C = \frac{1}{\sum P_x^2}$$

which takes values ranging between 1 (for no identification) to N (for perfect identification). This measure, known as the coefficient of specificity, can be thought of as the number of blocks of similar size that the identifier will split the population into. Table 1 presents the estimated identifying power of the standard health record identifiers based on a sample of health records drawn from the population of Northern Ireland.

Sex, an identifier of low power, splits the population into two similar-sized groups. Year of birth splits the population into the equivalent of 75.7 equal-sized groups, the approximate age range of persons in the population. Current address splits the population into the equivalent of 328,412.7 equal-sized groups (an average of $4\frac{1}{2}$ people/address since the population of Northern Ireland is about 1,500,000). In practice, of course, the groups are not of equal size and the actual number of groups into which the population is split is not exactly equal to C. However, C is clearly an intuitively meaningful summary of an identifier's power. We must, however, be somewhat careful when comparing identifiers in different populations. In large populations such as the 220,000,000 of the United States, the number of addresses and consequently the value of C will increase, although since the size of the problem has also increased, its usefulness will remain roughly the same. The value of C for sex will remain at 2, and consequently the larger the population, the less its relative usefulness.

COMBINING IDENTIFIERS

The identifying power of several identifiers is multiplicative not additive (provided that they are independent of one another). This can be easily verified by considering the identifiers day of the month and month of the year, which form part of our date of birth. Month on its own splits the population into 12 groups, day of the month on its own splits the population into the equivalent of 30.5 equal-sized groups, whereas the

two taken together (i.e., day of the year) splits the population into 12 ×
30.5 = 366 groups. However, we instinctively find it much easier (and
more meaningful) to add values together to get an overall view rather
than multiply them. Consequently, the identifying power of an identifier
is usually expressed as $\log_2 C$, and hence the identifying power resulting
from the use of two identifiers can easily be obtained simply by adding
together their individual identifying powers. The combined identifying
power of age and sex, for instance, is given by 6.2 + 1.0 = 7.2.

A perfect identifier for a population such as Northern Ireland's
1,500,000 would, on this scale, have an identifying power of 20.4
($\log_2 1,500,000$). Of course, no such identifier exists, but by combining the
identifying power of existing identifiers, we can achieve results which are,
at least in theory, powerful enough to identify uniquely each person in
the population (using surname and date of birth gives an identifying
power of 9.3 + 14.7 = 24.0). In practice there would always be a few
persons (twins, etc.) who would share a surname and date of birth, but the
vast majority of the population would indeed be uniquely identified. This
approach is extremely useful in deciding just what identifiers we would
need to record to be able to have a realistic prospect of identifying the
health record file of any person in a given population. [The addition rule
only applies, however, if the identifiers are independent. Since most
people who share the same address have the same surname knowledge of
a person's surname will give very little additional identifying power if
their address is already known (i.e., identifying power of surname +
address ≠ 9.3 + 18.2).]

POWERFUL OR RELIABLE?

The concept of identifying power assumes that the identifiers are always
correctly recorded. This is naturally never the case in practice. Table 1
gives the estimated error rates (in terms of percent of records incorrectly
completed) for each of the most commonly employed personal identifiers, again based on a Northern Ireland sample of medical records for
which the authenticity of the various items could be verified. Ironically,
the greater an identifier's identifying power, the more likely it is to be
incorrectly recorded. This is a consequence of the fact that the more
complex an identifier, the easier it is to get it wrong. Sex, for instance, is
on average recorded incorrectly on only 1 record in every 1000, while a
very powerful identifier such as current address is recorded incorrectly on
1 record in 10, thus severely limiting its practical usefulness. Clearly, we
need to consider both an identifier's power and its accuracy when
assessing its merits. Howard Newcombe of British Columbia, the father

TABLE 1   Relative Merits of Personal Identifiers

Identifier	Coefficient of specificity C	Identifying power I	Error rate E (%)	Merit ratio I/E
Surname	658.9	9.3	4.0	2.3
Surname (Russell Soundex code)	266.0	8.0	0.8	10.0
First forename	123.4	6.9	12.0	0.6
Mother's maiden name	658.9	9.3	5.0	1.9
Current address	328,412.7	18.2	10.0	1.8
Date of birth	28,547.6	14.7	15.0	1.0
Year of birth only	75.7	6.2	6.3	1.0
Sex	2.0	1.0	0.1	10.0
Current marital status	2.1	1.1	0.1	11.0

of medical record linkage, has proposed a measure, the merit ratio, which combines identifying power and error rate in the form of a ratio (Newcombe, 1967). Inspection of Table 1 indicates that three identifiers,—sex, marital status, and surname (in the form of the Russell Soundex code)—have outstanding merit ratios.

Both our intuition and the evidence quoted in Table 1 tells us that surname is a powerful identifier ($I = 9.3$). Furthermore, it is reasonably accurately recorded in practice (errors on about 4 records in every 100). Most of these errors are, however, usually fairly trivial spelling mistakes, generally involving versions of the name which sound similar, either because of similar-sounding consonants (Hassard and Hazzard), interchanging of vowels (Hassard and Hasserd), or errors in repeating letters (Hassard and Hasard). A method of coding surnames, the Russell Soundex code, is available which eliminates practically all these trivial but common mistakes. Full details of the Russell Soundex code are given in the appendix to this chapter. In essence, however, it retains the first letter of a surname, eliminates all vowels, retains only one of any duplicated letters, and gives each of the remaining letters a one-digit code which ensures that easily confused letters (such as s and z) receive the same code. Only the first three codeable letters are coded, and if less than three codeable letters are left, then the code is made up to three digits with zeros. Thus all Russell Soundex-coded surnames consist of a letter followed by three digits.

Using the rules laid out in the appendix, you can easily verify that Hassard, Hazzard, Hasserd, and Hasard, or Hazard will all be recorded as H263 using the Russell Soundex code. Table 1 confirms that restructuring surnames in this way will greatly improve the reliability with which they are recorded (40 errors in every 1000 records for conventional surnames compared with only 8 errors in every 1000 using the Soundex code). However, it must also reduce its identifying power (I = 9.3 for conventional surnames compared with I = 8.0 for Soundex-coded surnames). After all, the coding procedure does lump together surnames which are potentially genuinely different. A glance at the respective merit ratios (an increase from 2.3 to 10.0 when we replace conventional surnames with Soundex codes) suggests that the increase in reliability much more than compensates for this small loss in identifying power. In computerized medical record linkage systems, surnames are routinely coded into their Russell Soundex code format.

The high merit ratios of sex and marital status are somewhat misleading in that they chiefly reflect thte very great accuracy with which these identifiers are recorded. Soundex surnames are, however, in many ways the ideal single identifier. They are extremely reliable and at the same time reduce a large population to a large number of relatively small groups or blocks of persons. Soundex codes therefore offer an ideal basis for carrying out the searching step which we outlined earlier as the initial step in any human-based process for linking a new health record with the correct health history. (Indeed, we intuitively suggested that a clerk would probably use the surname as a basis for searching through a large file of health histories.) In an automated medical record linkage system the main health history file is therefore held in order of Soundex codes and a new medical record is only compared in detail with health histories that share the same Soundex code.

MEASURING THE BALANCE OF THE ODDS

To automate the comparison step requires some way of replacing our intuitive grasp of the relative importance of various agreements and disagreements by numerical values which reflect the same ideas but are rather more easily handled by an unthinking machine. The idea of a "balance of odds" ("How unlikely would this be if the records belonged to the same person?" versus "How likely would this be if the records belonged to different people?"), which is how we construct our judgments on the importance of agreements, has an exact parallel in statistics, where it is known as a balance of probabilities or a *likelihood ratio*. The question "How likely would this be?" can be answered by the very simple and

TABLE 2   Weights of Comparison Outcomes

	Weight $\left(\log_2 \dfrac{FM}{FU}\right)$	Frequency/1000 records	
		Matched (FM)	Unmatched (FU)
Sex			
Disagree	−7.35	3	505
Agree			
Male	+1.00	494	246
Female	+1.01	503	249
Address (house number, street name)			
Complete disagreement	−2.42	180	972
Complete agreement	+7.84	715	3
Agree on street name only	+3.28	89	9
Agree on house number only	+0.00	16	16
Year of birth			
Disagree			
Ages differ by < 2.5%	+2.94	132	17
Ages differ by < 5.0%	+1.14	42	19
Ages differ by < 7.5%	−0.08	18	19
Ages differ by < 10.0%	−1.80	6	21
Ages differ by < 15.0%	−3.88	3	45
Ages differ by < 15.0%	−8.70	2	866
Agree	+5.90	797	13

direct expedient of actually collecting samples of records that we know from external evidence belong to the same persons (matched records) and to different persons (unmatched records) and observing just how frequently various agreements or disagreements occur. Table 2 contains evidence on several commonly used identifiers acquired by just such a procedure.

Disagreements on a person's sex, for example, occur only on about 3 in every 1000 pairs of genuinely matched records, whereas it occurs on around 500 in every 1000 pairs of genuinely unmatched records. This produces a balance of odds of 1:168 in favor of unmatched records. In other words, if the recorded sex disagrees on two records, it is 168 times more likely that the records relate to different people than that they relate to the same person. Using the same empirically derived evidence, we see that the balance of odds for agreement between recorded sex (either male or female) is 2:1 in favor of the records belonging to the same person. The principle of the balance of odds or likelihood ratio therefore offers us a means of deciding which conclusion the evidence supports and also how strongly individual pieces of evidence support this.

THE WEIGHT OF EVIDENCE

Just as the identifying power of two identifiers used together is given by the produce of their individual identifying powers, so the "balance of odds" suggested by two pieces of evidence is given by the product of their individual likelihood ratios (assuming that the pieces of evidence are independent of one another). Again, since we find it so much easier to obtain the overall picture by summing individual items of evidence together, it is much more convenient to work with the logarithms of the various ratios. The weight, or importance, to be attached to the outcome of a particular comparison is therefore defined as

$$\text{Weight} = \log_2 \frac{\text{frequency in matched records}}{\text{frequency in unmatched records}}$$

and the total weight of comparison of two records is obtained simply by adding together the weights of all the individual comparison outcomes (e.g., sex agrees as *M*, date of birth disagrees, etc.). Expressing these weights in logarithmic form has several other advantages; if the evidence favors the conclusion that the records refer to the same person, the weight is +*ve*; if it favors the conclusion of different people, then the weight is −*ve*; and the stronger the evidence either way, the larger the weight.

Carrying out the matching step automatically requires first, the calculation of the weights of the various possible agreements and disagreements for the range of identifiers to be used, based on evidence from samples of records. Each record sharing the same Soundex code is then compared with the new record, the various agreement and dis-agreements noted, and a total weight of comparison calculated for each record by summing the appropriate individual weights (see Figure 3).

REACHING A FINAL DECISION

The health history with the largest +*ve* total comparison weight is, of course, the record that most closely resembles the new record. However, this is not the same thing as saying that the two records must refer to the

(a)

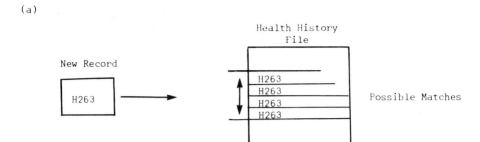

(b)

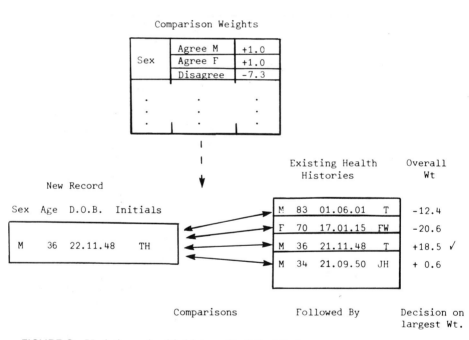

FIGURE 3 Updating a health history: Statistical linkage.
(a) Searching step. (b) Matching step.

(a)

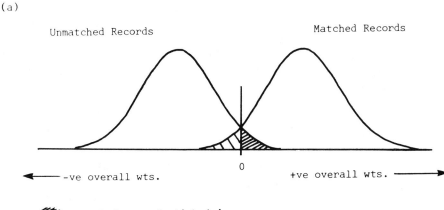

(b)

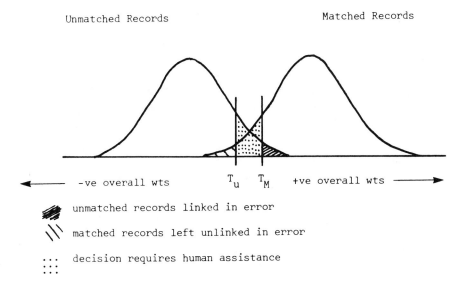

FIGURE 4 Distribution of overall weights of comparison. (a) Using O as a decision value. (b) Using T_u and T_m as decision values.

same person. Although any +ve total comparison weight is evidence in favor of linking two records, it is of course possible, although unusual, for two records relating to different people to show an overall +ve weight. This could easily happen for same-sexed twins, for example. Similarly, two records belonging to the same person might very occasionally show a --ve overall weight if the address had changed. Because of this possible overlap between the total comparison weights for matched and unmatched records and the risk of making the incorrect decision in some cases (see Figure 4a) it would seem advisable to demand some stronger evidence when deciding if two records relate to the same person or not. Most linkage systems therefore link two records only if the overall weight exceeds some value, T_M, and leave two records unlinked if the overall weight is less than some value, T_u. If the weight falls between these values, no decision is taken and human assistance is called for (see Figure 4b). The larger the gap between T_u and T_M, the fewer incorrect decisions will be made, but of course, the greater the number of situations in which human assistance is required.

The values of T_M and T_u can be set on a commonsense basis. For instance, values of $T_M = +10$ and $T_u = -10$ are often used, a value of $+10$ indicating odds of more than 1000 to 1 ($2^{10} = 1024$) in favor of the records belonging to the same person, and similarly -10 indicating odds of more than 1000 to 1 in favor of the records relating to different persons. These may seem impressive odds, but remember that when hundreds of thousands of records are being processed in a year a few mistakes will still happen (as, of course, they also will in a human-based system). Alternatively, the experience gained in running a record linkage system for a year or so can be used to modify the values of T_u and T_m if it is felt that the number of errors involved is too large or that too many decisions require human intervention.

RECORD LINKAGE IN ACTION

The weights assigned to the individual comparison outcomes are the real key to the success of a record linkage system. It is these which cause an inanimate piece of machinery to mimic the experience and judgment of a human being. To do this successfully, the weights must be based on evidence gathered from the community in which they are to be used; otherwise, the subtle interpretations of information, which is the real power of record linkage, become meaningless. Ulan Bator may be a very unusual place of birth for residents of the United States, but it is likely to be less impressive when linking records in Outer Mongolia. Once a record linkage system is running, new weights can be regularly calculated by checking the frequency of occurrence of the various agreements and disagreements among pairs of records that have been linked or left unlinked. This regular recalibration parallels the changes in our view of the world as we grow older and wiser and ensures that our "statistical clerk" evolves along with the community it seeks to serve.

Record linkage systems similar to the one described here are now in operation in a number of centers throughout the world [Oxford, England (Acheson, 1968); Belfast, Northern Ireland (Cheeseman, 1968); Vancouver, Canada (Newcombe, 1967)]. The information they are gathering together is being used, among other things, to help pinpoint possible causes of cancer (Newcombe, 1973) and to locate mothers who are at risk of producing handicapped children (Newcombe, 1966). The potential usefulness of the information generated by a record linkage system is probably incalculable. The ideas of statistics and the power of the computer may well have opened the door to a healthier future for many of our children, a door which otherwise would have had to remain firmly locked.

APPENDIX: THE RUSSELL SOUNDEX CODE

The Russell Soundex code seeks to improve the reliability of surname recording by setting aside unreliable surname information. Surname errors are generally caused by (1) vowel substitution; (2) consonant substitution; (3) doubling of single letters, or vice versa; and (4) substitution of familiar for unfamiliar forms. The Soundex code aims, as far as is possible, to eliminate these sources of error.

The rules for surname coding are as follows:

1. The first letter of the surname is used as the prefix letter of the code. The remaining letters of the surname are coded to numerical digits using the following rules.
2. The letters A, E, I, O, U, Y (regarded as vowels), W, and H are not coded.
3. The remaining letters are codeable and are coded in the following way:

B, P, F, V	= 1
C, G, J, K, Q, S, X, Z	= 2
D, T	= 3
L	= 4
M, N	= 5
R	= 6

4. A codable letter is not coded if it would yield the same digit code as the immediately preceding codable letter in the surname unless the two codable letters are by a vowel (as defined in rule 2).

5. The coding stops when three digits have been obtained. If the coding yields fewer than three digits, zeros are used to complete the code, e.g.:

BROWN, BROWNE, BRAUN   = B650
McGINNIS, MacGUINNES, McINNES = M252

As an exercise you are invited to construct possibly confusing alternative versions of your own surname and check if the Russell Soundex code allows for them. It rarely misses.

REFERENCES

Acheson, E. D. (1968). The Oxford Record Linkage Study—the first five years. In *Record Linkage in Medicine* (E. D. Acheson, ed.) E. & S. Livingstone, Edinburgh, pp. 40–49.

Cheeseman, E. A. (1968). Medical record linkage in Northern Ireland. In *Record Linkage in Medicine* (E. D. Acheson, ed.) E. & S. Livingston, Edinburgh, pp. 70–76.

Dunn, H. L. (1946). Record linkage. *Am. J. Public Health 36*, 1412–1416.

Newcombe, H. B. (1966). Familial tendencies in diseases of children. *Br. J. Prev. Soc. Med. 20*, 49–57.

Newcombe, H. B. (1967). Record linkage: the design of efficient systems for linking records into individual and family histories. *Am. J. Hum. Genet. 19*, 335–359.

Newcombe, H. B. (1973). Record linkage for studies of environmental carcinogenesis. In *Proceedings of the 10th Canadian Cancer Conference.* University of Toronto Press, Toronto, pp. 49–64.

FURTHER READING

The technical aspects of record linkage are very readably discussed in Newcombe, 1967 (referenced above).

A good general introduction to the whole field of medical record linkage is given in

Acheson, E. D. (1967). *Medical Record Linkage.* Oxford University Press, London.

A wide variety of applications of record linkage techniques in medicine are discussed in

Acheson, E. D. ed. (1968). *Record Linkage in Medicine.* E. & S. Livingstone, Edinburgh, p. 339.

II
Condensing Complex Data

Human beings possess a remarkable capacity for assessing and integrating vast quantities of complex, interrelated information and drawing very subtle conclusions from it. One person's perception of this information, however, may be quite different from that of someone else. It would be helpful to have access to methods that would allow the data to be approached objectively and from different viewpoints. This is the field of multivariate statistics. It is not only ideally suited to the study of humans and their world, personalities, sociology, and health, but also is highly relevant to any field when the objects of study are complex and require a wide variety of measures to describe them adequately.

The chapters in Part II explore both the techniques of multivariate statistics and the ideas which underly them. In his paper on Cluster Analysis, Brian Everitt discusses what we mean by similarity and how we can construct numerical measures of it based on information from a number of variables. He then shows how such measures of similarity can be used to detect groups of inherently similar individuals.

David Hand explores a related problem. Suppose we know in advance that particular groups of people are different. They may, for instance, have been diagnosed as suffering from different diseases. Can we effectively describe how the groups differ, that is, what particular pattern of variables characterizes a particular disease? This pattern recognition problem mimics the diagnostic skills of the physician and indeed has a major role in training and aiding physicians in diagnosis.

One of our favorite and most effective devices for displaying and describing complex information is the diagram. Multivariate statistics draws very frequently on diagrams. Visual displays are the ideal medium for conveying complex results on similarities in a way that we find very easy to comprehend. One technique of presenting the relative similarities of groups of individuals in diagrammatic form is known as multidimensional scaling. Alan Tyree uses this in the field of law to investigate the similarities of a number of key legal cases dealing with the subject of lost property.

A slightly different approach to multidimensional scaling is discussed by Allan Anderson in the context of market research. Formally measuring similarity in the way described by Brian Everitt and Alan Tyree means initially deciding on a set of variables to measure. This in itself can be a major problem. Frequently we know how similar things are (in our view) but find it difficult to define those variables which make them similar. Instead of measuring similarity, nonmetric multidimensional scaling utilizes the less exact, but frequently more practical, strategy of simply ranking objects (in this case, ice cream confectionaries) in terms of their relative similarities and then visually reconstructing these relationships.

In her chapter on management training, Judy Brook takes another approach to the problem of which variables to measure. When studying human subjects, why not let them decide which variables describe their personal situation best? After all, who is better qualified to make such a decision? This ability for the participants in a study to construct a framework of descriptors tailor-made to their personal circumstances is central to the technique know as "Kelly's repertory grid."

In Phil Gendall's investigation of the differences among soft drinks as perceived by their potential purchasers, the reasons why purchasers view soft drinks as very similar or very different is informative to market researchers. This chapter reinforces this point by summarizing the similarities between drinks and the variables that buyers used to describe the drinks, on a single diagram displaying a large amount of interrelated information. This is a classic example of multivariate analysis at work.

In many fields of study, especially that of human personality, the patterns of response to a series of questions can provide fascinating glimpses into the way we view our world. Very positive responses to questions such as "Do you enjoy parties?" and "Would you address a large audience, if invited to?" might be indicative of an outgoing or extroverted personality. These underlying themes are known as factors and the scanning of large number of variables for groups that show similar patterns of response, followed by the identification of the factors involved, is known as factor analysis. Bill Stiles demonstrates this technique in action, and one of its leading exponents, Hans Eysenck, gives his perspective on some of the controversies that have surrounded factor analysis.

4

Numerical Approaches to Classification

Brian S. Everitt
University of London
London, England

CLASSIFICATION

The need to classify objects or people is fundamental both for everyday living and, more specifically, in the development of many areas of science. In its most general form classification is the process of giving names to collections of things which are thought to be similar to each other in some respect. The ability to sort similar things into categories is clearly a primitive one since if nothing else, early man must have been able to realize that many individual objects shared certain properties, such as being edible, or poisonous, or ferocious, and so on. Indeed, without this ability to classify it is difficult to see how language could have developed since a new word would have to be found for each separate object or event encountered. However, by using the ability to classify, we can name, for example, animals as cats, dogs, or horses, and such a name collects individuals into groups.

Classification has also played an important role in the development of many areas of science. Most notable of course, has been its contribution to biology and zoology leading eventually to Darwin's theory of evolution. It has, however, also played a central part in other fields. For example, the classification of chemical elements in the periodic table, produced in its most complete form by Mendeleyev in the 1860s, has had a profound influence on the understanding of the

structure of the atom. Again in astronomy the classification of stars into dwarf stars and giant stars using the Hertsprung-Russell plot of temperature against luminosity (see Struve and Zebergs, 1962) has strongly affected theories of stellar evolution.

The classification process consists essentially of two separate but related steps. The first is the construction of a sensible and informative classification of an initially unclassified set of objects. The second involves the derivation of rules for allocating objects to one of a number of previously defined categories. The first stage can be illustrated by the example of the field worker in archaeology who finds large numbers of objects, such as stone tools, funeral objects, pieces of pottery, ceremonial statues, and skulls; the worker would like to produce a classification of these objects since this might aid in discovering whether they arose from a number of different civilizations. A further example is provided by psychiatry; diseases of the mind are more elusive than diseases of the body, and the classification of such diseases is in an uncertain state. A psychiatrist may collect a large amount of information on a sample of mentally ill patients and use this to try to determine whether a classification of the patients can be produced which has implications for etiology and treatment. More will be said about these and further examples later.

The second stage of the classification process, rules for allocating objects or individuals to a priori defined classes, can be illustrated by the example of a disease that can be diagnosed without error only by means of a postmortem examination. The physician would clearly like to develop a rule for allocating suspected cases to either the disease class or the no disease class while the patient is still alive, so that if necessary, appropriate action could be taken. A futher example is provided by research into the question of authorship. Most people have heard of the Shakespeare-Bacon-Marlow controversy over who wrote the great plays usually attributed to Shakespeare, and other work has dealt with the authorship of a number of Christian religious writings called the Paulines, and *The Federalist* papers published anonymously in 1787–1788 by Alexander Hamilton, John Jay, and James Madison to persuade the citizens of the state of New York to ratify the Constitution. Here we wish to assign essays, and so on, to the classes, which in this case are the various possible authors.

Numerical methods and statistics have much to contribute to both these aspects of classification. Methods applicable to the first stage have developed primarily over the last two decades and are known collectively as *cluster analysis techniques* or methods for *unsupervised pattern recognition*. It is these techniques that we shall be primarily interested in in this chapter.

Methods applicable to the second stage were first developed in the 1930s and are known as *discriminant function techniques* or methods for *supervised pattern recognition*. These techniques are the subject of Chapter 5.

CLUSTER ANALYSIS

Cluster analysis is a generic term for a large number of techniques designed to construct a classification of a set of objects given a description of each object in terms of a number of numerical variables. Many of these methods have appeared only during the last two decades or so, their development following that of the electronic computer, on which they critically depend to perform the prodigious amount of arithmetic generally involved.

To illustrate in a little more detail what the techniques of cluster analysis are attempting to do, let us consider a problem in which we have a set of essays considered by most historians to have been written by a single author. However, the eminent historian, Professor Aardvark

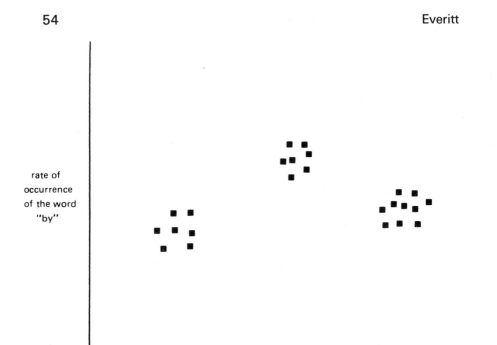

mean sentence length

FIGURE 1   Plot of mean sentence length and rate of occurrence of the word "by" for a number of essays.

believes otherwise, and to investigate further records for each essay two measures, mean sentence length and the rate of occurrence of the word "by." He then constructs a picture of the data by plotting the two variable values of each essay. Let us suppose that he arrives at the diagram given in Figure 1. Such a picture would imply that the essays are naturally classified into three distinct classes. One possible explanation of such a finding is that the essays were written by three different authors. On the other hand, had the plot looked like Figure 2, a single author would probably be indicated.

Here the method of cluster analysis used has been very simple, involving simply a visual examination of a two-dimensional plot. In essence what we appear to do to identify the groups of points is to examine the relative *distances* between points and place in the same group those that lie close together. If all data sets consisted of only two measurements for each of the objects to be classified, this method would provide a very effective clustering procedure since the human eye appears to be an excellent pattern recognition device in two dimensions. Indeed, many clustering techniques may be thought of as attempts to imitate this

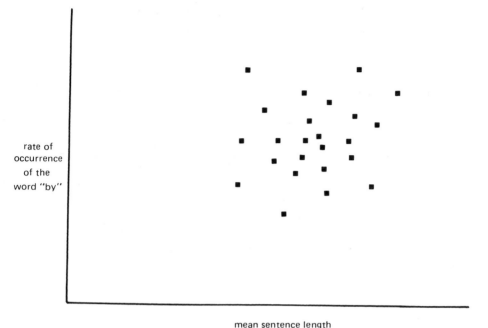

mean sentence length

FIGURE 2   Plot of mean sentence length and rate of occurrence of the word "by" for a number of essays.

procedure on data consisting of many more than two variables for each object or individual under consideration; this implies that it will be necessary to consider how we should measure the distance between pairs of objects given a set of variable values common to both.

DISTANCE MEASURES

The most familiar distance measure is, of course, *Euclidean*; it is familiar because we live in a locally at least, Euclidean universe, and when we talk of the distance between point A and B in everyday conversation, we generally mean Euclidean distance. To introduce this measure, consider Figure 3. The Euclidean distance between the two points is indicated and from Pythagoras's theorem it is clear that it takes the following value:

$$d_{AB} = \sqrt{(2.5 - 1.5)^2 + (2.5 - 1.5)^2} = 1.41$$

In general if we have two points with coordinates (x_i, y_i) and (x_j, y_j), the Euclidean distance between them, which we shall represent as d_{ij}, is given by the expression

$$d_{ij} = \sqrt{(x_i - x_j)^2 + (y_i - y_j)^2}$$

This expression is easily extended to the situation where we have, say, p variable values for each of the objects under study:

$$d_{ij} = \sqrt{(x_{i1} - x_{j1})^2 + (x_{i2} - x_{j2})^2 + \cdots + (x_{ip} - x_{jp})^2}$$

where x_{ik} and x_{jk} for $k = 1,...,p$, represent the variable values for objects i and j, respectively. This expression may look rather fearsome, but it will become less so if we consider a small numerical example. So let us suppose that we have made measurements of four variables on each of six individuals with the following results:

Individual	Variable 1	Variable 2	Variable 3	Variable 4
1	6	3	4	5
2	2	3	5	4
3	5	4	6	3
4	9	1	1	8
5	8	2	0	9
6	8	0	1	8

The Euclidean distance between individuals 1 and 2 is given by

$$d_{12} = \sqrt{(6 - 2)^2 + (3 - 3)^2 + (4 - 5)^2 + (5 - 4)^2}$$
$$= 4.24$$

and between individuals 1 and 3 by

$$d_{13} = \sqrt{(6 - 5)^2 + (3 - 4)^2 + (4 - 6)^2 + (5 - 3)^2}$$
$$= 3.16$$

and so on. The distances for all pairs of individuals may be displayed in the form of a *distance matrix*, **D**.

$$
\mathbf{D} = \begin{array}{c} \\ 1 \\ 2 \\ 3 \\ 4 \\ 5 \\ 6 \end{array}
\begin{array}{c}
\begin{array}{cccccc} 1 & 2 & 3 & 4 & 5 & 6 \end{array} \\
\left[\begin{array}{cccccc}
0.00 & 4.24 & 3.16 & 5.57 & 6.08 & 5.57 \\
4.24 & 0.00 & 3.46 & 9.22 & 9.33 & 8.77 \\
3.16 & 3.46 & 0.00 & 8.66 & 9.22 & 8.66 \\
5.57 & 9.22 & 8.66 & 0.00 & 2.00 & 1.41 \\
6.08 & 9.33 & 9.22 & 2.00 & 0.00 & 1.41 \\
5.57 & 8.77 & 8.66 & 1.41 & 1.41 & 0.00
\end{array}\right]
\end{array}
$$

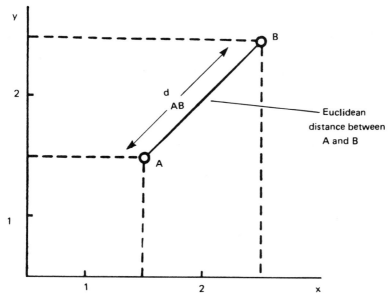

FIGURE 3   Euclidean distance in two dimensions.

Two important points to notice about this matrix are (1) that it is *symmetric* (i.e., the distance d_{ij} is equal to d_{ji}), and (2) the entries in the main diagonal are zero (i.e., the distance of an individual from himself or herself is zero). Such distance matrices are generally the first step in the development of a classification for a set of objects or individuals, as we shall see in the next section.

Although Euclidean is the most commonly used distance measure in clustering applications, other measures are occasionally used, for example, the so-called *city block distance*. This measure is illustrated in Figure 4, and in general is given by the expression

$$d_{ij} = |x_{i1} - x_{j1}| + |x_{i2} - x_{j2}| + \cdots + |x_{ip} - x_{jp}|$$

where $|x_{ik} - x_{jk}|$ means the absolute value of $x_{ik} - x_{jk}$ (i.e., the numerical value ignoring the sign).

FROM DISTANCE MATRIX TO CLUSTERS

Now we have to consider how we can construct a set of clusters for our objects or individuals from the information contained in the distance matrix. A large number of methods are available, but here we shall

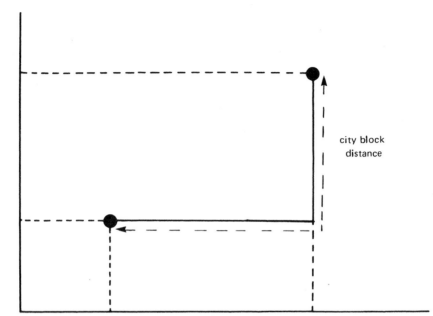

FIGURE 4   City block distance in two dimensions.

concentrate on just two, one of which is known as *group average clustering* and the other as *k-means clustering*.

The first of these is an example of a large class of clustering methods known collectively as *agglomerative hierarchical techniques*. These methods proceed by a series of steps in which objects or groups of objects are successively clustered together into larger and larger clusters; they begin with as many clusters as there are objects under investigation and end with a single cluster containing all objects. Investigators have then to choose which of the classification schemes produced best fits their data. At any particular stage the two objects or two groups of objects that are closest to each other are fused into a single cluster. It is the different possibilities for defining distance between groups that gives rise to the different methods of this class. The one we shall discuss, group average clustering, defines this distance as the average of all the distances between pairs of individuals, one from one group and one from the other. This is illustrated in Figure 5.

As an example of the operation of group average clustering we shall apply it to the distance matrix met in the preceding section. The first step is to examine this distance matrix for its smallest entry and form a group

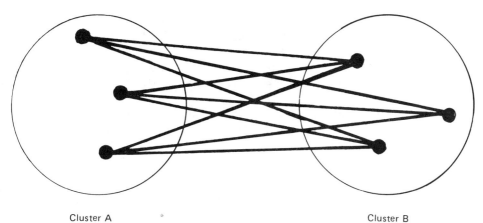

Cluster A   Cluster B

FIGURE 5   Group average distance for two groups. Group average distance between clusters A and B is the average of the pairwise distances shown.

of the two individuals or objects concerned. Examination of **D** shows that the smallest entries are d_{46} and d_{56}, both of which equal 1.41. Either of these pairs might be taken to form a group, so let us take individuals 5 and 6. Consequently, at the end of this stage we have five clusters, four of which contain a single individual and one of which contains individuals 5 and 6.

Now we need to construct a new distance matrix showing inter-individual distances and individual-to-cluster distances. For example, the group average distance between individual 1 and the cluster containing individuals 5 and 6 is given by

$$d_{1(56)} = \tfrac{1}{2}(d_{15} + d_{16}) = 5.83$$

Consequently, our new distance matrix is

$$\mathbf{D}_1 = \begin{array}{c} \\ 1 \\ 2 \\ 3 \\ 4 \\ (56) \end{array} \begin{array}{ccccc} 1 & 2 & 3 & 4 & (56) \\ \begin{bmatrix} 0.00 & 4.24 & 3.16 & 5.56 & 5.83 \\ 4.24 & 0.00 & 3.46 & 9.22 & 9.05 \\ 3.16 & 3.46 & 0.00 & 8.66 & 8.94 \\ 5.55 & 9.22 & 8.66 & 0.00 & 1.71 \\ 5.82 & 9.05 & 8.94 & 1.71 & 0.00 \end{bmatrix} \end{array}$$

The smallest entry in this matrix is 1.71, which is the distance between individual 4 and the cluster containing individuals 5 and 6. So at this stage individual 4 joins this cluster and a further distance matrix, $\mathbf{D}_2$, is computed:

$$\mathbf{D}_2 = \begin{array}{c} \\ 1 \\ 2 \\ 3 \\ (456) \end{array} \begin{array}{cccc} 1 & 2 & 3 & (456) \\ \left[\begin{array}{cccc} 0.00 & 4.24 & 3.16 & 5.74 \\ 4.24 & 0.00 & 3.46 & 9.11 \\ 3.16 & 3.46 & 0.00 & 8.85 \\ 5.74 & 9.11 & 8.85 & 0.00 \end{array}\right] \end{array}$$

The smallest entry in $\mathbf{D}_2$ is 3.16, the distance between individuals 1 and 3, so these are now placed together in a cluster. At this stage, therefore we have three clusters:

	Cluster 1	Cluster 2	Cluster 3
Members	2	1,3	4,5,6

A new distance matrix, $\mathbf{D}_3$, is now calculated. One of the entries in this matrix will be the distance between clusters 2 and 3. According to the definition used by group average clustering, this will take the value

$$d_{(13)(456)} = \tfrac{1}{6}(d_{14} + d_{15} + d_{16} + d_{34} + d_{35} + d_{36})$$
$$= 7.29$$

$\mathbf{D}_3$ is therefore as follows:

$$\mathbf{D}_3 = \begin{array}{c} \\ 2 \\ (13) \\ (456) \end{array} \begin{array}{ccc} 2 & (13) & (456) \\ \left[\begin{array}{ccc} 0.00 & 3.85 & 9.11 \\ 3.85 & 0.00 & 7.29 \\ 9.22 & 7.29 & 0.00 \end{array}\right] \end{array}$$

The smallest entry in $\mathbf{D}_3$ is 3.85, which is the distance between individual 2 and the cluster containing individuals 1 and 3. Therefore, individual 2 joins this cluster and at this stage we have the following two clusters:

	Cluster 1	Cluster 2
Members	1,2,3	4,5,6

Finally, these two clusters are joined.

The series of steps in this procedure may be represented in the form of a *dendrogram*, which is a diagram showing which objects and groups of objects have been joined together at each stage. The dendrogram for this example is shown in Figure 6. Examination of this diagram is sometimes helpful in deciding how many groups an investigator should have in his or her classification scheme. Here the data appear to divide fairly naturally into two groups, individuals 1, 2, and 3 and individuals 4, 5, and 6, since there is a relatively large change in the dendrogram on joining these two groups compared to fusions made earlier.

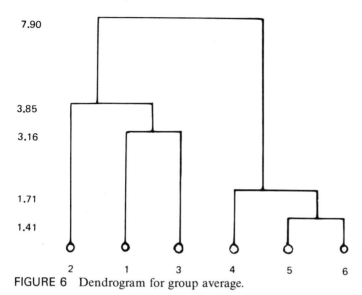

7.90

3.85

3.16

1.71

1.41

2   1   3   4   5   6

FIGURE 6   Dendrogram for group average.

The k-means clustering method does not produce a series of solutions in the same way as group average clustering and other hierarchical techniques. Instead, an investigator specifies the number of groups required in a classification scheme and the k-means method divides the individuals into this number of groups in such a way as to minimize the average within group distance between pairs of individuals. The idea behind this approach is, of course, to ensure that individuals within the same group are relatively close to each other. The operation of the k-means approach can be illustrated using the distance matrix of the preceding section, by supposing that we wish to divide the six individuals into two groups. The simplest method of choosing the two groups which leads to the lowest value of the average within-group distance would be simply to examine its value for every possible division of the six individuals into two groups. For example, a division $(1,2), (3,4,5,6)$ gives a value of

$$\frac{4.24 + 8.66 + 9.22 + 8.66 + 2.00 + 1.41 + 1.41}{7} = 5.09$$

whereas a division (1) $(2,3,4,5,6)$ gives a value 6.21. For a small-scale example such as presented here, this approach is quite possible. However, in general it is quite impracticable even with the fastest available or projected computers, since the number of possible divisions is, for even a moderate number of objects, extremely large. For example, for 19 objects there are a total of 1,709,003,480 divisions into eight groups. To overcome this problem the k-means algorithm starts with an arbitrary division of the data into the required number of groups and then considers whether moving an object from the group it is in to another group would cause a decrease in the average within-group distance. If so, the object is moved. Each object is considered in turn and the procedure ends when no move of a single object causes an improvement. For our small example, let us suppose that we begin with the division (1,2), (3,4,5,6); we know from the above that for this division the average within-group distance is 5.09. Consider moving object 2 from the first to the second group, to give the division (1), (2,3,4,5,6); this has an average within-group distance of 6.21. Since this is larger than the previous value object, 2 is not moved. Now consider moving object 3 from the second to the first group; the average within-group distance becomes

$$\frac{4.24 + 3.16 + 3.46 + 2.00 + 1.41 + 1.41}{6} = 2.61$$

Consequently, the division (1,2,3), (4,5,6) represents a distinct improvement and the procedure would now continue from here.

For more details of both group average clustering and the k-means algorithm, readers are referred to Everitt (1980).

EXAMPLES OF CLUSTERING APPLICATION

The increase in interest in cluster analysis methods over the last 10 to 15 years has been dramatic. The methods have now been used in the analysis of problems as diverse as classifying puberty rites of North American Indians, studying the penile morphology of New Guinea rodents, investigating the process by which cockroaches behaviorally recover from cold stress, and assessing extracts from Plato and Jane Austen. Clearly, the ability to find a use for cluster analysis in a particular research study is limited only by the imagination of the researcher. (This is not to imply, of course, that all such uses would be either very sensible or very informative.) Here we shall describe in a little more detail a number of applications of cluster analysis techniques.

Classification of Suicide Attempters

From a number of clinical studies it is evident that people who attempt suicide are very heterogeneous, spanning a range of severity of attempt, apparent motivation, previous history, and other phenomena. Clearly, the treatment of such patients and investigations of the cause of attempted suicide would be greatly enhanced if a reasonable classification of the people involved could be produced. With this in mind, Paykel and Rassaby (1978) studied 236 suicide attempters presented at the main emergency service of one city in the United States. From the pool of available variables, 14 were selected as particularly relevant to classification and used in the analysis. The variables included age, number of previous suicide attempts, severity of depression and hostility, and a rating of the overall severity of the attempt in terms of medical consequences and intention to end life; in addition, a number of demographic characteristics were recorded. A number of clustering techniques were applied to the data and the final classification was one with three groups, having the following general characteristics.

Group 1: Patients taking overdoses, on the whole showing less risk to life, less psychiatric disturbance, and more evidence of interpersonal rather than self-destructive motivation.

Group 2: Patients in this group made more severe attempts, with more self-destructive motivation, and by more violent methods than overdose.

Group 3: Patients in this group had a previous history of many attempts and gestures, their recent attempt was relatively mild, and they were overtly hostile, engendering reciprocal hostility in the psychiatrist treating them.

Such a classification might prove extremely valuable as a basis for future studies into the causes and treatment of attempted suicide. Readers interested in other applications of cluster analysis in psychiatry are referred to Paykel (1971, 1972).

Planning the Needs of the Handicapped

The second example, described by Jones (1979), involves a study in which the aim was to devise a planned program of identification and assessment of individual handicapped people in the city of Birmingham in the United Kingdom that would enable their needs to be met gradually, over a specified period of years. The method used to identify the groups of individuals with particular forms of handicap was that of cluster analysis. The goal was to classify the handicapped people into a reasonable number of groups, in such a way that all the people in the same group would have similar characteristics, and people in different groups would have significantly different characteristics. Once the clusters had been determined, the needs of the typical representative of each group could then be assessed. It was then possible to predict the total service requirements and costs involved by scaling up the requirements of the representatives according to the number of people within a group.

CONCLUSION

Classification is a fundamental process in many areas of science, in particular in those disciplines that are still at the stage of developing theories and explanations for observed phenomena. Consequently, numerical methods that lead to classifications have proved extremely popular in such areas as sociology, psychiatry, and psychology, and many applications of the methods have been reported. However, it should be remembered that many problems remain unsolved, and a somewhat cautious approach is needed at this stage to avoid misleading and unhelpful results which could have the effect of throwing potentially very useful methods into disrepute.

REFERENCES

Everitt, B. S. (1980). *Cluster Analysis.* 2nd ed. Heinemann Educational Books, London.

Jones, B. (1979). Cluster analysis of some social survey data. *Bull. App. Stat. 6*, 25-56.

Paykel, E.S. (1971). Classification of depressed patients; a cluster analysis derived grouping. *Brit. J. Psych . 118*, 275-288.

Paykel, E.S. (1972) Depressed typologies and response to amitriptyline. *Brit. J. Psych. 120*, 147-156.

Paykel, E.S. and Rassaby, E. (1978). Classification of suicide attempters by cluster analysis. *Brit. J. Psych. 133*, 45-52.

Struve, O. and Zebergs, V. (1962). *Astronomy of the Twentieth Century.* MacMillan, New York.

5

Pattern Recognition, or How to Tell It's One of Those

David J. Hand
University of London
London, England

What do the following exercises have in common?

1. Recognizing spoken words
2. Deciding from which ancient civilization a pottery fragment originates
3. Diagnosing a disease
4. Choosing the most suitable subject for a student to study
5. Classifying crops from satellite pictures
6. Detecting particular types of subatomic particles in bubble chamber photographs
7. Target recognition
8. Fingerprint identification
9. Recognizing printed words

Perhaps the most striking observation is not that they have something in common, but that they have a great many differences. The objects are different in each case (words, potsherds, diseases, etc.); some cases involve only a few classes (2 and 5, perhaps), while others involve a large number (1 and 8); for some of them speed is vital (1 and 7) while for others it does not matter (2 and 4). However, despite these differences they do have a fundamental underlying common nature: they all involve classifying something into a number of classes. This abstract operation of classification is the same, even if it masquerades under names such as

recognition, diagnosis, or identification. *Pattern recognition* is the collective term for a set of formal techniques for carrying out such classifications. Methods of pattern recognition enable us to train computers to carry out the exercises listed above.

Why might we wish to do that? In some cases it might be because we can get more reliable and consistent results using formal methods—the subjectivity of a human's decisions is removed. This is the case with medical diagnosis, for example. In other cases it might be because a human being cannot carry out the classifications quickly enough (as in classifying each cell of a 1000 by 1000 cell satellite photograph—with a new picture arriving every minute or so) or to avoid intrinsically boring operations with their attendant high risk of error (sorting through tens of thousands of bubble chamber photographs or fingerprints). Other situations involve straightforward commercial considerations: computers that can understand the spoken word will have a tremendous range of applications.

Pattern recognition techniques can be divided into two types (one might say that there are two classes of pattern recognition techniques): *unsupervised* pattern recognition and *supervised* pattern recognition. Unsupervised pattern recognition is also called *cluster analysis*. In cluster analysis nothing is assumed known beforehand about the class structure. The problem is to see if such a structure exists and, if it does, to identify the distinguishing characteristics of the classes. This is a very interesting and very demanding type of problem, opening up all sorts of questions, both methodological and deeply philosophical. If one is to use cluster analytic methods effectively, one is forced to think very carefully about the precise nature of the area one is studying.

Supervised pattern recognition, on the other hand, confronts those problems for which some class structure is known to exist. In the examples above, we know that there are different words and a human being could tell us what a particular word was; we know that there are different diseases and a physician could identify a particular instance; and so on. One consequence of this is that we can find examples of objects from each class. We can then use these examples to obtain a more formal description of the class structure. This description will, in turn, enable us to classify new objects. In what follows the discussion will be concerned exclusively with supervised pattern recognition (which henceforth will simply be called pattern recognition).

Although we have identified a common aim to the above, it is obvious that there are fundamental differences. How can we use the same kinds of methods to classify words, diseases, or archaeological specimens?

We might begin to try to answer this question by speculating about what subconscious methods a human being uses. Perhaps the first thing done is that certain features of the object to be classified are identified. So, for example, in the case of recognizing friends' faces, the features will be facial characteristics such as shape of nose, eye color, and hair length. For speech recognition the features will be characteristics of frequency, intensity, and duration of the words. For potsherds they might be color and texture, and for diseases such things as blood pressure and biochemical indicators. Presumably the human brain then in some way compares these features with the corresponding features of objects with known class memberships. The new object is then assigned to the same class as the objects it most closely resembles, where "resemblance" is in terms of the chosen features. Thus a spoken word is recognized as being a particular instance of enunciations of that word. A face is classified into the class of views of a particular friend. A student is identified as being similar to other students who did well in a particular subject.

Whether or not this is the way the brain works, this is the basic principle behind pattern recognition methods. We illustrate by taking a concrete example [the data below are reconstructed from Figure 6 of Strandjord et al. (1973)].

The aim is to produce a simple diagnostic rule that will enable us to

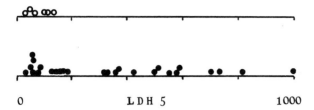

0   L D H 5   1000

o   hepatitis secondary to infectious mononucleosis

•   acute infectious hepatitis

FIGURE 1   LDH5 activity for the classified patients.

distinguish between two diseases: acute infectious hepatitis and hepatitis
secondary to infectious mononucleosis. We have a sample of patients
from each disease class available, and measurements can be taken of
whatever features we think might be appropriate.

As a first feature we consider a measure of the activity of lactate
dehydrogenase isoenzyme-5 (LDH5) as a possible characteristic that
distinguishes between the classes. Figure 1 shows a plot of the values of
LDH5 for the patients in our samples.

How can this be used to diagnose new patients based on a
measurement of their LDH5 activity? One thing is clear—that anyone
with an LDH5 score greater than 200 is unlikely to be suffering from
secondary hepatitis. Thus if a patient scores greater than 200, we can
diagnose him or her as suffering from acute infectious hepatitis.

For patients who score less than 200, however, things are not so clear.
The lower their score, the more confident we will be that they have
secondary hepatitis, but we cannot by 100% certain. (Neither, as it
happens, can we be 100% confident that they have acute infectious
hepatitis if they score above 200. Our initial sample is, after all, only a
sample and there might be occasional unusual patients with secondary
hepatitis who score higher than 200. By chance our sample has not got
any of these.)

Perhaps we can find some other feature that does better than LDH5.
Figure 2 shows an analogous situation using a measure of the activity of
lactate dehydrogenase isoenzyme-3 (LDH3). This does, indeed, seem
better.

We could go on like this, searching for new features in the hope that
we will find one that is very good—but there is a better way.

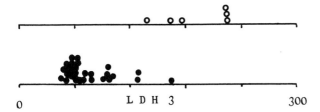

0   L D H 3   300

FIGURE 2   LDH3 activity for the classified patients.

The features, such as LDH5 and LDH3 in this example, are different. They may have a lot in common, but they each also have something the other does not have. In other words, they each contain extra information relevant to the diagnosis problem—informational additional to that contained by the other. So maybe we can use them together.

Figure 3 shows just this. There is a slight complication. Whereas before we diagnosed future patients according to whether their score was above or below a particular value—that is, according to on which side of a point they fell in Figures 1 and 2—now we must see on which side of a line they fall. Using our Figure 3, future patients whose points fall above the line will be diagnosed as having secondary hepatitis, and those with points below the line will be diagnosed as having acute infectious hepatitis.

Note that none of the initial sample will be misclassified by this rule. This is satisfying, but it does not happen very often. It can arise for two reasons: first, because the two populations from which the initial sample comes do not overlap in any way and can be perfectly separated by a straight line (this is extremely rare and is unlikely to occur except in contrived situations), and second, because the initial sample just happened to come out in that way. If we had not been so lucky, we might have ended up with a situation such as that shown in Figure 4. Here the sample does not permit us to find a line that perfectly separates the two classes. The best we can do is as shown, giving a rule that misclassifies one of the initial sample points.

It is also important to note that even if the initial sample is perfectly classified, this does not mean that future patients will all be correctly classified. The initial sample is, after all, only a *sample*—it will not describe all aspects of the two population distributions perfectly. More generally, following from this, we can expect the classification perform-ance on future patients to be less good than that on the initial sample, however many or few of the initial sample are misclassified.

The larger the initial sample (which, we can now reveal, is called the *design* set or *training* set because the classifier is designed or trained using

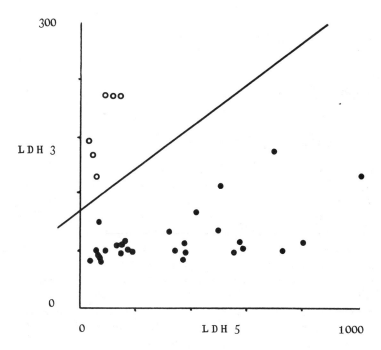

FIGURE 3   Using LDH5 and LDH3 simultaneously.

it), the more closely it will represent the true distributions of the features and the better will be the classifier. We can go on adding more and more features to reduce the number of design set elements misclassified. If the design set is representative of future objects to be classified, a smaller design set misclassification rate can be expected to yield fewer misclassations on future objects.

There are complications, however, and this is where pattern recognition becomes interesting.

Figures 1 and 3 show simple cases with only one or two features. But how can we draw a figure and select the best separating surface when there are 10 or 100 features? We cannot, of course, and we have to resort to more formal mathematical approaches. There are two broad types of these. The first type decides on the general shape of the separating surface (is it to be a flat plane, a quadratically curved surface, or what? A curved line used in Figure 4 would lead to all of the design set points being correctly classified) and then estimates the precise position of the surface by minimizing some criterion. For example, an obvious criterion is to try to find the position that results in the fewest design set elements being misclassified, but there are many other possibilities. The second type is

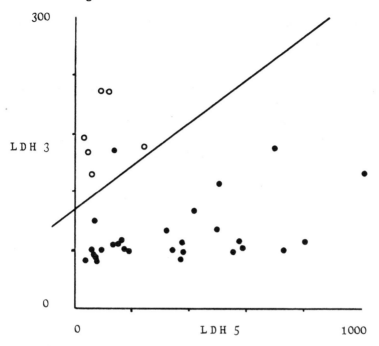

FIGURE 4   Using LDH5 and LDH3 simultaneously.

based on the idea of assigning a new object to the same class as the design set object that is nearest to it—where distance is measured in terms of the features. To a certain extent the two types of method have complementary properties and are suitable for different problems. One of the exciting areas of application of pattern recogntion methods is deciding what kind of technique is appropriate.

Another more subtle complication arises when we consider the choice of features to use. It is true that adding new features cannot increase the misclassification rate of design set objects. But our concern is with future objects. Unfortunately, it has been found that for a fixed design set, as the number of features increases, the future misclassification rate at first decreases but then begins to increase. This is a subtle phenomenon, involving several different factors, but the basic reason is that as the number of features increases, the fixed size design set describes the distribution of possible future objects less and less well. When the description is being degraded faster than information is being added by extra features, the misclassification rate starts to increase. In addition to this, there are simple practical reasons, such as cost and classifier speed, for wishing to avoid using too many features.

Given that we do not want to use too many features, the question naturally arises as to which we should choose. Feature selection occupies a very large area within pattern recognition and is still the focus of much concentrated research. Important new discoveries are still being made.

Other areas within pattern recognition also raise interesting questions. For example, how should we estimate future misclassification rate? We have already commented that the design set misclassification rate tends to underestimate the future rate. Should we always force a classification, or can we reduce the number misclassified by sometimes delaying a decision ? Are there special problems associated with some subjects—for example, with psychological data, which are mostly measured only on ordinal scales? What do we do if the design set classifications are not 100% correct? And so on.

Pattern recognition has been applied in a colossal number of areas—as the diversity of fields listed at the beginning of this chapter illustrates. Moreover, new applications are being found all the time—the scope is limited only by the imagination. Some applications, especially, such as computer recognition of speech and writing and computer-assisted medical diagnosis, will have a major impact on our lives in the near future. It is an exciting area for research, in both its applications and methodological issues. The latter, the methodology, changes as new tools—bigger and faster computers—become available. Pattern recognition is one of the areas of statistics that has been most drastically affected by computers. Indeed, apart from a few early applications of classical discriminant analysis, it would not be stretching things to say that its development has depended almost completely on the advent of the computer. Conversely, in many areas in which it has been applied, the need was barely perceived until computers became available. Its existence demonstrates once again the mutual interrelationship of science and technology: each advances using the latest ideas and developments of the other as a stepping-stone.

FURTHER READING

Hand, D. J. (1981). *Discrimination and Classification*, Wiley, Chichester, West Sussex, England.

Strandjord, P. E., K. J. Clayson, and R. J. Roby (1973). Computer assisted pattern recognition and the diagnosis of liver disease. *Hum. Pathol. 4*, 67–77.

6

Finders Keepers

Alan L. Tyree
University of Sydney
Sydney, Australia

A RECENT CASE

On November 15, 1978, Alan George Parker was an outgoing passenger on a British Airways flight out of Heathrow Airport in London. [1]. While awaiting his flight in the international executive lounge, he found a gold bracelet lying on the floor. Being an honest man, Mr. Parker handed the bracelet to a British Airways official together with a note of his name and address and a request that the bracelet be returned to him in the event that the owner of the bracelet could not be found. Although the bracelet was never claimed by the true owner, British Airways did not return it to Mr. Parker but sold it and kept the proceeds, which amounted to some 850 pounds sterling. Mr. Parker thus involuntarily became the most recent participant in a long list of persons who have contested the right to an object that neither party owns but which both claim in the absence of the rightful owner.

In this chapter a series of these so-called "finders cases" will be analyzed using quantitative methods. Since it is not expected that the majority of readers will have had legal training, it is necessary to include a brief description of the way in which law is made and develops in a legal system such as ours, which is inherited from the common law system of England.

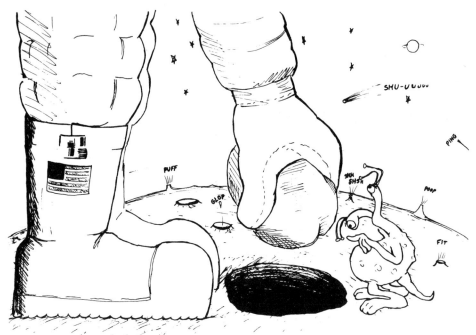

HOW THE LAW IS MADE

The Common Law is not, as widely supposed by nonlawyers, a set of rules that are applied to fact situations which arise from time to time * The development of the law is on a case-by-case basis, a procedure that is far more inductive than deductive. Although statutes passed by legislatures may appear to contradict this, even in this circumstance the development is the same, although the analysis becomes somewhat more complicated.

Common law development proceeds through the mechanism known as "the doctrine of precedent." This doctrine requires that lower courts apply *as law* certain propositions and findings of courts above them in the judicial hierarchy. The "certain propositions" are carefully (and sometimes confusingly) circumscribed so that superior courts are not unrestrained lawmakers. The chief restrictions are that the judgments of the higher courts are binding only when the "relevant" facts of the case before the lower court are "sufficiently" analogous to those which were

---

*"Common Law" as used here means the system of law derived from England in which the doctrine of precedent operates. In short, it refers to "judge-made" law.

before the higher court and that the proposition of the higher court which is being invoked was "necessary" for the decision reached by the higher court. The part of the judgment that is binding as law is known as the *ratio decidendi* of the case. All else that the court may say is *obiter dicta*.

It is clear that the words in quotation marks leave a wide scope for interpretation. This problem of interpretation is part of the task that confronts a practicing lawyer when faced with a new dispute. The facts will seldom, if ever, be identical with cases that have been decided by courts. The lawyer proceeds by analogy, arguing that the facts of the dispute are similar to case X (which was decided in a way favorable to the interest of his client) and different from case Y (which was decided in a way inimical to the interests of his client). This last type of argument is known in the trade as "distinguishing" a case.

To prevent the entire system from becoming a hopeless tangle, those with a more theoretical bent attempt to extract fundamental principles from a body of case law. If this attempt is successful, order may be introduced into the area, with the result that novel problems may be more readily analyzed and resolved without the need for expensive and prolonged litigation.

SIMILARITY IN THE COMMON LAW

The way in which the common law develops and the way in which lawyers talk about cases both suggest that there is a notion of similarity among cases. Indeed, one of the leading textbooks on the matter, Cross, *Precedent in English Law* [2] begins with the proposition: "It is a basic principle of the administration of justice that like cases should be decided alike,"

It also seems natural to define measures of similarity in terms of the factual content of cases. Mention has already been made of the importance which lawyers place on factual distinctions when attempting to "distinguish" cases in arguments. There is also theoretical support for the use of factual content as a basis for defining a measure of similarity. Thus, in discussing the doctrine of precedent, Cross [2] cautions that "judgments must be read in the light of the facts of the cases in which they are delivered." Some writers have gone even further and have attempted to define the binding part of this judgment, the *ratio decidendi*, solely in terms of factual content and its relation to the decision in the case.

Later in the chapter we describe the construction of a quantitative measure of similarity on a small body of case law and then use that measure in an analysis of the law. It would be possible for the interested reader to turn directly to those sections, but for a full understanding of the method it is desirable to have some acquaintance with the factual and legal context. Consequently, the next section is devoted to a brief discussion of problems involving the legal rights of finders.

FINDERS CASES

The process of common law law-making and the way in which it leads to complications may be illustrated by a series of cases which are concerned with the rights of various parties to movable property where the "true owner" is not known. A typical example of such a situation is when personal property is lost by the owner and found by another person. Because of this, the line of cases is commonly referred to as "finders cases," although it will be noted that in the cases discussed the article would not necessarily be "lost" in common language.

The earliest finder case is *Armory* v. *Delamirie*, 1 Strange 505, decided in 1721. The plaintiff was a chimney sweep who found a jewel. Although the report does not say, most readers presume that the jewel was found in a chimney while the plaintiff was at work. Although this might or might not be true, it is important to keep in mind that the defendant was neither the owner of the jewel nor the owner of the chimney, but was a jeweler to whom the plaintiff had taken the jewel for cleaning and/or appraisement.

The jeweler removed the stone from the mounting and refused to return it to the chimney sweep, perhaps salving his conscience by offering that unfortunate gentleman three half-pence for the stone, take it or leave it. The court ordered that the jeweler return the stone or, alternatively, that the jury should listen to evidence to prove "what a jewel of the finest water that would fit this socket would be worth" and ordered the jeweler to pay that sum in compensation.

The case also provides an example of statements made by a court which would not bind later courts, for it was said that "the finder of a jewel, though he does not by such finding acquire an absolute property of ownership, yet he has such property as will enable him to keep it against all but the rightful owner . . . ." Notice that the jeweler in this case had no claim whatsoever to the jewel, that he merely grabbed it. It was thoroughly unnecessary for the court to make any comment concerning the finder's rights against other parties, parties that might have some legitimate claim on the jewel which might fall short of complete ownership. The words quoted must be *obiter dicta*, since they are quite unnecessary to resolve the dispute before the Court.

The point may be illustrated by a consideration of the facts in *Hannah* v. *Peel* [1945] 1 KB 509. Major Peel acquired a house in Shropshire which he left unoccupied until near the end of 1939, when it was requisitioned by the government for army use. In 1940 Lance Corporal Hannah was stationed in the house. Hannah was adjusting the blackout curtains when he felt something lying on the top of a window frame which he first thought to be a piece of dirt or loose plaster. In fact, the report tells us that he dropped the object on an outside window ledge and that it was not until the following morning that he learned that it was a brooch covered with dirt and cobwebs.

On the advice of his commanding officer, Hannah handed the brooch over to the police, receiving a receipt for it, for the purposes of finding the true owner. No owner had been found by mid-1942 when the police, for reasons not explained, turned the brooch over to Major Peel, who promptly sold it for £66 to a London jeweler, who resold it the following month for £88. Peel had offered Hannah a reward which was declined, Hannah insisting at all times that he was entitled to the brooch.

Notice how different this case is from *Armory* v. *Delamirie*, and how inappropriate it would be to resolve the dispute between Hannah and Peel on the basis of the words quoted above, words spoken when the mind of the court was directed toward quite different circumstances. Whereas the jeweler in *Armory's* case had a claim scarcely better than that of a thief, the justice in *Hannah's* case is considerably harder to perceive.

In the event, the Court found in favor of Hannah, the finder, and against Peel, the owner (but not the occupier) of the premises. Notice that Hannah was not a trespasser; he had every right to be in the room. Further notice that the brooch was not fastened or fixed to the premises in any way. Any of these variations are thought to make the finder's case weaker.

Indeed, a case that is at the other extreme from *Armory* v. *Delamirie* is worth a brief consideration. In *Moffatt* v. *Kazana* [1969] 2 QB 152, a man named Russell purchased a house and lived in it from early 1951 to early 1961, when the house was sold to Kazana. Three years later, Kazana was having some restoration work done on the house. Some bricks were dislodged from the chimney, exposing a tin which was found to contain nearly £200 in £1 notes. The tin was taken by the workman who discovered it to the police, who ultimately returned it to Kazana.

On the basis of the evidence of the trial, the court came to the conclusion that the tin had been deposited in the bungalow by Russell.

Unfortunately, Russell's direct evidence was unavailable since he had died before the trial, but the court did accept that Russell had simply forgotten about the tin and that the buyers had never known of its existence.

Under these circumstances, it was easy for the court to find that the money still belonged to Russell. He had merely forgotten about it, a lapse which in no way disentitled him to any of his property. He was, in the language of the earlier courts, the "true owner" of the tin and its contents.

It will be seen that there are many variations on this theme which are possible and, in many cases, there is no clear sense of justice to guide a court to a decision. The contest is often between two parties whose claim to the goods is based on circumstances alone. Both are seeking a windfall which is quite fortuitous. Yet a decision must be made, for in the common law there is seldom any power for the court to order a division of the windfall.

If a new case were to arise, what principles might be used to predict the outcome or to advise the disputants as to the likelihood of success should they wish to litigate their claim? That the task of identifying such principles may not be an easy one may be gathered from the following description of the cases which was given by an experienced judge [3]: "These cases have long been the delight of professors and text writers, whose task it is to reconcile the irreconcilable."

IS THERE A COMMON NOTION OF SIMILARITY?

Notice that the quotation from Cross above presupposes the existence of a common notion of similarity between cases, at least common to all lawyers. This assumption does not appear unreasonable in view of the relatively uniform cultural and educational background of modern lawyers.

In 1976, I conducted a rudimentary experiment to test this presupposition. Four subjects who were at the time law students who had completed an introductory course on the legal system were asked to order 28 pairs of finders cases according to. their own notions of similarity among case pairs.* The cases used were eight finders cases which had been used as course material. The students were required to be thoroughly familiar with the cases and to have formulated their views as

---

*The subjects were students who had completed LAWS101: *The Legal System* at Victoria University of Wellington (New Zealand) during the 1975 session.

TABLE 1   Cases

A	*Armory* v. *Delamirie* (1721) 1 Strange 505
B	*Bridges* v. *Hawkesworth* (1851) 21 LJQB 75
C	*Elwes* v. *Briggs Gas Co* (1886) 33 Ch D 562
D	*Hannah* v. *Peel* [1945] 1 KB 509
E	*Corporation of London* v. *Yorkwin* [1963] 1 WLR 982
F	*Moffatt* v. *Kazana* [1969] 2 QB 152
G	*South Staffordshire Water Co* v. *Sharman* [1896] 2 QB 44
H	*Yorkwin* v. *Appleyard* [1963] 1 WLR 982

to the law represented by them. Each pair of cases was written on a separate card, the cards thoroughly shuffled, and instructions issued to order the deck so that the two most similar cases were on the first card, the two least similar on the final.

The cases used are listed in Table 1; the rankings generated by the four subjects are listed in Table 2, together with a ranking generated by a measure of distance between cases which is defined in the next section. For convenience, all the cases are represented by letters as indicated in Table 1.

The level of agreement exhibited by the four subjects is measured at $W = 0.83$ using Kendall's coefficient of concordance and is significant at the 1% level (see [4]). Closely related to that measure, Table 3 shows the Spearman rank correlation coefficient for each pair of subjects, including the ranking generated by the distance measure. Again, all are significant at the 1% level.

This "experiment" is not claimed to be rigorous or conclusive, but it does provide an indication that there is a common notion of similarity, at least among this very restricted group of subjects. Although this experiment must be considered to be very preliminary, when combined with the prevailing jurisprudential notions of similarity it provides a basis to continue to the next stage of the program, namely, attempting to define a measure of similarity on the cases which will reflect the intuitive notion.

QUANTIFICATION

Rather than construct a measure of similarity, a distance measure is defined on the body of case law through the use of yes-no responses to questions which represent the factual content of the cases [5]. The factual content of the cases is elicited by means of a series of questions which reflect the facts which have been considered as relevant in various

TABLE 2   Rankings of the 28 Pairs of Cases by the Four Subjects and
the Ranking Generated by the Metric

Case	Subjects				
pairs	1	2	3	4	5
A–B	24	16	21	23	9
A–C	26	22	28	25	27
A–D	21	15	23	24	15
A–E	25	24	15	26	23
A–F	28	28	27	28	28
A–G	27	20	26	27	24
A–H	23	23	25	19	19
B–C	5	11	9	11	11
B–D	2	1	3	3	2½
B–E	3	21	11	14	16
B–F	17	26	20	22	22
B–G	7	8	6	9	7
B–H	20	13	16	21	11
C–D	8	9	5	6	5½
C–E	4	6	2	1	2½
C–F	11	17	18	20	20½
C–G	1	2	8	5	8
C–H	10	3	12	10	13
D–E	14	12	7	8	11
D–F	16	27	14	15	18
D–G	9	7	4	4	4
D–H	19	10	10	12	5
E–F	12	25	24	16	26
E–G	13	5	13	7	14
E–H	22	14	17	13	17
F–G	15	19	22	18	25
F–H	18	18	19	17	20½
G–H	6	4	1	2	1

TABLE 3 Spearman Rank Correlation Coefficients

	1	2	3	4	d
1	1	0.65	0.80	0.83	0.63
2		1	0.77	0.80	0.84
3			1	0.90	0.88
4				1	0.80
d					1

judicial and academic writings (see [6].) Two points about this process should be noted. First, it would be scarcely possible to proceed in this way without the legal background necessary to identify the salient facts. Secondly, it seems right to include as many questions as possible at this preliminary stage and then to use various mathematical techniques to eliminate fact variables or to redefine questions in an attempt to achieve a more economical description.

The questions used to define the fact variables are:

Q1: Did the finder control the real estate where the chattel was found?
Q2: Was the chattel attached to the real estate?
Q3: Did the other claimant have title to the real estate?
Q4: Did the other claimant have title to the chattel?
Q5: Was there a prior legal agreement concerning the chattel?
Q6: Did either party rely on a lease?
Q7: Is there a master/servant relationship between the other claimant and the finder?
Q8: Was the chattel in a position so as to be difficult to find?
Q9: Was there an attempt by the finder to locate the true owner of the chattel?
Q10: Had either party knowledge of the chattel prior to the finding?

The coded answers are displayed on Table 4. The answers were determined by three law students working independently in the first instance. When the results were compared, it was found that there was some disagreement which resulted from the slight ambiguity of some of the questions; these differences were resolved by discussions among the coders.

It might have been better in a preliminary study to stay with a simple measure of distance in which all variables are given equal importance, but it was thought desirable to use one that reflects the relative importance of the facts. In legal argument, a fact that is "unusual" is likely to be of more importance than others. The distance chosen

TABLE 4   Response to Questions

	Q 1	Q 2	Q 3	Q 4	Q 5	Q 6	Q 7	Q 8	Q 9	Q 10
A	0	0	0	0	1	0	0	0	0	0
B	0	0	1	0	0	0	0	0	1	0
C	1	1	1	0	0	1	0	1	1	0
D	0	0	1	0	0	0	0	1	1	0
E	1	1	1	0	1	1	0	1	1	0
F	1	0	0	1	0	0	0	1	1	1
G	0	1	1	0	0	0	1	1	1	0
H	0	1	0	0	0	0	1	1	1	0

attempts to build in this legal bias by weighting facts by the inverse of the variance of each of the fact variables. Thus the measure chosen is defined by

$$d(x,y) = \frac{\sum (x_i - y_i)^2}{\mathrm{VAR}_i}$$

where x_i is the coded response to the i-th question for case x.

Applying this measure to the set of eight finders cases produces a distance table (Table 5). The distance measure was used to rank the 28 case pairs with the results shown in Tables 2 and 3. The substantial agreement with each of the four subjects indicates that the distance has captured much of the intuitive notion of legal similarity between cases.

In addition to the agreement with the human subjects, the distance exhibits a characteristic which might be called the nearest-neighbor rule: each case has been decided in a manner compatible with its nearest neighbor (i.e., the finder wins case X if and only if the finder wins in the case nearest to X). Note that the distance incorporates no direct information concerning who won or lost the case.

This nearest-neighbor rule may be applied to the *Parker* case mentioned in the introduction [1]. The case nearest to *Parker* is *Bridges* v. *Hawksworth*, with a distance of 9.6. The second nearest is *Hannah* v. *Peel* at a distance of 14.9. The finder won in both cases, a result that agrees with the English Court of Appeal finding in *Parker* itself.

THE GEOMETRIC REPRESENTATION OF A BODY OF CASE LAW

In one sense, the finders cases now have a geometric representation, but a 10-dimensional representation is not the sort of thing that is too easy to

TABLE 5   Table of Distances

	A	B	C	D	E	F	G	H
A	0	18.74	37.68	24.08	32.34	42.36	33.71	29.14
B		0	18.93	5.33	24.27	32.15	14.67	18.93
C			0	13.60	5.33	31.89	14.93	19.20
D				0	18.93	26.82	9.33	13.60
E					0	37.22	20.27	24.53
F						0	36.15	31.89
G							0	4.27
H								0

visualize. What is desired is to represent the cases in a small number of dimensions with a view toward discovering legal information from the geometric representation.

An exploratory technique known as multidimensional scaling has been found to be of considerable value in various disciplines [7,8]. The purpose of the technique is to represent objects (finders cases in this application) as points in a geometric space. Consideration of the geometric properties (e.g., dimension, clustering, etc.) then suggests inferences concerning the real properties of the objects represented. It is important to emphasize (particularly to people in a discipline such as law, who are likely to be unfamiliar with quantitative methods) that the technique is exploratory and suggestive only. It does not purport to *prove* the existence of relationships, but merely suggests directions which may then be followed by more traditional methods of legal research. In spite of these comments and the lack of any firm logical basis, scaling has led often enough to the discovery of real relationships so as to inspire a certain degree of confidence in its usefulness.

The fundamental ideas of multidimensional scaling are most easily understood by means of the following example.* Suppose that a road map has become separated from the table of distances between cities. Further suppose that the map has somehow been lost but that the table of distances has been retained. Is it possible to reconstruct accurately the road map from the distance table? The answer is a qualified "yes": if the roads are relatively straight, a good approximation to the original map may be reconstructed except that the orientation may be lost and the scale of the reconstructed map is arbitrary.

---

*The explanation here follows Kendall [9].

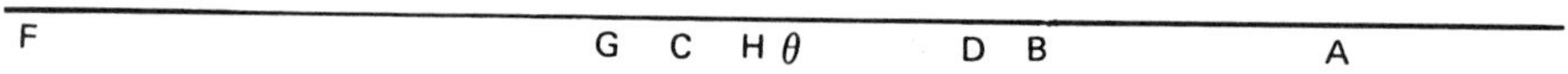

F								
		G	C	H θ		D	B	A

FIGURE 1   One-dimensional plot; coordinates from Bell Laboratories KYST; stress = 20.1%.

Now consider a slightly more complicated problem. It is not known whether the distance table relates to a two-dimensional configuration (e.g., a road map) or to a three-dimensional system (e.g., a map of a galactic system). Is it possible to determine the dimensionality from the table of distances? Again the answer is "yes." For any such table, it is possible to calculate a number known as the "stress" for each dimension. The stress is a measure of how well the distance table may be represented by a configuration of points in the given dimension. If the representation is exact, the stress will be zero. If the stress is nonzero, but still small, the geometric representation is thought of as an approximate reflection of the objects that are to be studied.

There is, of course, no a priori reason to suppose that a distance table such as that of Table 5 corresponds to a "map" in any dimension smaller than 10. Experience has shown, however, that such a representation is frequently possible and that drawing the map will provide meaningful information concerning the real relationships between the objects. Aside from the geometric considerations already mentioned, there is a simple but powerful model which is used in scaling: if the data will "fit" in N dimensions, the explanation of that data will require precisely N factors or concepts. Again, for the benefit of those unfamiliar with quantitative techniques, it should be stated that no logical argument to support the method is known, but its usefulness has been demonstrated in many different fields provided only that it is used with caution and in combination with other established research methods.

Scaling the cases in one dimension produces the configuration shown in Figure 1.* The stress is 20.1%, a rather high figure which indicates that the one-dimensional plot is not very suitable for the finders cases. According to the scaling paradigm mentioned above, this suggests that a single concept will be incapable of explaining or rationalizing this series of cases and that any attempt to do so will probably result in a concept which either does not adequately explain the cases or which is difficult or impossible to apply to actual factual situations.

---

*The program used was the Bell Laboratories KYST, based on the algorithm of Kruskal [10].

Indeed, one attempt by an academic writer to define a single concept to explain the cases will illustrate the point. It has been suggested that all of the finders cases may be decided by the notion of "constructive possession" [11]. In legal terminology, the word "constructive" is used to denote those circumstances when the law will consider a fact to be true even when it is not. Thus "constructive possession" refers to those circumstances when the law will deem that some person is in possession of the object even though that person would not necessarily be in possession in the ordinary use of the language. Thus it is claimed that a person should be in "constructive possession" of the lost object if that person owns or occupies the land on which the object is found *and* that object is somehow "attached" to the land.

Aside from the fact that this kind of analysis seems very artificial to the nonlawyer (and to many of us who are lawyers), the analysis also fails to help us solve novel cases which might arise. Who is in possession (constructive or otherwise) of a ring found by workmen at the bottom of a swimming pool? Or of a bundle of money found in a desk recently purchased? The concept of "constructive possession" turns out to be highly *nonconstructive* in the logical and mathematical sense; that is there would not appear to be any algorithm which will enable us to determine if any particular person is in "constructive possession" in any particular circumstance. Furthermore, attempting to define "constructive possession" by reference to factual situations is merely starting over and implicitly admitting that the concept is not a workable one-dimensional one.

On the other hand, the cases do exhibit some one-dimensional characteristics. The cases are properly ordered in the sense that those lying to the left of the point marked 0 were all lost by the finder, while those lying to the right were all won by the finder.

This brief analysis of the one-dimensional plot may provide some clue as to the popularity of these cases among academic writers. The cases are not properly one-dimensional because of the high stress, but they nevertheless exhibit certain one-dimensional characteristics, such as the proper ordering. These properties are the geometric reflection of the fact that the cases are difficult to rationalize yet still seem to "hang together," a combination that might be supposed to be very attractive to academic inquirers.

Scaling the cases in two dimensions produces the results shown in Figure 2. The stress is a comfortable 8.3%, suggesting that the cases may be adequately rationalized through the use of two concepts. If this research were aimed at a complete explanation of the finders cases, a search would be made for legal concepts that could be used to identify the

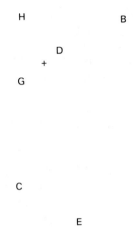

FIGURE 2   Two-dimensional plot; coordinates from Bell Laboratories KYST; stress = 8.3%.

two dimensions in Figure 2. Assuming that such purely legal research produced any results, the next step would be to attempt to apply the new concept to further finders cases to test for validity.

However, the two-dimensional plot suggests a different approach and it is interesting to note that this alternative approach is one which *has* been followed in legal research. Two of the cases, *Armory* v. *Delamirie* and *Moffatt* v. *Kazana*, are "outliers," well separated from the rest of the cases along the horizontal axis. Rather than attempting to explain all of the cases with a two-dimensional formulation, an alternative approach is to declare that these two cases are not really finders cases at all. These cases would then be discarded, the remaining cases rescaled, and the analytic process repeated [12,13].

This approach, as mentioned, accords reasonably well with current legal thought on the subject. In *Armory's* case the jeweler was a complete stranger to the entire sequence of events; in *Moffatt's* case, the true owner was one of the litigants. If cases of this type are excluded, we are left with cases where one of the litigants is the finder and the other is the owner or occupier of the land on which the chattel is found. Some of the higher dimensionality may be circumvented by restricting the domain of discourse, by redefining the meaning of "finders cases."

We are not in a position to pursue this second line of inquiry for a technical reason: eight cases already represent a dangerously small number of objects for reliable scaling in two dimensions. The results obtained from the program using only six cases could possibly be quite meaningless.

Furthermore, even though this second approach does correspond with current legal thought, a consideration of the two-dimensional plot suggests that the banishment of *Armory* and of *Moffatt* may have been premature. If it is possible to reconcile all of the cases by the use of two concepts, which is suggested by the relatively low two-dimensional stress, the resulting theory would be of a much wider application and richness than the one-dimensional theory that is available for the restricted class of "finders cases."

CONCLUSION

The finders cases constitute a well-known and well-researched body of case law which exhibits an intriguing complexity. But even with this small number of well-known cases, scaling has shed some light on their complexity and structure. This result has been achieved using a relatively elementary quantification, although the computer programs used are of some sophistication. At the risk of being repetitive, I should like to emphasize once again that these methods are, at the present state of development, of value only when used in conjunction with the more traditional research methods. As such, the quantitative methods can provide a useful guide to the more efficient use of the traditional methods of legal research.

REFERENCES

1. *Parker* v. *British Airways Board* [1982] 1 All ER 834, English Court of Appeal. Leave was granted to appeal to the House of Lords.

2. Cross, *Precedent in English Law*, p. 39.

3. *Hibbert* v. *McKiernan* [1948] 2 KB 142, 149 per Lord Goddard.

4. Kendall, *Rank Correlation Methods*, 2nd ed. (London: 1955).

5. Kelley, *General Topology*. (New York: 1955).

6. Harris, The Concept of possession in English law, in *Oxford Essays in Jurisprudence* G. G. Guest, eds.(Oxford: Oxford University Press, 1961).

7. Shepard, Rommy, and Nerlove, eds., *Multi-dimensional Scaling: Theory and Application in the Behavioral Sciences* (New York: 1972).

8. Green, and Carmone, *Multi-dimensional Scaling and Related Techniques in Marketing Analysis* (Boston: 1970).

9. D. G. Kendall, Maps from marriages: an application of non-metric multi-dimensional scaling to parish register data, in *Anglo-Romainian Conference on Mathematics in the Archeological and Historical Sciences*, Mamaia, 1970, Hodson, Kendall, and Tauto, eds. (Edinburgh: 1971).

10. Kruskal, Non-metric multi-dimensional scaling: a numerical method, *Psychometrika 29* 1964, 115.

11. Tay, Possession and the modern law of finding, *Sydney Law Rev. 4*, 1964, 383.

12. Cohen, The finders cases revisited, *Tex. Law Rev. 48*, 1970, 1001.

13. Reisman, Possession and the law of finders, *Harv. Law Rev. 52*, 1939, 1105.

7

How to Assess Management Training

Judith A. Brook
Massey University
Palmerston North, New Zealand

INTRODUCTION

A major concern of our bureaucratic society is to find men and women who can assume a leadership role within our organizations, and for this reason training and development schemes for managers have proliferated in recent years. One of the goals of such training is to enable people to deal more effectively with the management problems they encounter in their daily work. This usually includes developing better relationships with subordinates, superiors, and colleagues. It has become clear to social scientists working in this area that while some people naturally seem to make better leaders than others, there is plenty of evidence that leadership skills can be improved with training. Moreover, the particular style of leadership adopted depends to some extent on individual taste, but this may be tempered by the characteristics of the followers, the relationships existing between the parties, the type of working environment, the amount of power the leader possesses by virtue of his or her position, and the types of problems to be solved, among other things.

ASSESSING MANAGERIAL BEHAVIOR

If, as we have suggested, it is desirable and possible to train men and women for management, the next question is: How can changes in

managerial behavior be measured? How do we measure success, and how do we evaluate the effectiveness of such training? Furthermore, since managerial performance depends to a large extent on the characteristics of the situation, including the quality of the relationship between leader and followers, the method we choose to assess change should include both the situational and the interpersonal aspects of the job. Often, in the past, when an evaluation of a management course has been conducted, it has been left up to the judgment of the trainers, the employees' superiors, or the trainees themselves to decide whether or not positive changes have occurred as a result of training. These judgments are usually highly subjective and simply reflect the opinions and biases of the persons concerned. If a free-response style is used, it is very difficult to quantify the information so that it can be analyzed in an objective way. On the other hand, there is a danger that an interview or questionnaire designed by and couched in the language of the evaluator may restrict the respondent to a perspective and a framework that is inappropriate and unfamiliar to him or her.

There is one method of data collection which has become increasingly popular. This method attempts to gather evaluative informa-

1	3	5	1	2	2	1	3	1	1	Scientific approach to problems	Emotional approach to problems	1
1	4	2	3	3	3	2	4	2	1	Research oriented	— Administration oriented	2
2	5	7	2	3	7	3	2	2	2	Not conservation minded	— Conservation minded	3
7	1	7	3	3	3	2	1	4	5	Nonlaboratory worker	— Laboratory Worker	4
6	3	1	3	2	4	7	3	3	3	Nonacademic attitudes	— Academic attitudes	5
1	1	7	1	1	1	1	1	1	1	Professional	— Nonprofessional	6
5	5	3	2	3	4	5	5	2	3	Flexible	— Inflexible	7
1	3	7	2	3	2	1	4	1	1	Interested in science	— More interested in other matters	8
2	7	3	1	3	1	3	6	1	1	Peers	— Nonpeers	9
1	3	7	2	2	2	1	3	1	1	Intellectual	— Nonintellectual	10

Personal constructs (Case Study A)

Elements:

1. A time when I delegated an important task to a coworker
2. The time when I actively opposed the ideas of my controlling officer (or someone in authority)
3. A time I had to deal with a problem brought to me by a member of my staff
4. A time I had to make an important decision concerning my research (or other work)
5. A time when I had a professional association with some outside organization (business, industry, etc.)
6. The occasion when I made (or proposed) changes in the running and conduct of section meetings or other procedures of a similar nature
7. An occasion when I felt most satisfied with my work performance
8. An occasion when I felt least satisfied with my work performance
9. My professional self *now*
10. My professional self *a year ago*

FIGURE 1 The repertory grid form, consisting of 10 rows (constructs) and 10 columns (elements).

tion as seen through the eyes of the respondent and yet is indirect enough to reduce the natural tendency for the person to give the "socially desirable" answer. It has the added advantage of producing quantitative data that lend themselves to a type of statistical analysis which simulates to some extent the way in which the respondent categorizes and organizes his or her own ideas. The method, known as the repertory grid technique, is described in the next section.

THE REPERTORY GRID TECHNIQUE

A group of managers who had completed a training course 12 months earlier were presented with a list of 10 of the most important inter-personal situations encountered in their day-to-day work. These 10 "elements" were written at the top of an empty grid form and cor-responded to the 10 columns, as shown in Figure 1. The managers were then asked to think carefully about these 10 familiar work roles and to make a series of 10 judgments, comparing three of them at a time (triads). For each comparison, they had to write down the most important similarity between two of the interpersonal situations which distin-guished them from the third. The particular triads of elements had been chosen in advance by the evaluator with the aim of providing the most interesting and varied set of comparisons possible. These similarities and differences were recorded beside the empty rows of the grid form (Figure 1). They are called personal constructs because they are the individual's personal constructions of reality in relation to his or her work. An example of one such set of personal constructs is illustrated in Figure 1.

As each verbal construct was elicited, the manager was asked to rate each of the 10 elements, in turn, on that particular construct using a 7-point scale. For example, each of the 10 interpersonal situations listed at the head of the columns was rated on the construct "Scientfic approach to problems—Emotional approach to problems," where 1 refers to "scientific approach" and 7 refers to the "emotional approach" extreme. The manager recorded his or her ratings in the appropriate cell of row 1 of the grid (Figure 1).

After the 10 constructs had been elicited and ratings of elements completed, the resulting 10 × 10 raw data matrix consisting of numerals between 1 and 7 was analyzed using a single linkage cluster analysis procedure similar to that described by Everitt (Chapter 4). In this procedure the raw data are first converted into a similarity matrix based on the Euclidean distances between data points. Similarity between points is measured on a scale of 0 to 100, where 0 represents minimum similarity (greatest distance) and 100 represents perfect similarity (the

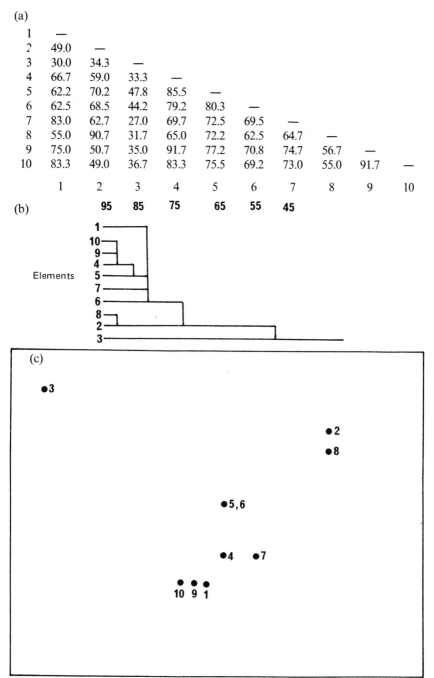

(a)

	1	2	3	4	5	6	7	8	9	10
1	—									
2	49.0	—								
3	30.0	34.3	—							
4	66.7	59.0	33.3	—						
5	62.2	70.2	47.8	85.5	—					
6	62.5	68.5	44.2	79.2	80.3	—				
7	83.0	62.7	27.0	69.7	72.5	69.5	—			
8	55.0	90.7	31.7	65.0	72.2	62.5	64.7	—		
9	75.0	50.7	35.0	91.7	77.2	70.8	74.7	56.7	—	
10	83.3	49.0	36.7	83.3	75.5	69.2	73.0	55.0	91.7	—

(b)

(c)

FIGURE 2 Cluster analysis and principal components analysis results of repertory grid element data for case study A consisting of similarity matrix (a), dendrogram (b), and two-dimensional plot of elements (c).

points coincide). Two such similarity matrices are reproduced in Figures 2a and 3a.

Since we were interested in the way in which both the elements and the constructs clustered, the first similarity matrix derived from each data set reflected distances between element columns (Figure 2a) and the second similarity matrix reflected construct row distances (Figure 3a). For example, any two constructs which were applied by the respondent in a very similar manner over all elements will appear close together in the classification system produced by the analysis, as will any two elements which were rated in very similar ways over a number of constructs. In forming the clusters from a similarity matrix, single linkage cluster analysis involves locating the two points which are closest together to form the nucleus of the first cluster. Subsequently, other points are added to existing clusters or linked together to form new clusters according to their proximity to other points. The computer program that was used to perform these operations displays the linkages first by means of a minimum spanning tree. This depicts diagramatically the sequence and position of the points incorporated into the treelike structure, according to the values derived from the similarity matrix. After the clusters have been so formed, the resulting hierarchical arrangements of points (elements or constructs) at successive linkage levels are displayed by means of a dendrogram (see Figures 2b and 3b). We assume that this hierarchical arrangement of constructs and elements reflects the similarities and differences as perceived by the respondent. This simulates, at least to some degree, the structuring of the various subsystems within the person's personal construct system.

Finally, a principal components analysis is performed which allows the data points to be represented on a graphical spatial plot illustrating the approximate positions of all the points on a two-dimensional space (see Figures 2c, 3c, 4, 5, 6, and 7). Starting with the same similarity matrices the principal components analysis uses the information concerning the relationships of the elements with one another to map their dispersion on the axes of the two major components. A similar procedure is followed for the relationships between the constructs. From this we can see, at a glance, the pattern of clustering effects and the degree of "tightness" within and between adjacent clusters.

The results of these analyses provided a way of showing how similar elements or constructs were grouped together while others were more loosely related or linked together to form other clusters. Moreover, in the present context, where the sets of elements included Self now and Self a year ago, it was possible to observe important changes that had occurred from before to after training.

(a)

	1	2	3	4	5	6	7	8	9	10
1	—									
2	69.7	—								
3	62.2	68.5	—							
4	41.0	54.0	55.5	—						
5	50.0	63.0	50.5	57.7	—					
6	82.7	52.3	55.5	45.0	32.7	—				
7	54.3	71.3	71.5	52.0	71.0	37.0	—			
8	84.7	78.3	73.5	52.3	44.7	74.0	63.0	—		
9	72.3	69.3	61.2	30.0	39.0	58.3	68.0	71.0	—	
10	91.7	71.3	70.5	49.3	51.7	81.0	56.0	93.0	64.0	—

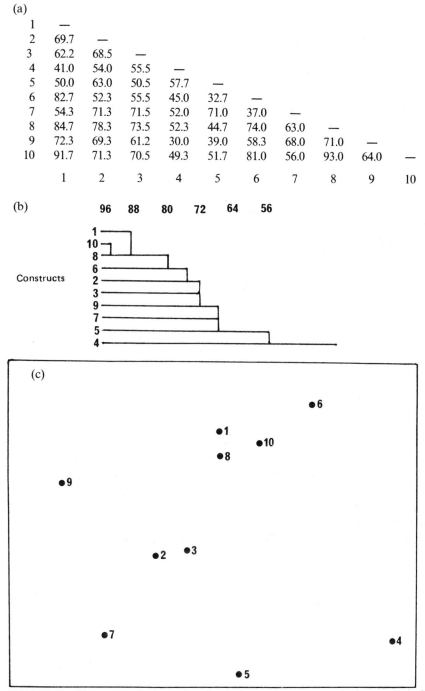

FIGURE 3 Cluster analysis and principal components analysis results of repertory grid construct data for case study A consisting of similarity matrix (a), dendrogram (b), and two-dimensional plot of constructs (c).

The three examples that follow were selected to illustrate the application of this technique to the evaluation of a particular management course. This course had been designed to train a group of research scientists in the managerial skills needed to successfully lead their work teams, consisting of other scientists and technical staff from within their organization. The names and personal details of the people involved are fictitious.

CASE STUDIES

Case Study A: Graham Stewart

Graham Stewart had been employed as a scientist by the organization for 15 years. For the last nine months before the training course he had been occupying the position of group leader in charge of younger scientists and technicians. Twelve months after training he completed a repertory grid similar to the one shown in Figure 1 and the results were analyzed using the analyses described. The dendrogram and graph reproduced in Figure 2 display the levels at which each of his 10 elements are linked and the distances between elements. Both are derived from the similarity matrix that accompanies them.

Since Self now, labeled (9) in Figure 2, and Self a year ago (10) are very close together, it appears from his repertory grid that little change has taken place in the months since the training course. Both elements are closer to Most satisfied (7) than to Least satisfied (8). Furthermore, there is quite a strong relationship between Most satisfied (7) and Contact with outside organization (5) suggesting that this is one rewarding aspect of his job. In fact, most of the elements are associated with Contact with outside organization, indicating that this is an important part of his work. There are many other interesting linkages between elements which could be explored, depending on one's purpose but we will note one final example. The element Opposed someone in authority (2) is closely related to Least satisfied (8), which may have something to say about Graham Stewart's attitude to conflict.

Even a preliminary glance at his 10 personal constructs elicited by this method is informative (see Figure 1). They indicate Graham Stewart's general attitudes toward his work and the underlying constructs which he tends to use in his perceptions about himself and his coworkers.

The similarity matrix data for his constructs (see Figure 3) link Scientific approach to problems (1) to Professional (6) to Interested in science (8) and to Intellectual (10). There is an interesting link between

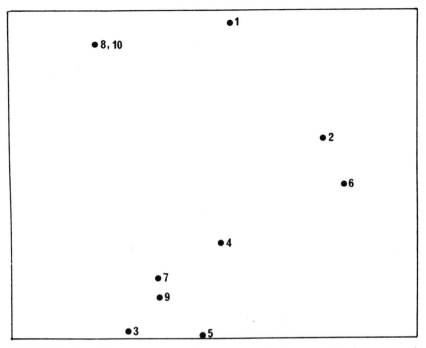

FIGURE 4   Two-dimensional plot of repertory grid element data for case study B.

Not conservation minded (3) and Research oriented (2). It would appear that Graham Stewart does not find some of his fellow scientists and peers particularly open to conservationalist attitudes. The construct, Non-laboratory worker (4) is at some distance from all other constructs except Nonacademic attitudes (5). This suggests that his perceptions about nonlaboratory workers are not highly elaborated and probably less relevant to him as far as his work is concerned.

Case Study B: John Parker

John Parker is another management course trainee in his mid to late thirties with less than 10 years on the staff and a little less than five years as a section leader at the time of training. The elements in his repertory grid (see Figure 4) show that Self now (9) and Self a year ago (10) are some distance apart, and over the months since training he has moved closer to Most satisfied (7) and farther from Least satisfied (8). The elements Self now (9) and Most satisfied (7) are both related to Problems from staff (3), perhaps reflecting his current ability to deal with such problems. Self now

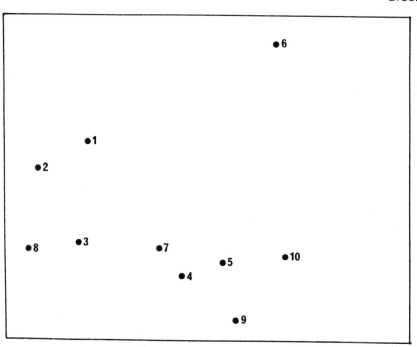

FIGURE 5   Two-dimensional plot of repertory grid construct data for case study B.

(9) is also related to Important decision (4) and Contact with outside organization (5). This cluster of related elements seems to indicate a change in his work emphasis toward an increased interest in working with outside organizations.

John Parker's constructs were:

Emergent pole	Implicit pole
1. Successful	Confusion
2. Disciplined	Undisciplined
3. Organizing/efficient	Cussed/awkward
4. Harmony	Conflict
5. Satisfied	Dissatisfied
6. Official interactions/impersonal	Unofficial/personal
7. Controlled	Uncontrolled
8. Rational	Irrational
9. Unselfish	Selfish
10. Friendly	Unfriendly

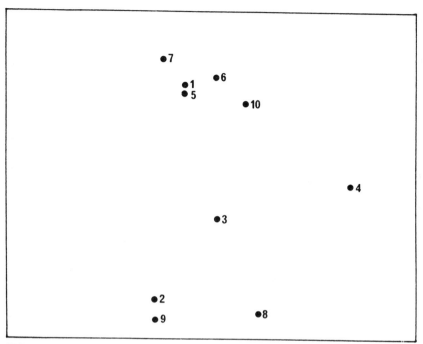

FIGURE 6   Two-dimensional plot of repertory grid element data for case study C.

His first construct in the preceding list contrasts Successful with Confusion. This is a rather unusual contrast to make, but he appears to treat it as a single dimension. Successful (1) is perceived as somewhat similar to Disciplined (2). Not surprisingly, Organizing/efficient (3) is associated with Rational (8). John Parker also relates Harmony (4), Satisfied (5), Friendly (10), Unselfish (9), and Controlled (7) together as opposed to Conflict, Dissatisfied, Unfriendly, Selfish, and Uncontrolled. This provides additional insight into his value system. The remaining construct, Official/impersonal (6), in contrast to Unofficial/personal, stands, more or less, apart suggesting that this dimension is less central to his thinking.

Case Study C: David Winter

In contrast to the two previous studies, David Winter was a member of a matched control group of scientists who had not taken part in the management training course. He completed the repertory grid at the same time as the ex-trainees and was presented with exactly the same list

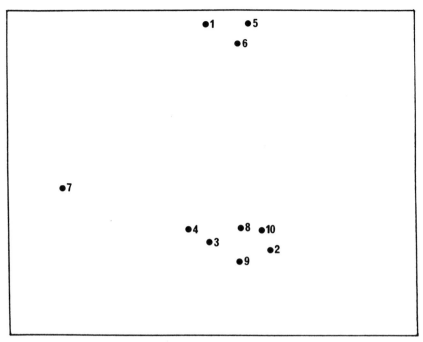

FIGURE 7   Two-dimensional plot of repertory grid construct data for case study C.

of elements. This subject was in his early thirties with less than 10 years experience with the organization and less than 5 years in a leadership position.

His elements Self now (9) and Self a year ago (10) are widely separated (see Figure 6). Here is a case where Self now (9) is much closer to Least satisfied (8) and farther from Most satisfied (7). A year ago the situation was reversed, with Self a year ago (10) being closer to Most satisfied (7) and farther from Least satisfied. The element Least satisfied (8) is associated with Opposed someone in authority (2), which suggests some conflict in this area of his work. Moreover, it is interesting to note that Self a year ago (10) is fairly close to Delegated an important task (1), Contact with outside organization (5), and Changes in work procedures (6), whereas none of these elements are related to Self now (9). Problems from staff (3) and Important decision (4) seem to stand in more-or-less isolated positions. Perhaps David Winter does not see these as a major aspect of his work.

His list of construct choices reflects a certain amount of conflict in his work situation.

Emergent pole	Implicit pole
1. Annoyed	Satisfied, knowing that you have done a good job
2. Idealistic	Cynical
3. Beneficial	Destructive
4. Stimulating	Depressing
5. Disinterested	Interested
6. Disruptive	Supportive
7. Decisive	Indecisive
8. Relieved	Discontented
9. Pleased	Disappointed
10. Cooperation	Disagreement

His construct Annoyed (1) is related to Disinterested (5) and to Disruptive (6), while Idealistic (2), Cooperation (10), Pleased (9), Relieved (8), Stimulating (4), and Beneficial (3) are linked together (see Figure 7). Decisive (7) as opposed to Indecisive appears to be a peripheral construct. The deterioration in his job satisfaction over the past year may be associated with events at work which have been disruptive and a source of anger and frustration to him. This could be a valuable lead to follow up in subsequent discussion with this employee.

PERSONAL CONSTRUCT THEORY

For those readers who would like to know a little more about the ideas underlying the repertory grid technique, there follows a brief description of personal construct theory.

G. A. Kelly (1955), a teacher and counselor, developed a theory to explain how people interact with their surroundings and, in particular, with other people in their environment. Kelly's personal construct theory maintains that men and women come to terms with their everyday lives by acting in a manner similar to scientists investigating a problem. When confronted with a new or unfamiliar situation they formulate hypotheses about it on the basis of their former experiences, then test and consequently modify or extend their previous theories in the light of the new information. A person's personal construct system is thus the whole set of representations or models that develop as he or she interacts with the world and the people in it. Parts of this system are shared, at least to some extent, with other people, but other parts are quite unique to the person. The important thing is that the personal construct system is not immutably fixed but is in a constant state of change and development as new events and experiences impinge.

Each person's personal construct system is built up of interrelated constructs. These are simply bipolar dimensions such as outgoing-reserved, flexible-inflexible, mean-generous, which enable the person to discriminate between elements belonging to a particular class. The elements in a class may be people, things, events, or concepts, the only restriction being that they can be compared and contrasted in some way. Because the relationship between constructs is relatively structured and systematic, if one could gather the necessary information about the relationships between a person's constructs and elements, it should be possible to analyze the data to produce a plan or representation of that part of the person's personal construct system. Kelly went on to develop the method known as the repertory grid technique to gather this information, and he used it in his counseling practice to help clients explore and make sense of their interpersonal worlds. Since then the technique has been developed not only for the clinical setting but more recently for business and industry, to help people explore their work environment. Even here it may be used for a wide variety of purposes, for example, to develop training needs, in market research, to investigate motivation at work, to aid in the counseling of employees or, as in the present case, to evaluate the effectiveness of a management training course.

Kelly's original analyses were carried out by hand using a laborious visual-scanning technique, but the more recent developments of statistical packages suitable for this type of multivariate analysis has meant that clients can now receive immediate feedback on their grid performance and on the functioning of their personnel construct systems.

REFERENCE

Kelly, G. A. (1955). *The Psychology of Personal Constructs*, Vol 1. Norton, New York.

8

Multidimensional Scaling in Product Development

Allan M. Anderson
Salmond Industries, Ltd.
Palmerston North, New Zealand

INTRODUCTION

Most companies producing goods rely very heavily on new product development for continuing profits. The product development function has therefore become important to most companies as new and improved products must be efficiently and economically developed on an ongoing basis.

Above all, product development must seek to define new products which are demanded by the consumer and which are in areas of low competition from other producers. Market research for product development therefore involves searching for gaps in the market place. This *gap analysis* can be either over a whole range of products or within a specific product line.

Gap analysis implies looking at a pictorial representation of a selection of products already in the market place, assessing their relative positionings, and defining areas where voids exist and potential products could be developed. Multidimensional scaling can be used to produce such a pictorial representation or product map, which can then be used as a basis for locating potential new areas which might be suitable for product development.

In this chapter we show how multidimensional scaling can be used to formulate a new ice cream novelty and then, after it has been made, to test how well it fits the gap in the market.

WHAT IS MULTIDIMENSIONAL SCALING?

Multidimensional scaling is a statistical technique which is used to translate numerical measures of product similarity (see Chapter 4 on cluster analysis for a discussion of the measurement of similarity) into a pictorial representation of similarity. It aims to derive the basic pattern or structure underlying a set of data and reproduce it in a form which is more acceptable to the human eye—namely, a geometrical model. (This concept of creating a picture to illustrate how similar certain things are is utilized in several chapters herein, notably Alan Tyree's study of the similarity of law cases in Chapter 6 and Phil Gendall's study of the soft drink market in Chapter 9.)

The similarity of two particular products is, of course, a matter of personal opinion. The simplest and most direct approach is simply to ask a group of consumers to rank a particular pair of products on a scale of "very similar" to "very dissimilar." Repeating this exercise for all possible pairs of products involved in a study yields a matrix which numerically describes the relative similarities of the various products.

The pictorial representation created by multidimensional scaling will seek to have distances between the points (products) which faithfully reproduce these similarities (small distance corresponding to high similarity). An exact reproduction is, of course, unlikely and the objective, in practice, is to find a useful pictorial representation which reproduces the similarities acceptably well.

The graphical representation created by multidimensional scaling can have any dimensionality, 1 (a line), 2 (a plane—the basic pictorial model), 3 (normal space), and 4 or more (possible in theory but not much use in practice). The more dimensions utilized, the more flexibility the technique has and the more exactly it can reproduce the original pattern of relative similarity. The amount of disagreement between the pictorial distances and the numerical definitions of similarity is referred to as the stress value for a particular graphical representation. Stress must decrease as more dimensions are used to create the picture. The object of multidimensional scaling is to obtain a representation of the products under study in as low a dimensionality (and therefore as simple) as possible consistent with maintaining a low stress value (and therefore high accuracy).

AN EXAMPLE OF MULTIDIMENSIONAL SCALING IN PRODUCT DEVELOPMENT

The objective was to define and develop a new ice cream novelty product to appeal to the teenage market. A list of new product ideas was obtained by brainstorming, literature review, and product morphology techniques. This list was screened down to three product ideas on the basis of technical feasibility, cost, and anticipated consumer acceptance. Concept descriptions for the three product ideas are as follows. Each is a novelty food sold on a wooden stick.

Concept 1: a honey-flavored ice cream with nuts throughout and coated with chocolate

Concept 2: plain ice cream with candied fruit, nuts and spices throughout, coated with chocolate

Concept 3: ripples of green peppermint and brown chocolate ice cream coated with chocolate

Seven products, already on the market, were chosen to be representative of the total market offering and their names and descriptions are given below. All except Chocbar are sold on wooden sticks.

A. Jelly Tip: vanilla ice cream with a strawberry-jelly tip and coated with chocolate

B. Chocbar: solid chocolate center surrounded by vanilla ice cream and coated with chocolate roasted coconut

TABLE 1   Similarity Matrix for the Original Analysis

	A	B	C	D	E	F	G	1	2	3
A	—									
B	2.05	—								
C	1.75	2.48	—							
D	2.71	3.37	3.04	—						
E	2.32	2.75	2.10	2.68	—					
F	3.00	2.62	2.71	2.73	2.50	—				
G	4.52	4.36	3.41	2.95	3.59	3.14	—			
1	2.55	2.73	3.33	3.60	3.00	2.95	3.29	—		
2	3.05	2.82	2.43	3.29	2.65	3.04	3.67	3.24	—	
3	2.10	2.90	2.10	2.82	2.09	2.41	2.81	2.75	2.50	—

C. Topsy: vanilla ice cream coated with chocolate
D. Fruju: Fruit-flavored ice block with true-to-fruit flavors
E. Toppa: vanilla ice cream coated with a fruit-flavored water ice
F. Milkshake: ripple vanilla and chocolate-flavored milk ice
G. Popsicle: flavored ice block

The seven existing and three concepts were combined into a single list of 10 products. Similarities data were obtained for the 10 products and these were analyzed using multidimensional scaling.

The Consumers

A group of potential consumers, with ages ranging from 13 to 17 years, were selected from a local school. A representation of both male and female students and all ages was obtained.

The Similarities Data

Each pair of products was compared on the basis of substitutability as a measure of similarity. The students were asked the following question:

If you went into a shop and asked for the first ice cream in the pair and it was not available, how good would the second ice cream be as a replacement for the first?

The judgments were made on a 5-point scale ranging from 1 for an excellent replacement to 5 for not at all suitable as a replacement. The question and method of judgment were carefully explained to all students.

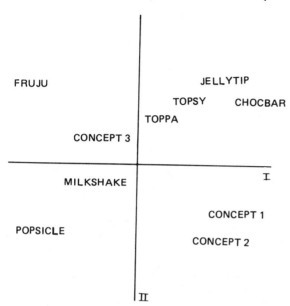

FIGURE 1 Product map showing the position of the three new product concepts.

Data Collection

The presentation of all pairs of the 10 products to each student required a total of 45 judgments. This was considered too many judgments, particularly for the younger students. Each student was therefore presented with only 11 pairs of products to judge. The pairs were selected in such a way that, over all the students, each pair of products was judged a total of 22 times. The scores for each pair were averaged and a similarity (more exactly, dissimilarity) matrix was derived, as shown in Table 1.

Multidimensional Scaling Analysis

The data in Table 1 were analyzed using a multidimensional scaling computer program called KYST. A two-dimensional product map was chosen as the lowest dimensionality with a reasonable stress value. This product map is shown in Figure 1.

The map appears to reflect the customers' distinction between ice block–based products (Fruju and Popsicle) and ice cream–based products (Toppa, Topsy, Jelly Tip, and Chocbar). The milk-flavored ice product is viewed as being intermediate between these two types.

TABLE 2   Similarity Matrix for Final Analysis

	A	B	C	D	E	F	G	1
A	—							
B	2.21	—						
C	1.58	1.80	—					
D	3.90	4.35	3.90	—				
E	2.50	2.90	1.75	3.26	—			
F	3.58	3.32	3.26	3.35	2.65	—		
G	3.83	4.65	3.89	1.70	3.42	3.20	—	
1	2.75	3.75	2.90	4.25	3.20	3.53	3.95	—

The map also shows that concept 3 is located in the center and close to the existing ice cream–based products. It is assumed that the close proximity of concept 3 to the existing products would reduce its probability of success in the market because of the intense competition.. In contrast, concepts 1 and 2 are located in a quadrant not occupied by existing products. They fell into a "gap" or "hole" in the product space and it would be expected that they would experience very little direct competition from existing products. Concept 1 was selected for further development on the basis of its projected ease of development and of economics.

Product Development and Production

Concept 1 was formulated and processed on pilot plant equipment. Regular taste panels were conducted to guide the formulation toward that outlined in the initial concept description. When an acceptable product had been formulated to fit the concept description, a small-scale production run was made under normal factory conditions. The product samples were used for consumer testing and multidimensional scaling analysis to compare the positioning of the formulated product to the positioning of the initial concept description.

Testing the Formulated Product

Samples of the made-up product based on concept 1 were taken to the school from which the original similarity data were obtained. Fifty-six students between 13 and 17 years were each given one of the ice creams to eat. The seven existing products and the new product were combined to form a list of eight products. This gave a total of 28 possible pairs of products. Each of the 56 students judged 11 pairs of products using the

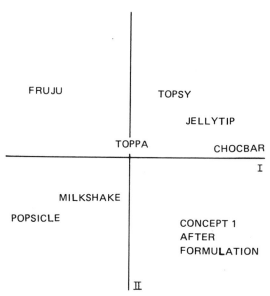

FIGURE 2   Product map showing the position of the formulated concept 1.

same procedure as for the original data collection. This again gave a total of 22 judgments for each pair. These data were averaged for each pair to derive the similarity matrix shown in Table 2. These data were used to derive a two-dimensional product map, as shown in Figure 2. A comparison of Figures 1 and 2 shows that the product based on concept 1 has been correctly formulated to fit the gap in the existing product map.

DISCUSSION

Although this study resulted in a close fit of the formulated product to its concept description, there are a number of aspects of the approach which require careful execution and where problems might arise:

1. There is a heavy reliance on the concept description. This description must not only be good enough to create a clear and realistic product image in the potential consumer's mind, but it must also provide an adequate basis for product design. The result should be a real product with attributes very near those envisaged by the consumer through the concept description.

2. The selection of products used to define the existing product space must be representative of the total market offering. Where the total

number of products in the market is large and the range diverse, a compromise may have to be reached between a reasonable number for adequate representation and a number which will not result in the need for excessive similarity judgments. In this study only seven products were used. This is considered to be at the lower limit of the number of products required for adequate representation using multidimensional scaling. The use of 15 to 20 products would not only provide a more accurate representation, but would also provide the basis for a more stable product space, less subject to changes in attitude toward one or two products.

3. The subjects chosen should be representative of the segment of potential consumers. Standard statistical sampling procedures should be used. As the technique relies on an aggregate of attitudes from the sample of subjects, it is important that these attitudes be representative of the total market of potential consumers.

4. The multidimensional scaling analysis could be used during formulation of the product. In this way it would be possible to note the changes in product positioning resulting from formulation changes and thereby move the product progressively toward the desired gap in the product space.

5. The consumers' attitudes toward existing products are a result of the interaction of all market forces, of which tangible product attributes are only a part. These forces include price, promotion, and packaging. The approach described in this chapter cannot take account of all the diverse market forces in the positioning of the concepts. It must be assumed that in making judgments on substitutability, the subjects build up their own image of the new product. It should then be possible to direct the factors of the marketing mix to create an image for the new product according to its position in the product space. The positioning of the new product in the total product space should therefore form a basis for its market planning.

9

The Positioning of Zap: An Application of Discriminant Analysis

Philip J. Gendall
Massey University
Palmerston North, New Zealand

BACKGROUND

Boy: Come on, Maureen?
Girl: No, Wayne.
Boy: Oh, come on!
Girl: I don't want to.
Boy: Oh, Maureen, you said you would.
Girl: I know but, oh, gee, Wayne.
Boy: You'll like it, Maureen, you really will.
Girl: But Wayne.
Boy: Trust me, Maureen.
Girl: Oh . . . Wayne.
Boy: There you are, I told you, didn't I?
Girl: Oh, Wayne, you were right. Big strawberry M is every bit as good as big chocolate M. You open new doors for me, Wayne.

Yea! Mmmmm, flavored milk.
Mmmmm, Big M.

This radio commercial,* together with heavy television and print media advertising, launched the Victorian Dairy Industry's Big M campaign in January 1978. The campaign focused on the theme "Milk will never be

---

*Reported in Rados and Gilmour (1981).

the same again," and promoted fresh, flavored milk in cartons as an alternative to soft drinks. It was such a success that other Australian state dairy authorities rapidly developed their own versions of Big M, with equally impressive results.

The success of these flavored milk products was attributed to the change in target market from the traditional milk market to the highly competitive soft-drink market and to their positioning alongside established products in the teenage segment of this market. In other words, Big M and its imitators were aimed at teenagers and were promoted as an exciting, enjoyable alternative to Coca-Cola and Fanta, rather than as a substitute for ordinary white milk (which was generally regarded as a nutritious, boring "food" for children or old people).

On the other side of the Tasman, the New Zealand Dairy Board watched these developments with interest. In 1977 the board had finalized plans for an ultra-high-temperature (UHT) milk processing and packaging plant at Takanini, South Auckland. As its name suggests, UHT milk is produced by heating milk to a very high temperature, then cooling it quickly. This process does not affect the nutritional value of milk, but sterilizes it and gives it a shelf life of several months when aseptically packed.

Although this new product was intended primarily for export markets in the South Pacific, it also provided the Dairy Board with an opportunity to gain production and marketing experience closer to home. Consequently, the Board set about developing a range of flavored UHT milk in cartons for the New Zealand market under the brand name Zap. On the basis of what it had observed in Australia, the Board's strategy for Zap was to "position" it as a soft drink that appealed to teenagers.

THE SOFT-DRINK MARKET

The soft-drink market in New Zealand is very competitive; this is reflected in the fact that the market leaders spend up to $500,000 a year on television advertising alone. The market is characterized by heavy consumption among teenagers and young adults, who have strong psychological needs for fun, activity, social acceptance, emancipation, and social mobility. Success in this market depends on creating the right image for a product, mainly through advertising.

Clearly, the Dairy Board faced a major task in its attempt to penetrate this market with Zap. For a start, existing brands of soft drinks were already well established, and brand names like Coca-Cola and Fanta immediately evoked mental images of fun, excitement and peer-group acceptance among teenagers. By contrast, milk was regarded as nutritious, healthful, and dull. The heaviest consumers of milk were young children—teenagers drank very little; for them, milk was definitely not a preferred drink. So convincing teenagers that Zap was an alternative to Coke and Fanta rather than simply a substitute for ordinary white milk was not going to be easy. But the Board believed that it could be done.

Zap was launched in November 1979 with a promotion campaign based on the theme, "Zap . . . Get it and you've got it." Heavy use was made of television and radio to convey the ideas of refreshment and group activity and to reflect the life-style and aspirations of teenagers. The objective of the campaign was to position Zap in the minds of teenagers as a soft drink and not as a flavored milk.

THE STUDY

To see whether this positioning strategy had been successful, a small-scale study was carried out at Massey University in August 1981, just under two years after the initial launch of Zap.* In this study teenagers

---

*This study is reported in Fearon (1982).

were asked to rate seven drinks, including Zap, on 18 features, or attributes. Then discriminant analysis, a form of multivariate statistical analysis, was used to produce a "map" of the soft drink market from these ratings. The success of the Dairy Board's positioning strategy was then deduced by comparing the location of Zap with that of the other products.

The seven products included in the study were: Coca-Cola, Fresh Up, Lemon & Paeroa, Fanta, Zap, lemonade, and milk. Coca-Cola, Fanta, Lemon & Paeroa, and lemonade are well-known carbonated soft drinks, Fresh Up is a pure apple (or apple and orange) juice. All of these products, together with white milk, were readily available competitors for Zap.

Each of the 105 teenagers who took part in the study rated three products on 18 attributes which had previously been identified as relevant to soft-drink buyers. These attributes were:

Fizzy/not fizzy
Exciting/unexciting
Refreshing/not refreshing
Popular/unpopular
Modern/old fashioned
Drunk mainly in a group/drunk mainly on your own
Drunk mainly in hot weather/drunk mainly in cold weather
Nourishing/not nourishing
Healthful/unhealthful
Filling/not filling
High in energy value/low in energy value
Natural taste/artificial taste
Strong flavor/weak flavor
Sweet/not sweet
A lot of added sugar/no added sugar
High in additives/low in additives
Luxury purchase/everyday purchase
Good value for money/poor value for money

Each sample member rated his or her three products* on a 7-point scale—one scale for each attribute. So if a particular product was considered to be very fizzy, it was given a score of 1; if it was not considered to be fizzy at all, it was given a score of 7. Similarly, if a

---

*There are 35 possible combinations of three objects from a group of seven. Each of the seven objects is included in 15 of these combinations, and each combination was rated by three different people, giving a total of 45 observations per product.

TABLE 1 Average Ratings for Each Product in Each Attribute

	Zap	Coca-Cola	Fanta	Lemonade	Lemon & Paeroa	Fresh Up	Milk
Fizzy	6.3	1.9	2.8	2.4	2.8	6.3	6.4
Exciting	4.6	3.7	4.2	4.1	3.3	4.6	5.6
Refreshing	4.1	3.2	3.1	2.8	2.6	2.6	3.8
Popular	4.4	1.9	3.4	3.0	3.2	3.3	3.2
Modern	2.6	3.4	4.0	4.4	3.1	3.5	5.9
Drunk mainly in a group	4.9	2.6	3.7	3.4	3.2	4.3	5.7
Drunk mainly in hot weather	3.6	2.4	2.4	2.4	2.2	2.7	4.4
Nourishing	3.3	5.4	5.4	5.1	5.1	2.6	1.8
Healthful	3.4	6.0	5.7	5.4	5.2	2.0	1.3
Filling	3.1	3.6	3.8	3.9	3.9	3.6	2.5
High energy value	3.4	4.0	4.6	4.4	4.5	2.7	2.2
Natural taste	4.3	5.6	5.3	4.5	4.4	1.9	1.6
Strong flavor	3.3	2.7	3.0	4.1	3.8	2.8	4.8
Sweet	3.2	2.6	2.0	2.6	2.7	3.6	5.9
A lot of added sugar	4.8	2.5	2.1	2.7	3.2	5.9	6.6
High in additives	3.4	2.2	2.0	2.4	2.9	5.2	6.2
Luxury purchase	3.4	3.7	3.7	3.8	3.6	3.3	6.7
Good value for money	5.3	4.5	4.8	4.4	4.7	3.8	1.7

Source: Adapted from Fearon (1982).

product was thought to be exciting, it was given a score of 1 or 2 on this attribute; if it was unexciting, its score would be 6 or 7. And so on for each of the 18 attributes.

The average ratings for each product on each attribute are shown in Table 1. From these average ratings we could establish the image of each product, and by comparing these images we could tell how teenagers in the study saw Zap in relation to the other six drinks. If the Dairy Board's positioning strategy had succeeded, the image of Zap would be similar to the image of Coca-Cola and the other carbonated soft drinks and well away from that of ordinary milk.

The attributes that best described Coca-Cola, Fanta, lemonade, and Lemon & Paeroa were "fizzy," "refreshing," "popular," "drunk mainly in a group and in hot weather," "high in additives," "sweet," and "containing a lot of added sugar." White milk, on the other hand, was regarded as unexciting and old fashioned. but nourishing, healthful, high in energy value, natural tasting, and free from additives. Fresh Up was seen as similar to milk, but not quite so natural and nourishing.

These observations confirmed what we already suspected about the images of these products. But what about the image of Zap? Well, Zap appeared to fall somewhere between the carbonated soft drinks and the natural products (milk and Fresh Up). Zap was regarded as more modern than Coca-Cola, Fanta, lemonade, and Lemon & Paeroa, but less exciting, less popular, less refreshing, and less likely to be drunk in a group or in hot weather. Zap was considered to be more nourishing and healthful, higher in energy value, more natural tasting, and lower in artificial additives and sugar than the carbonated drinks, although it did not rate as well on these characteristics as milk or Fresh Up.

Thus we concluded that for this group of teenagers, Zap was more closely associated with Fresh Up and ordinary milk than with Coca-Cola and the other "fizzy" drinks. In other words, the Board's attempt to position it as a soft drink had failed. But by how much had it failed, and just where was Zap positioned in relation to other drinks? To answer these questions required a more sophisticated form of analysis than our simple comparison of average ratings. What we needed was a statistical method which would allow us to compare the seven products *simultaneously* along the 18 attributes—a "multivariate" method. The method we used was discriminant analysis.

DISCRIMINANT ANALYSIS

One way of regarding discriminant analysis is as a form of multidimensional scaling. Multidimensional scaling methods are concerned

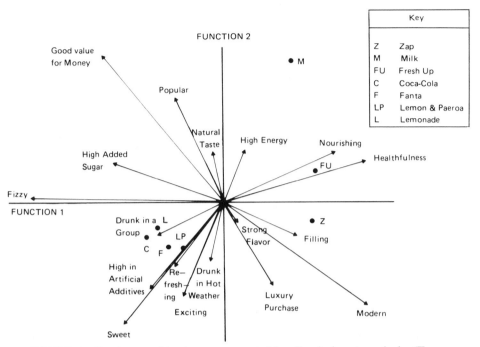

FIGURE 1   Product positioning map created by discriminant analysis. (From Fearon, 1982).

with the problem of expressing the relationships among a set of objects. For example, consider the similarity or dissimilarity among a set of products as perceived by consumers. We can assume that consumers see a group of products as being more or less similar on a number of dimensions rather than only one. But it is often impossible for consumers to say exactly what these dimensions are or, even if they could, to say where each product is located in relation to these dimensions.

However, multidimensional scaling methods such as discriminant analysis can be used to *infer* the number and type of dimensions which underlie perceived similarities or dissimilarities. Multidimensional scaling also develops a geometric representation of the "product space" such that products that are perceived as similar are positioned near one another and products that are seen as dissimilar are far apart. The objective of the exercise is to find the minimum number of dimensions of the set of objects compared and then to locate the objects in relation to these dimensions (and to each other).

Thus all multidimensional scaling methods have two features in common. First they try to discover the criteria used by people when comparing a set of objects, then they locate, or position, each object in relation to these criteria. The result is a two-dimensional or three-dimensional "map" with the criteria as its axes and the objects as points in the spaces between the axes. Discriminant analysis is only one of a number of multidimensional scaling methods, but it does have the advantage that a statistical test can be used to determine if observed perceptual differences between objects could have occurred by chance.

In this particular study, discriminant analysis of the attribute ratings for the seven products involved produced a two-dimensional map of the soft-drink market. This map is illustrated in Figure 1. In addition to the two axes representing the underlying combinations of attributes that best distinguish between the products, and the positions of the products themselves, the map displays the 18 product attributes. They are shown as lines radiating out from the intersection of the axes; the longer the line, the more important the attribute is in differentiating one product from another, and the direction of one line relative to another indicates the degree of association between the attributes they represent.

From the product positioning map created by discriminant analysis the relationship between Zap and other drinks is immediately obvious. Coca-Cola, Fanta, lemonade, and Lemon & Paeroa are closely grouped together in an area of the map which is described by the attributes "sweet," "exciting," "refreshing," and "drunk in a group and in hot weather." Milk and Fresh Up are found in the opposite direction, in an area described by attributes such as "nourishing" and "healthful." Zap is positioned closer to Coca-Cola than to milk, but is obviously not regarded as a direct competitor of the "fizzy" drinks. This suggests that the Dairy Board may have been successful in differentiating Zap from ordinary milk on some attributes, but that overall it was still regarded by teenagers as a flavored milk rather than a soft drink.

By looking at the attributes that are closest to the axes of the product positioning map, we can gain an insight into the underlying features which are most important when teenagers judge the differences among drinks. In fact, with a bit of deduction and intuition we can describe each of these axes (or discriminant functions, as they are called) by giving it a label that summarizes the attributes which characterize it. What is more, discriminant analysis tells us which of these axes is most important in differentiating among the drinks in our study.

The horizontal axis in Figure 1 is the more important of the two. This axis is closely associated with attributes related to health; at one end with nourishment and healthfulness, at the other end with high added sugar

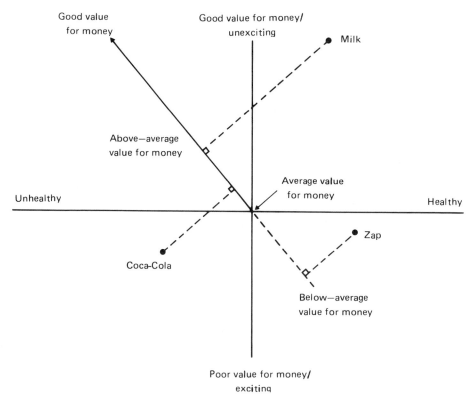

FIGURE 2   Relationship between products and attributes.

and fizziness (which we know is equated with unhealthy). We labeled this axis "healthfulness" and concluded that the major factor that distinguishes this group of drinks from each other in the eyes of teenagers is how healthy each drink is perceived to be.

The vertical axis, the second discriminant function, contrasts attributes such as "good value for money," "popular," and "natural tasting" with "sweet," "high in artificial additives," "exciting," "modern," and "a luxury purchase." This axis is not so easy to describe concisely, but we labeled it "value for money" after noting that teenagers appear to regard good-value-for-money drinks as unexciting.

Examination of the length and direction of the attributes displayed on the product positioning map also allows us to draw a number of conclusions about the relative importance of these attributes and the relationships between them. For example, because the line representing "fizzy" is relatively long, we conclude that this attribute is more important in differentiating between the drinks studied than "strong flavor," which has a very short attribute line.

From the fact that the lines representing "nourishing" and "health-fulness" are close together and point in the same direction, we deduce that teenagers see these attributes as very similar. On the other hand, "good value for money" and "refreshing" are at right angles, which indicates that there is no relationship between them; and "good value for money" and "luxury purchase" point in completely different directions, indicating that they are direct opposites as far as teenagers are concerned.

By drawing a perpendicular line from a particular drink to a particular attribute, it is possible to see how that drink compares with other drinks on that attribute. This is illustrated in Figure 2. From this diagram we can see that milk, for example, is regarded by teenagers as a better value for money than Coca-Cola, which in turn is considered better value for money than Zap. However, while both milk and Coca-Cola are perceived as being above-average value for money, Zap is seen as below-average value for money (which is not surprising given that, at the time, a 500-ml carton of Zap was about 10 cents more expensive than a 333-ml can of Coca-Cola and more than twice as expensive as a 600-ml bottle of milk.

CONCLUSIONS

The main conclusion of this study was that the Dairy Board's attempt to position Zap as a soft drink had failed. In retrospect there were several reasons why this happened, several of them completely beyond the Board's control. As the first cartoned liquid product sold in New Zealand, Zap inevitably generated a great deal of controversy. This controversy was fueled by arguments over the advantages and disadvantages of cartons, heightened by the fact that Zap was a milk-based product and hence subject to the political attention of the entire dairy industry.

Critics of Zap attacked it on the grounds that it was overpriced, created litter, caused hyperactivity in children, and used imported packaging materials and hence would damage the New Zealand glass manufacturing industry. The milk vendors claimed that Zap was the thin end of the wedge of Dairy Board plans, which would ultimately lead to the demise of home milk delivery service. The adverse publicity generated by these groups opposed to Zap focused attention on the fact that Zap was a milk product. This made it very difficult for the Board to convince consumers that Zap was an alternative soft drink rather than simply another flavored milk, and the successful positioning of Zap depended on achieving this objective.

The town milk industry responded to the launching of Zap by conducting its own "Flavour it Instead" campaign, which promoted milk and milk flavorings and created further confusion in consumers' minds. In addition, the type of packaging used for Zap, the aseptic Tetra Brick, was totally unknown in New Zealand, and this accentuated the problems of introducing a new product onto the beverage market in competition with soft drinks. So, while the Dairy Board had no illusions about the difficulty of competing in the soft-drink market, it could not have anticipated the magnitude of the task it had set itself.

Despite these problems it was apparent from the product positioning map that the Dairy Board had achieved some success with Zap. The Board had clearly differentiated Zap from milk and positioned it toward the exciting end of the soft-drink spectrum with no loss in perceived healthfulness compared to milk. On the basis of this observation and the other information which Figure 1 provides about teenagers' perceptions of the soft-drink market, we concluded that Zap should not try to compete directly with the carbonated soft drinks. In our opinion a better strategy was to seek a unique position in the market for Zap as an exciting and healthy drink.

There are many practical situations in which the sort of product mapping illustrated in this study can help to provide answers to important questions. For example, by including an hypothetical "ideal" brand in the group of products rated, a firm may be able to identify a new product opportunity if none of the existing products is close to the theoretical ideal. Similarly, by observing the location of its product in relation to others in the market a firm can see which product features or advertising elements it should change to shift its product toward or away from its competitors.

Thus discriminant analysis not only allows simultaneous comparison of objects on a number of attributes, it also enables the locations of these objects to be portrayed in the form of a map, or picture (and, incidentally, it allows various statistical tests to be carried out on the positions of the objects and the significance of the attributes). This picture which discriminant analysis creates may not be worth a thousand words, but as an aid in visualizing complex relationships among objects it is often worth two or three hundred at least.

REFERENCES

Fearon, D. C. (1982). A product positioning study using both metric and nonmetric multidimensional scaling techniques. Unpublished research report, Massey University, Palmerston North, New Zealand.

Rados, D. L., and P. Gilmour (1981). Big M. In *Australian Marketing Casebook*, University of Queensland Press, Queensland, Australia.

10

Understanding Patient-Physician Communication

William B. Stiles
Miami University
Oxford, Ohio

Samuel M. Putnam
St. Mary's Hospital
Rochester, New York

Mary Casey Jacob
University of London
London, England

INTRODUCTION

Much of science consists of taking things apart and putting them back together again. Being able to disassemble and reassemble something is one way of showing that we understand it. Sometimes, particularly when the thing to be understood is a process such as a conversation between two people, the taking apart and putting back together must be conceptual rather than physical.

A process that we have been studying is the medical interview—the verbal interaction between physician and patient, in which the physician's expertise is brought to bear on the patient's medical problems, to relieve the patient's worry and suffering. In this chapter we explain how we have conceptually disassembled and reassembled one portion of the medical interview, medical history-taking.

The disassembly has been accomplished by classifying each physician and patient utterance according to a verbal interaction coding system. Coding "atomizes" the conversations into frequency counts of various categories of utterances. The reassembly has been by the statistical procedure of *factor analysis*, which groups categories that tend to occur together in the same interviews.

As usually conducted, a medical interview consists of three more-or-less distinct segments, (1) a *medical history*, in which the patient presents the problem(s) that prompted the visit and the physician gathers information about the background and circumstances of the problem; (2) a *physical examination*, in which the physician examines the patient's body and conducts procedures designed to test alternative diagnoses; and (3) a *conclusion segment*, in which the physician offers diagnoses and explanations, gives directions for further tests, prescribes treatment, and so on. We have chosen to concentrate on the medical history segment in this chapter, but the same procedures for disassembling and reassembling have been applied to the other segments and could be applied to any type of verbal encounter.

TAKING IT APART

To take the history apart, we have used a coding system that focuses on one aspect of the verbal interaction, the speech acts. A speech act is what is *done*, as opposed to what is *said*, when someone says something. For example, in saying "What brought you to the clinic today?" the physician *asks a question*; in saying "I have pains in my leg," the patient makes a *self-disclosure*. Questioning and disclosing are categories of speech acts.

The speech act coding system that we have used is a taxonomy of verbal response modes (VRMs), which includes eight basic categories, question (Q), acknowledgment (K), interpretation (I), reflection (R), disclosure (D), edification (E), advisement (A), and confirmation (C). According to the VRM system, each utterance is classified twice, once with respect to its grammatical *form* or literal meaning, and once with respect to its communicative *intent* or occasion meaning (the meaning intended by the speaker on that occasion). Category summaries are given in Table 1.

An utterance's form and intent may be the same or different. To illustrate, "Would you open your mouth?" would be coded Q(A)—question in form (inverted subject-verb order) with advisement intent (i.e., the speaker means to give a polite directive, not to ask a question). The VRM system thus allows 64 different classifications—8 "pure modes," in which form and intent coincide, and 56 "mixed modes," in whic form and intent differ. However, in any one type of conversation, such as the history segment of a medical interview, most utterances fall in a relatively small number of categories. (Different categories are prominent in different types of conversations.)

Of course, speech act coding looks at only one aspect of human

TABLE 1   Verbal Response Modes[a]

Mode		Grammatical form	Communicative intent
Question	(Q)	Interrogative, with inverted subject-verb order or interrogative words	Requests information or guidance
Acknowledgment	(K)	Nonlexical or contentless utterances; terms of address and salutation	Conveys receipt of or receiptive to other's communication; simple acceptance, salutations
Interpretation	(I)	Second person ("you"); verb implies an attribute or ability of the other; terms of evaluation	Explains or labels the other; judgments or evaluations of other's experience or behavior
Reflection	(R)	Second person; verb implies internal experience or volitional action	Puts other's experience into words; repetitions, restatements, clarifications
Disclosure	(D)	Declarative, first-person singular ("I") or first person plural ("we"), where other is not a referent	Reveals thoughts, feelings, perceptions, intentions
Edification	(E)	Declarative, third person (e.g., "he," "she," "it")	States objective information
Advisement	(A)	Imperative; or second person with verb of permission, prohibition, or obligation	Attempts to guide behavior; suggestions, commands, permission, prohibition
Confirmation	(C)	First-person plural ("we") where referent includes other	Compares speaker's experience with other's; agreement, disagreement, shared experience, or belief

[a]Both the form and intent of each utterance are coded. Form is written first, intent second in parentheses; thus Q(A) means question form with advisement intent.

interaction, albeit an important one. It would also be possible to code medical interviews according to the content of what was said, the nonverbal display of emotion, or any of a variety of other aspects.

Table 2 shows the *VRM profiles*, the mean percentage of the modes used most commonly by physicians and patients in medical histories. These data are from a study of 115 interviews with adult female patients, tape recorded with patients' permission in a primary care clinic in Chapel Hill, North Carolina. The interviews were all first visits to the clinic for that episode of illness. Presenting problems were highly varied and typical of a primary care practice. [For more details, see Stiles et al.

TABLE 2   Verbal Response Mode Profiles: Medical History[a]

	Mean percentage of utterances	
Mode[b]	Physicians	Patients
Q(Q)	36.7	0.8
K(K)	15.5	3.6
I(K)	9.8	0.6
R(R)	5.9	0.1
D(D)	2.9	14.8
E(D)	0.3	13.6
K(D)	0.4	5.4
E(E)	4.0	15.2
D(E)	2.1	19.2
K(E)	0.1	10.6
Other	16.3	8.6
Unscored	6.0	7.5
Total	100.0	100.0
Mean number of utterances	104.9	134.5

[a]N = 115 interviews.
[b]Q, question; K, acknowledgment; I, interpretation; R, reflection; D, disclosure; E, edification. Form is written first, intent second in parentheses. Other, modes used for less than 5% of utterances by physicians and patients; unscored, inaudible, incomprehensible, or coder disagreement.

(1982).] Each interview was transcribed verbatim and then coded independently by three trained VRM coders. A final "composite" set of codes for each transcript was compiled on the basis of two-out-of-three intercoder agreement on each utterance.

Not surprisingly, as Table 2 clearly shows, physicians and patients play very different roles in medical histories. Most physician utterances can be characterized as attentive—oriented toward eliciting and receiving information from patients. These include questions ("When did the pain start?", "Did you have a cough?"), reflections ("Then you stayed home anyway," "So you've been feeling well generally"), and acknowledgments—both in pure form ("mm-hm," "yeah"), coded K(K), and in the form of evaluative words ("right," "okay"), coded I(K) (see Table 1). Most patient utterances were informative—oriented toward giving information to the physician. The information was both subjective (disclosure intent) and objective (edification intent), and it was cast in the forms of disclosure (first person), edification (third person), or acknowledgment ("yes," "no," "mm-hm").

I have a sharp pain in my leg. D(D)
My leg hurts. E(D)
(Does your leg hurt?) Yes. K(D)
My ankle is swollen. E(E)
I have a swollen ankle. D(E)
(Is your ankle swollen?) Yes. K(E)

PUTTING IT BACK TOGETHER:
THE VERBAL EXCHANGE STRUCTURE

Our strategy for reassembling the medical history was to fit the VRM pieces together into *verbal exchanges*—associated speech act categories that have functional relationships with each other. For example, questions and answers might form one kind of exchange. Describing the history's verbal exchange structure includes identifying what types of exchanges are usually present and specifying which speech act categories make up each type of exchange.

In this chapter we focus on only the major exchanges—those involving modes used for an average of 5% or more of either physicians' or patients' utterances. As Table 2 shows, 10 categories met this criterion: physician Q(Q), K(K), I(K), and R(R) and patient D(D), E(D), K(D), E(E), D(E), and K(E). Although this inclusion criterion is in terms of percentages, we will use mode *frequencies* (i.e., the actual number of utterances in each mode) for reconstructing the verbal exchange structure. The concept of verbal exchanges implies that an interview might have high levels of all types of exchanges or low levels of all types of exchanges, whereas having a high *percentage* of some modes would necessitate low percentages of others (since percentages must sum to 100). In fact, these medical histories did vary greatly in length, ranging from 16 to 288 physician utterances (average, 104.9) and from 5 to 701 patient utterances (average, 134.5).

We used factor analysis to determine how modes were grouped in these medical histories. The notion underlying factor analysis is that a large number of measures (such as mode frequencies) can often be understood as representing a much smaller number of common, underlying factors (such as verbal exchanges). If several physician and patient modes are regularly used together as an exchange, interviews in which that sort of exchange is common will tend to have high frequencies of *all* of the constituent modes, whereas those interviews may or may not have high frequencies of other modes. Consequently, modes that constitute a particular type of exchange should "go together." If one of

TABLE 3 Intercorrelations of Mode Frequencies[a]

	Physician modes				Patient modes					
	Q(Q)	K(K)	I(K)	R(R)	D(D)	E(D)	K(D)	E(E)	D(E)	K(E)
Physician modes										
Q(Q)	1.00									
K(K)	0.41	1.00								
I(K)	0.58	0.18	1.00							
R(R)	0.60	0.53	0.55	1.00						
Patient modes										
D(D)	0.45	0.65	0.32	0.45	1.05					
E(D)	0.45	0.50	0.26	0.35	0.67	1.05				
K(D)	0.60	0.43	0.44	0.40	0.40	0.54	1.00			
E(E)	0.45	0.61	0.27	0.37	0.82	0.56	0.42	1.00		
D(E)	0.48	0.62	0.32	0.46	0.86	0.57	0.38	0.88	1.00	
K(E)	0.79	0.39	0.58	0.62	0.36	0.26	0.57	0.42	0.43	1.00

[a]N = 115 interviews.

TABLE 4   Principal-Axis (Unrotated) Factor Matrix for Medical History[a]

		Factor I	Factor II
Physician modes	Q(Q)	0.78	0.42
	K(K)	0.73	−0.29
	I(K)	0.58	0.55
	R(R)	0.71	0.33
Patient modes	D(D)	0.83	−0.43
	E(D)	0.70	−0.29
	K(D)	0.69	−0.25
	E(E)	0.80	−0.42
	D(E)	0.83	−0.39
	K(E)	0.72	0.52

[a]N = 115 interviews.

them is common in particular interview, the others will tend to be common in that interview; if one of them is rare, the others will tend to be rare, too. The frequencies of modes from different exchanges should not be so predictable from each other. Thus by seeing which modes "go together," it may be possible to identify groups of modes that make up distinct types of exchanges.

The statistical procedure of factor analysis includes three steps: (1) intercorrelating the variables (in this case, the mode frequencies), (2) extracting the common factors, and (3) rotating the factor axes to obtain the best conceptual fit. We will first present the results of these three steps and then return to interpret each step's meaning for verbal interaction in the medical history.

Table 3 gives the intercorrelations of the frequencies of the 10 major modes across the 115 interviews. A *correlation coefficient* is a measure of association between two variables that can range from −1.00 to +1.00. Positive values indicate a direct relationship—interviews high on one mode tend to be high on other positively correlated modes—whereas negative values would indicate an inverse relationship—interviews high on one mode would tend to be low on negatively correlated modes. A correlation of 0.00 would indicate that the frequency of one mode gives no information about the frequency of the other. (Note that only half of the table is given. The other half would be a mirror image, giving the same information. Correlations on the major diagonal are all 1.00, since the frequency of each mode is perfectly correlated with itself.)

The second step in finding association among the major VRM categories is to extract the common factors. As explained in texts on

TABLE 5   Varimax-Rotated Factor Matrix for Medical History[a]

		Factor 1 Exposition	Factor 2 Closed question
Physician modes	Q(Q)	0.31	0.83[b]
	K(K)	0.74[b]	0.26
	I(K)	0.08	0.80[b]
	R(R)	0.32	0.71[b]
Patient modes	D(D)	0.91[b]	0.22
	E(D)	0.72[b]	0.24
	K(D)	0.36	0.64[b]
	E(E)	0.88[b]	0.22
	D(E)	0.88[b]	0.25
	K(E)	0.20	0.87[b]

[a]N = 115 interviews.
[b]Factor loading ⩾ 0.6.

factor analysis (e.g., Harman, 1967), there are a variety of mathematical techniques for extracting factors from a matrix of correlations, and a variety of considerations and controversies surrounding the choice of which to use. We have used the principal-axis method, a commonly used approach. (Fortunately, for data in which the factor structure is fairly clear and simple, different methods produce very similar results.)

Table 4 shows the results of extracting two common factors. Each extracted factor can be considered as an imaginary measure of the interview, a sort of composite index made up of some weighted combination of mode frequencies. The entries in Table 4 are *factor loadings*, which may be understood as the correlations of each VRM measure with the (imaginary) common factors. The first principal factor (factor I in Table 4) is the one "best" index of all the variables in the analysis; that is, a dyad's "score" on this imaginary, composite measure would be the best single predictor of all 10 major mode frequencies. As another way to say the same thing, the first principal factor explains the maximum amount of variance on the 10 variables entered in the analysis. The second extracted factor (factor II, Table 4) is the best *additional* index; in conjunction with factor I, it adds most to one's ability to predict the mode frequencies (or, equivalently, it explains the most additional variance).

The two extracted factors together accounted for 71% of the total variance of the variables represented in the correlation matrix. (Note, however, that the variables were standardized so that each one's variance

Stiles, Putnam, Jacob

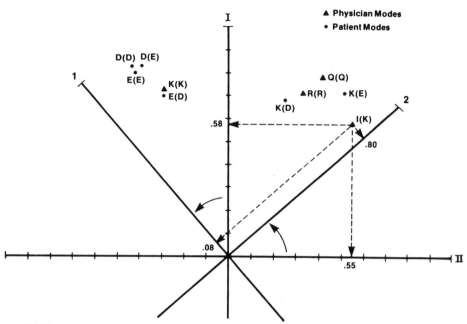

FIGURE 1   Spatial representation of factor matrices; patient and modes plotted with respect to the principal-axis extracted factors, I and II, and the varimax-rotated factors 1 and 2.

was 1.0, whereas the actual variances of the mode frequencies differed across modes.) It would have been possible to consider additional factors, but they each accounted for less than 10% of the total variance, whereas each of the 10 modes accounted for 10%. Since the purpose of factor analysis is to *reduce* the number of factors needed to understand the interview, it makes sense to consider only factors that "explain" more variance than would be explained by one of the original measures.

The final step is *rotating* the factors—converting the unrotated matrix of extracted factors (Table 4) into an equivalent matrix that has certain conceptually desirable features. The rotated factor matrix is given in Table 5.

The term *rotating* is based on a spatial representation of the factor matrix. This is illustrated in Figure 1, in which the 10 major modes are plotted with respect to *both* sets of factor axes, the extracted factors I and II and the rotated factors 1 and 2. As the figure shows, the sets of loadings in Tables 4 and 5 describe the same relationships among the modes, but they give coordinates with respect to sets of axes that are "rotated" with respect to each other. From a mathematical viewpoint, these points could be equally well described by any pair of axes in the plane. However, there are conceptual advantages to the rotated axes, 1 and 2. The points in Figure 1 can be described, approximately, as being in two clusters, one consisting of physician K(K) and patient D(D), D(E), E(E), and E(D), and the other consisting of physician Q(Q), R(R), and I(K) and patient K(E) and K(D). In the rotated version, axis 1 passes as close to the first cluster, and axis 2 as close to the second cluster, as they can if the axes are to remain at right angles to each other. This geometrical situation is represented in Table 5 by factor 1 having high loadings for modes in the first cluster and relatively low loadings for modes in the second cluster, and vice versa for factor 2. This is an approximation to *simple structure*; each factor represents (approximately) a distinct cluster of variables and may be interpreted as representing conceptually the core of meaning that is common to variables in the cluster.

There are many alternative criteria for deciding exactly how factor axes should be rotated, and the choice is at least partly a matter of taste, since rotations are mathematically equivalent. We have used a popular method, called *varimax*, which *max*imizes the *vari*ance among the factor loadings, that is, by finding a rotation in which each factor has many high loadings and low loadings, but few intermediate loadings, an approximation to simple structure.

The varimax rotation constrains the axes to remain *orthogonal* (i.e., at right angles to each other). However, inspection of Figure 1 shows that the clusters could be even better represented by and *oblique* rotation, one in which the axes were inclined slightly toward each other, so as to pass through the centers of the two clusters. We have not done this because oblique rotations are mathematically much more complex and because the orthogonal varimax solution is adequate for our purposes.

INTERPRETING THE RESULTS

With the data presented, we can now return to consider each step's contribution to understanding medical history-taking. Table 3's most striking feature is that all the correlations are positive. This indicates that longer interviews tended to include more of all of the major modes than one would expect if each mode were used independently of the others. However, the correlations were not perfect, and they varied considerably. To illustrate, a patient who used a great many *D(E)*s was very likely to have used many *E(E)*s (correlation = 0.88), but she was only moderately likely to have used many *K(D)*s (correlation = 0.38). Note also that the correlation coefficients are independent of the actual numbers of utterances (cf Table 2), so that "many" must be interpreted as relative to the average for each mode; for example, "many" physician questions would be a much larger number than "many" physician reflections because questions are much more common than reflections.

The first extracted factor in Table 4 is the one "best" index of all major modes used in the interviews. It could be interpreted as a "length of interview" factor, since all of the major modes load highly and positively. The second extracted factor (factor II, Table 4), the best *additional* index (in conjuction with factor I), has both positive and negative loadings. It could be interpreted as a qualitative polarity representing *how* participants spend their time, given the interview's length as represented by factor I. As shown in Figure 1, the positively and negatively loading modes fall into fairly well-defined clusters which correspond to the clusters identified in the rotation.

The rotated factors (Table 5) indicate clearly that there are two major types of verbal exchange used in these medical histories. They represent two distinctly different ways in which physicians gather information from patients. The first, which we call *exposition*, consists of physician $K(K)$—pure acknowledgments—and four patient modes, $D(D), E(D), E(E)$, and $D(E)$ These patient modes consist of first-person (disclosure form) and third-person (edification form) declarative statements that convey subjective information (disclosure intent) or objective information (edification intent).

Pt: Last Friday I felt like I had some phlegm right here in my throat. $D(D)$
Dr: Uh-huh. $K(K)$
Pt: And when I spit it, I noticed it wasn't white. $D(E)$
      It was kind of brownish-looking. $E(E)$
Dr: Mm-hm. $K(K)$
Pt: I don't have any other complaints. $D(D)$
      So I called $D(E)$
      and they said, "Well, you have to go to the screening clinic before we can send you to . . . whatever." $E(E)$
Dr: Mm-hm. $K(K)$
Pt: And so that's why I'm here. $E(D)$.

The second kind of exchange we call *closed question*. In this type of exchange, physicians ask questions, coded $Q(Q)$, while patients respond with "yes" or "no" answers, coded $K(D)$ (subjective information) or $K(E)$ (objective information). Physicians also sometimes acknowledge receipt of information using evaluative words such as "good," "right," or "okay," coded *I(K)*, or they repeat or summarize the patient's communication, coded *R(R)*.

Dr: So your arm doesn't get better during day. *E(R)*.
By the time you're home in the evening, you actually feel worse. *R(R)*
Pt: Yes. *K(C)*
Dr: When you first noticed it hurting, did it come on suddenly? *Q(Q)*
Were you doing anything that might have aggravated it? *Q(Q)*
Pt: No. *K(E)*
Dr: Okay. *I(K)*
Do you have any pain when you move your neck? *Q(Q)*
Pt: No. *K(D)*
Dr: Okay. *I(K)*
Do you think you arm has lost any muscle mass? *Q(Q)*
Is it still the same size as it was? *Q(Q)*
Pt: Yeah, *K(E)*
I think it's the same size. *D(E)*

[*Note*: This excerpt includes examples of two modes not included in Table 2, that is, modes used less than 5% by both participants. As indicated in Table 1, utterances that use third-person forms to restate the other's communication are coded *E(R)*. Utterances that use acknowledgment forms to agree or disagree with the other's presumptions are coded *K(C)*.]

Exposition and *closed question* represent two distinct ways in which physicians gather information from patients. In *exposition* exchanges, patients tell their story "in their own words" (i.e., in declarative sentences) whereas in *closed-question* exchanges, information is transmitted in "yes" or "no" answers to questions—in effect, "in the physician's words." Both types of information gathering are used in most medical interviews. However, a particular interview may have a lot or a little of each type.

The alternative pairs of factors given in tables 4 and 5, illustrated in Figure 1, give equivalent descriptions of these histories. That is, histories may be considered as varying in (1) overall quantity and (2) quality—exposition versus closed question—or as varying in (1) *exposition* exchanges and (2) *closed-question* exchanges. The choice of which to use may be made on the basis of taste or conceptual preference. Our preference has been to use the rotated factors.

Our research has also shown that the amount of *exposition,* but not *closed question,* is related to the amount of satisfaction and rapport that patients report (Stiles et al., 1979). To asses this, we constructed indexes of *expostion* and *closed question* and correlated these with patients' scores on a paper-and-pencil satisfaction questionnaire that they completed immediately after their visit. (The exchange indexes can be constructed in several ways. Perhaps the simplest is the sum of the frequencies of the modes in each of the two clusters—those indicated by footnote b in Table 5. Other methods take into account the different loadings of each mode on each factor. Similar results are obtained in either case.) Patients with higher *expostion* indexes were more likely to agree with such questionnaire items at "The doctor seemed interested in me as a person," "The doctor seemed warm and friendly to me," and, "I really felt understood by this doctor." It is interesting that these items refer to the *doctor's* warmth, interest, and understanding, whereas the modes in the exposition exhange are mainly *patient* modes, and the physician's verbal contributions are restricted to simple acknowledgments.

Finding a correlation between *exposition* and patient satisfaction does not prove that exposition *causes* patients to be satisfied. Perhaps patients who are predisposed to report high rapport on questionnaires are also likely to talk more freely to their physicians. Nevertheless, the correlation is at least consistent with the hypothesis that physicians might improve their patients' feelings of rapport by being more attentive listeners—by allowing patients to tell their story in their own words. In our current research, we are testing this possibility by training physicians to encourage more *exposition* in history-taking, to see if there are then impovements in patient satisfaction and cooperation with their medical treatment.

CONCLUSIONS

If understanding can be shown by taking something apart and putting it back together, the verbal exchange approach shows one way to understand medical interviews. Interviews can be "atomized" by classifying each utterance according to what kind of speech act it represents and then reassembled by combining classes of speech acts into verbal exchanges that accomplish the interview's purposes, through the use of factor analysis. The major factor-based exchanges in the medical history segment of interviews include *exposition* exchanges, in which patients tell their story in their own words, and *closed question* exchanges, in which physicians probe for specific information and patients answer "yes" or "no." Patients seem more satisfied (they report greater feelings of rapport) if they engage in more *exposition*, whereas the amount of *closed question* seems unrelated to patient satisfaction.

ACKNOWLEDGMENT

The research on which this chapter is based was supported by Grant HS 03040 from the National Center for Health Services Research, OASH.

REFERENCES

Harman, H. H. (1967). *Modern Factor Analysis.* University of Chicago Press, Chicago.

Stiles, W. G., S. M. Putnam, M. H. Wolf, S. A. James, (1979). Interaction exchange structure and patient satisfaction with medical interviews. *Med. Care 17*, 667–681.

Stiles, W. B., S. M. Putnam, M. C. Jacob (1982). Verbal exchange structure of initial medical interivews. *Health Psychol. 1*, 315–336.

11

The Why and Wherefore of Factor Analysis

Hans J. Eysenck
University of London
London, England

Factor analysis is part and parcel of the general topic of correlational study; its main purpose is to analyze matrices of correlations between a large number of tests or test items, and reduce the tremendously complex sets of often thousands of intercorrelations to a few dimensions or factors a combination of which can account for all the nonerror correlations actually observed. It closely resembles in principle a Fourier analysis of waveforms, where complex waves are reduced to a combination of simple sine waves. The technique has been widely used in psychology and other social sciences, and also occasionally in physics and other "hard" sciences. Mathematically, it is perfectly acceptable, but is has also been criticized, not so much in terms of its mathematical derivations, but rather in terms of its application and the conclusions drawn from it.

Factors are hypothetical descriptive or causal entities which might be responsible for observed correlations between elements, whether these be items in a personality questionnaire, tests of intelligence, items in a social attitude questionnaire, or whatever. Let us assume that an investigator is interested in the social attitude dimension of conservatism-radicalism. He might write a series of items which he believes to be related to this dimension, or which earlier research has suggested to be relevant. Having administered the resulting questionnaire to a suitably chosen sample of subjects, he could then intercorrelate the items and inspect the resulting matrix. He should find that all the "radical" items should correlate

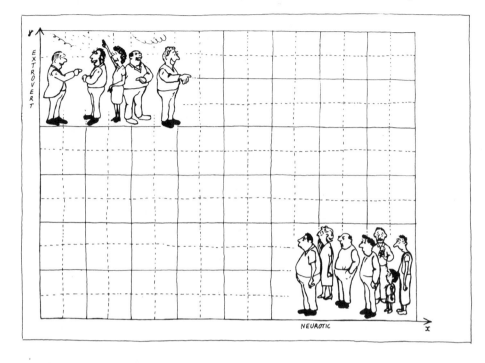

positively, the "conservative" items should correlate positively with each other, and the "radical" items should correlate negatively with the "conservative" items. Factor analysis would simply objectify such an eyeball inspection of the data; it would produce a "factor" of radicalism-conservatism, giving a "loading" to each item (positive or negative) indicating its correlation with radicalism-conservatism. This factor, and these loadings, would be derived from the observed intercorrelations, and if nothing but this single factor was active in producing the correlations, would completely account for them.

This seldom happens, of course, and the investigator might find that after he had derived the best possible estimate of a radicalism-conservatism factor, some residual correlations might be left over which might be high enough to call for further factors. Thus religious items might correlate together, over and above their conservative content, to form an additional "religiosity" factor. Items relating to unorthodox sexual behavior might correlate together, over and above their "radical" content, to define a factor of sexual expressiveness, and so on. A factor loading simply indicates that an item correlates to such and such an extent with the hypothetical factor, and a successful factor analysis results in a small number of factors the loadings on which would enable one to reconstruct all the intercorrelations between items within the limits of the sampling error. Mathematically, the method is simple, straightforward, and quite orthodox. There are of course technical problems, such as the significance of a factor statistically, the number of factors required, and so on; these need not detain us here as they are irrelevant to the logic of factor analysis.

Psychologists seem curiously ambivalent about factor analysis, being divided into those who use it very widely and consider it perhaps the most important multivariate technique of all, and those who consider it dangerous and misleading and prefer alternative techniques, such as analysis of variance. Mathematically, of course, the two techniques are not in any sense opposed to each other, but are more or less equivalent; the differences appear more readily in the type of question that is often asked. In analysis of variance we make an a priori decision as to the variables we regard as important, and in terms of which we dichotimize or otherwise divide our population into groups: male/female, old/young, extravert/introvert, and so on. We then seek to analyze the data to find out which of these divisions significantly influences the variance of the particular test or measure that is the focus of our interest.

In factor analysis, what is often done is to use the analysis not to test hypotheses, but to generate them. We may ask a large number of questions on social attitudes, say, and then look at their intercorrelations in order to derive some hypotheses about the way in which these attitudes can be organized into groups or factors ("what goes with what"). It is not necessary to proceed in this way, however, and my own approach has rather been to use factor analysis to test hypotheses. When this is done the technique is rather similar to analysis of variance, although perhaps more suitable to the kind of data involved, which are usually quantified in terms of correlations (Eysenck, 1952, 1953).

How can factor analysis be used for the study of specific hypotheses? Consider the following investigation (Eysenck, 1958). The study was based on a series of hypotheses which can briefly be stated as follows.

1. A trait of neuroticism is postulated which is characterized by anxiety, depression, moodiness, and a variety of other feeling and behavior patterns, some of which are expressed in question form as shown in table 1, keyed for "N."
2. Similarly, the personality trait of extraversion-introversion is postulated, which is made up of behavior patterns relating to sociability, activity, liveliness, and so on; items supposedly representing this trait are given in Table 1, keyed for "E."

TABLE 1   Six Neuroticism (N) and Six Extraversion (E) Questions

Questions	Key
A. Do you sometimes feel happy, sometimes depressed, without any apparent reason?	N
B. Do you prefer action to planning for action?	E
C. Do you have frequent ups and downs in mood, either with or without apparent cause?	N
D. Are you happiest when you get involved in some project that calls for rapid action?	E
E. Are you inclined to be moody?	N
F. Does your mind often wander while you are trying to concentrate?	N
G. Do you usually take the initiative in making new friends?	E
H. Are you inclined to be quick and sure in your actions?	E
I. Are you frequently "lost in thought" even when supposed to be taking part in a conservation?	N
J. Would you rate yourself as a lively individual?	E
K. Are you sometimes bubbling over with energy and sometimes very sluggish?	N
L. Would you be very unhappy if you were prevented from making numerous social contacts?	E

3. It is hypothesized that extraversion and neuroticism are entirely independent of each other (i.e., that factors of N and E would constitute orthagonal dimensions, that is, show zero correlation).

Correlations were run between the 12 questions, and the intercorrelations factor analyzed and two factors extracted. Three such analyses are shown in Table 2. One relates to the original sample used, constituting 200 men and 200 women, with the two sexes analyzed separately; a third sample, constituting a further male group of 200, was also used. It will be seen that for all three samples there are two very clear-cut factors, recognizable as E and N in terms of the items correlating highly with each factor (having high loadings). All the items that had been hypothesized to measure E have high loadings on one factor and practically zero loadings on the other; conversely, all the items hypothesized to constitute neuroticism have high loadings on the other factor, and practically zero loadings on the first. The independence of the two scales was demonstrated by the fact that their actual correlation was −0.05 in the third group, and only slightly larger in the two groups constituting the original sample. All these findings are exactly as predicted in terms of the hypothesis, and thus serve to support the general theory on which these predictions were based.

TABLE 2   Results of Three Factor Analytic Studies of the Personality
Inventory Given in Table 1

Item	Present sample			Original sample		
	E	N	E_m	N_m	E_f	N_f
A	0.01	0.75	−0.10	0.79	−0.05	0.72
B	0.48	0.01	0.70	−0.09	0.73	0.03
C	−0.06	0.74	0.03	0.82	0.08	0.58
D	0.59	0.04	0.59	0.12	0.66	0.10
E	−0.09	0.71	−0.04	0.75	0.04	0.69
F	0.02	0.58	−0.13	0.57	0.00	0.50
G	0.59	−0.06	0.72	−0.15	0.66	0.14
H	0.49	−0.04	0.58	−0.09	0.51	−0.04
I	−0.06	0.58	−0.06	0.67	−0.03	0.59
J	0.68	−0.02	0.87	−0.05	0.65	−0.16
K	0.09	0.63	0.23	0.55	0.17	0.43
L	0.64	0.09	0.67	0.03	0.58	−0.08

We can enter scores on these two short but reliable questionnaires in
an analysis of variance, incorporating such variables as sex, class, and
age; when this was done it was found that sex has a significant main
effect for both neuroticism and extraversion scores, men being more
extraverted but less neurotic. All these findings are merely intended to be
illustrative; they have been replicated many times on much larger
samples and using much longer questionnaires (Eysenck and Eysenck,
1969). Indeed, hundreds of studies in many different parts of the world
have shown that these two personality dimensions or factors emerge from
any large-scale study of personality questionnaires, rating studies, or
objective laboratory tests, wherever they are conducted; the factors have a
truly universal applicability (Eysenck and Eysenck, 1982).

Let us now consider Figure 1, which gives a diagrammatic representation of this two-factor structure. It illustrates dramatically the clustering
of the six E items, on the one hand, and the six N items on the other, as
well as the orthogonality of the two dimensions involved. The findings
are very clear-cut, but they should not be overinterpreted. Let us take the
question of independence of the two factors. Suppose that we had
incorporated two additional questions relating to impulsivity in our
questionnaire, had calculated their correlations with the other questions,
and had carried out a factor analysis of the new set of 14 questions; what
would have happened? Impulsivity items correlate both with E and with
N; as a consequence, they would have appeared in the now virginal first

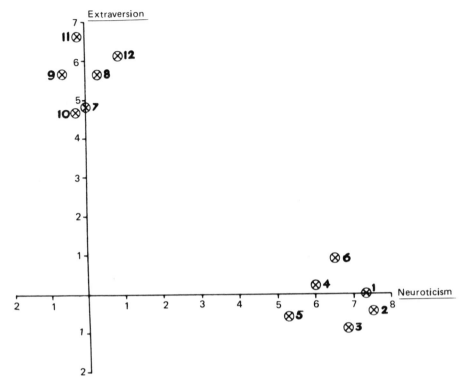

FIGURE 1   Relative position in a two-dimensional space of six neuroticism and six extraversion questionnaire items.

quadrant, and would have attracted the two axes closer to each other, giving a positive correlation between E and N! (The reason for this is, of course, that the axes are in fact located by joining the origin to the center of each cluster (hence the name "centroid" for this type of analysis and rotation of factors); and addition of new items may change the center of gravity of each factor and hence alter the relative position of the axes. (This can be avoided by insisting on retaining *orthogonal* axes rather than using *oblique* ones, but doing this imposes arbitrary assumptions on the analysis and does not properly represent observed reality.)

This raises an interesting question. What is the true correlation—that found in our original set of 12 items or that found in the enlarged set of 14 items? The answer to this question, of course, is that there is no "true" answer; E and N are concepts, not things occurring in real life, and scientific concepts are inevitably to some extent subjective and must be judged not in terms of "truth" or "falsehood," but in terms of the degree to which they mirror empirical reality. The subjectivity involved in the formulation of hypothesis, and the analysis of data, is not greater in psychology than in the "hard" sciences; this is a point often disregarded by psychological critics who are not well aware of the position of the "hard" sciences.

In science, measurement is often determined by agreement with theory. Consider the measurement of temperature in terms of liquid-in-glass thermometers. Clearly, water would not be a good liquid to use because it contracts from the ice point (0°C) to the temperature of maximum density (4°C), thus giving an illusory decline in temperature when actually the temperature is increasing. In actual fact the liquids most widely used (mercury and alcohol) were chosen in part because they fit in best with the kinetic theory of heat, which predicts that the final temperature reading of a fluid obtained by mixing two similar fluids of masses m_1 and m_2 at the initial temperatures t_1 and t_2 should be

$$t_f = \frac{m_1 t_1 + m_2 t_2}{m_1 + m_2}$$

The linseed oil thermometer was discarded because measurements made with the instrument did not tally with the predictions made by the kinetic theories; mercury and alcohol thermometers do tally. Thus the choice of the measuring instrument is based in part on its agreement with theory; the same is true of psychological measurement.

To apply this principle to our example, we would say that if a theory demands that E and N should be independent, the 12-item questionnaire is superior to the 14-item questionnaire. If we wanted to include questions on impulsivity, we should also include a similar number of questions on sociability. These correlate positively with E but negatively with N and thus appear in the fourth quadrant; they would serve to pull back the axes into orthogonality.

This may seem a very suspect way of working, but it is based on two essential findings which have been replicated many times. In the first place, many different groupings of questions relating to the general area of extraversion and neuroticism have given rise to axes which are essentially orthogonal (Royce, 1973). Some studies find a positive relationship, some find a negative relationship, but there is little systematic bias to all this. The best summary of all this work would therefore be that the hypothesis of orthogonality, being simplest, should be accepted until and unless there is good evidence that it is definitively contradicted by research findings. In evaluating all this evidence, one would bear in mind that for various reasons one would expect a negative correlation between the two variables in groups high on N, for reasons given elsewhere (Eysenck, 1967); high-N subjects are afraid of other people and hence tend to evince less sociability than would otherwise be expected. (In a similar way liquid-in-glass thermometers tend to give poor measurement in situations where the temperature is very high or very low, for obvious reasons.) Let me repeat, then, that the orthogonality of E and N is not "true" in any absolute sense; it is merely a convenient fiction, which summarizes a large number of observations and simplifies analysis and conceptualization.

Most factor analysts rest content with the desciptive analysis of the variables involved, furnished by the dimensions or factors emerging from the statistical manipulations involved. My own approach has always been that this is insufficient, that we must go on to a causal analysis of the dimensions involved, to get away from the inevitable subjectivity of correlational and rotational procedures involved in factor analysis. But before turning to a consideration of this point, on which I probably differ profoundly from most of the leading factor analysts, I would like to consider some objections that have been made to factor analysis by a variety of critics.

One of the most frequently heard criticisms is that the factor analytic procedure is essentially *tautological*. Thus we may be told that a trait of sociability is inferred from the intercorrelations between items which describe sociable behaviour, and that this trait of sociability, derived from these behovioral items, is then used to *explain* the occurrence of sociable behavior. This objection is similar to that which used to be made of instinct theories. We observe that people behave in a sociable manner, we then posit an instinct to account for that behavior; however, the instinct is posited because of the original observations which it is then supposed to explain.

We have already seen that this objection is unreasonable. Factor analysis attempts a more detailed and quantitative *description* of observable behavior; it does not attempt to make a causal analysis. Accurate observation and description are at the basis of all scientific endeavor; until such descriptions are available, it is difficult to posit any testable causal hypotheses or to investigate further the determinants of such behavior. The objection, therefore, turns out to be based on a misunderstanding of the purpose of factor analytic methods.

The second objection, also frequently encountered, is that you only get out of factor analysis what you originally put into it. In other words, when you put items measuring sociability into your scale, what you will get out is the factor of sociability. Up to a point this objection is of course correct; what you do not put into your tests, no factor analysis will be able to get out of them as a factor. However, the objection assumes that we already *know* what we are putting into the test items, but of course that is not so; we may have certain hypotheses or hunches as to the nature of the contents of our scales, and the factors to be expected, but there is no guarantee that what we thought we were doing is what actually happened. Factor analysis is required to confirm our hypotheses,and quite frequently it does not do so.

In essence there is a similarity to qualitative analysis in chemistry. It would be true there also that what we get out of the analysis (i.e., the composition of some alloy or material given to us) is only what is already contained in that alloy or material; indeed, this is self-evident. However, we would not have to undertake the analysis if we already knew what was in fact contained in the alloy or material; in this sense clearly the analysis reveals something new and not previously known (although it may previously have been suspected, or form the basis a of hunch or theory).

Several examples may usefully be given to illustrate this point, because it is a crucial one in the evaluation of the usefulness of factor analysis in psychological research, and because for many people it constitutes the major reason for rejection. Guilford and Guilford (1936) put forward a hypothesis that "social shyness" (i.e., the opposite of sociability) constituted a single general factor and constructed a scale to measure this trait. Eysenck (1956) put forward an alternative hypothesis. Based on the detailed study of social behavior of people of known degrees of neuroticism and introversion, he suggested that *introverted* social shyness is different in many ways from *neurotic* social shyness (Eysenck, 1956, p. 121).

> To put the hypothesis suggested here in a nutshell, we might say that the introvert does not care for people, would rather be alone, but if need be can effectively take part in social situations, whereas the neurotic is anxious and afraid when confronted with a social situation, seeks to avoid them in order to escape from this negative feeling, but frequently wishes that he could be more sociable. In other words the introvert does not *care* to be with other people; the neurotic is *afraid* of being with other people. If this hypothesis were true it seems likely that different items in the *S* (social shyness) scale would be chosen by introverts and neurotics respectively to express their non-sociable attitude.

In a correlational and factorial study Eysenck (1956) demonstrated that indeed the Guilford Scale broke into two quite independent halves, correlating with *E* and *N*, respectively. Here are some typical items that indicate the *introverted* type of social shyness:

1. Do you like to mix socially with people?
2. Are you usually a "good mixer"?
3. Are you inclined to keep in the background on social occasions?
4. Do you usually take the initiative in making new friends?
5. Do you like to have social engagements?
6. Are you inclined to limit your acquaintances to a select few?
7. Do you always have a "ready answer" for remarks directed at you?
8. Do you enjoy getting acquainted with most people?

(It will be obvious whether the answer "yes" or "no" is the appropriate one for a socially inhibited person.)

Here are some items that are indicative of *neurotic* social shyness:

1. Do you often experience periods of loneliness?
2. Are you troubled about being self-conscious?
3. Are you self-conscious in the presence of your superiors?
4. Are you usually well-poised in your social context?
5. Are you worried about being shy?
5. Do you often feel ill at ease with other people?
6. Are you troubled with feelings of inferiority?
7. After a critical moment is over, do you usually think of something you should have done but failed to do?
8. Do you worry over humiliating experiences longer than the average person?

These items are all from Guilford's "social shyness" scale, but it will be seen that they differ in content, and the study showed that they differ in their relation to the major factors of neuroticism and introversion, the items in the first group going with the introversion factor, the items in the second group going with the neuroticism factor. Clearly, Guilford attempted to select only items that would measure a general factor of sociability, and critics might have said that because all the items related to sociability, therefore it was tautological, and a foregone conclusion, that a single factor of sociability would emerge. However, clearly Guilford's hypothesis was erroneous, and he included two quite different types of sociability in his analysis; it required a factorial study to disentangle these two factors and to demonstrate that Guilford had been wrong.

Consider another example. It used to be customary in the literature on personality and social attitudes to posit a single factor of "suggestibility,' and in the literature many tests are given to measure this hypothetical factor. It might consequently be supposed that when a factor analysis is made of the intercorrelations between a whole sets of these tests, the discovery of a single factor of suggestibility was tautological and inevitable. Eysenck and Furneaux (1945) carried out such an analysis and found two major factors, which they called "primary" and "secondary" suggestibility. Primary suggestibility is of the ideomotor type, calling for motor movements following a suggestion that such a movement would take place. Secondary suggestibility is of a perceptual type, requiring agreement of perceptions with certain suggestions. There is also a suggestion of a tertiary type of suggestibility, of a more sociable nature, making people change their attitudes and opinions when confronted with a majority attitude or opinion. It is clear that the people who originated these various tests, assuming that they were all tests of a single trait of "suggestibility," were in fact wrong, and that again it needed a properly conducted factorial study to demonstrate the dimensionality of this particular universe of tests. Indeed, primary and secondary suggestibility are quite uncorrelated, and relate quite differently to external variables, such as hypnotizability. Primary suggestibilitiy is closely related to hypnotizability and predicts it with considerable accuracy; secondary suggestibility is not at all related to hypnotizability, nor does it predict it in any way. Thus clearly at the descriptive end, factor analysis is essential if we want to test descriptive hypotheses in the personality field.

We have seen in the case of sociability and suggestibility that where hypothesis and popular belief posit a single factor, factor analysis demonstrates this belief to be wrong and documents the existence of two or more factors. It is also possible that the original hypothesis of the existence of a particular factor is completely erroneous, and that *no* factor corresponding to the hypothesized dimension or trait exists at all. As an example, consider the hypothesis put forward by Kretschmer (1948), and widely accepted, of the existence of a fundamental personality variable named by him schizothymia-cyclothymia, which in its most extreme form is characterized by the distinction between schizophrenics and manic depressives. Kretschmer designed a number of objective laboratory tests which, according to him, measured this particular dimension, and this view was widely accepted. Kretschmer and his students never performed a factor analytic study to demonstrate that the test actually intercorrelated in the expected manner, giving rise to a "positive manifold" (i.e., positive correlations throughout), and defined a single general factor.

To test this hypothesis, Brengelmann (1952) gave to 100 normal subjects a number of objective behavior tests which, according to Kretschmer, were diagnostic of the schizothymia-cyclothymia dimension. The set of intercorrelations failed to show any tendency for the tests to intercorrelate positively or significantly, and we must conclude that the tests, all of which are declared by Kretschmer to measure the dimension schizothymia-cyclothymia in normal people, do not, in fact, produce a pattern of intercorrelations which would support the hypothesis. Hence factor analysis can be used to disprove a theory, and clearly what we get out of the analysis is not at all what Kritschmer thought was being put into it.

Even when a general factor is postulated and discovered, factor analysis can add a great deal to the detail of the final results. Let us take intelligence. Sir Francis Galton had postulated that all cognitive abilities share one common general factor of "intelligence." A. Binet had hypothesized that we are dealing rather with a set of separate abilities, with intelligence being nothing but the average of all these disparate entities. Factor analysis has shown that both were right. There is indeed a very strong general factor, giving rise to the "positive manifold" which is usually observed in intercorrelations between IQ tests, but in addition we have a number of minor factors or special abilities (verbal ability, numerical ability, visuospatial ability, memory, etc.) which add an important element to our picture to cognitive ability. Factor analysis thus showed what was right in both these divergent approaches and was able to reconcile them in a more complex hierarchical scheme (H.J. Eysenck, 1979, 1982).

Another argument often advanced against factor analysis is this. Factor analysis essentially discloses to us the position relative to each other of the tests or items intercorrelated in a multidimensional space determined by the rank of the matrix that is formed by the inter-correlations (rank 2 in the case of Figure 1). To put it in a slightly more mathematical fashion, if we take any two items in Figure 1, draw a line linking the position of the item in the two-dimensional space with the origin, and calculate the angle between the two lines, the scalar product is in fact the correlation between these two items. Factor analysis tells us, and this is an important contribution, that we can locate all the items in a two-dimensional space, in such a way that the scalar products correspond to their correlations within the limits of sampling errors. This is a very important item of knowledge, particularly when it agrees with the hypothesis on which the whole procedure was based. However, the procedure cannot identify the actual position of the dimensions or factors in this two-dimensional space; in the actual mathematical process of analysis these serve the purpose of a scaffolding which can be taken down once the building is completed. Clearly, we can put our two dimensions or factors in any position in this two dimensional space in Figure 1 without altering the relative postion of the dots identifying the test items. In this particular instance the two axes actually drawn do of course have a privileged position, as they clearly go through the centroids of the two groups of items, and indeed Thurstone made this the basis of a statistical rule for rotating factors.

But the position is not always, or even usually, so clear cut; it is clear cut in this case because the items were selected on the basis of a specific hypothesis, which in turn was based on a great deal of previous work, and consequently the results came out very clearly and precisely. This is not always so, and when it is not, the problem of rotation of factors becomes one of great complexity, often involving a considerable degree of subjectivity. To the degree that such subjectivity is involved (and note that no statistical rule, including Thurstone's rule of "simple structure," can guarantee the scientific and psychological meaningfulness of the factors extracted), we are departing from scientific rigor.

Here indeed we have a weakness of factor analysis, which becomes apparent most clearly when we are dealing with the use of factor analysis to suggest theories rather than to test theories. When theories are being tested, we know precisely what to expect and can express the deviation from such expectation in a fairly precise mathematical manner. When we do not know what to expect, however, subjectivity rules supreme, and we may choose arbitrarily between any number of possible and mathematically equivalent solutions. This, unfortunately, is the position in the great majority of factor analytic studies undertaken. The reason for this is often that such studies are undertaken with the minimum of foresight or theoretical consideration of the issues involved; Ph.D. students, and even advanced workers, often find themselves with a large group of data, possibly collected for a different purpose or for no specific purpose at all; they then resort to a correlational and factorial analysis of these data, in the hope of getting something out of this mass of data. This is clearly not an ideal way of doing research, and the fact that it is often done in this way has brought factor analysis into disrepute. All that one can say is that it is not the fault of the method that it is being misused in this fashion and that although occasionally useful suggestions may emerge from a factor analysis of such data, it would usually be better to throw the data away entirely and start again. The old adage "garbage in, garbage out" applies with equal force to factor analysis as to other types of statistical analysis.

The advantages and disadvantages of factor analysis have both to be recognized by the serious student. Confirmed factor analysts only seem to recognize the former, while habitual critics only recognize the latter. Both are wrong. Factor analysis fulfills an interesting and important function in the analysis of psychological data, but it does not furnish us with the strong kind of evidence that the experimentalist would like in order to fix firmly the variables with which he or she is dealing. To do that, or so I like to suggest, we have to go further and formulate causal hypotheses which can explain in a meaningful manner the nature of the observed correlations. Such causal hypotheses can then be investigated experimentally, along the lines of univariate analysis, and provided that the predictions are indeed borne out by the experimental or observational results, lend support to the general scheme. Without such support we will continue to be faced by alternative rotations of axes and interpretations of factors, and proliferation of explanatory concepts.

In the quest for such a causal explanation of, say, extraversion-introversion, we might be led first to ask the question of whether individual differences along this dimension are determined more by genetic or environmental factors. There is now a great deal of evidence on this point to indicate that genetics plays a significantly larger part than environment in causing these differences (Eaves and Eysenck, 1975; Fulker, 1981), while as far as the environmental contribution is concerned, within-family factors play an important part, whereas between-family factors are comparatively unimportant. Such findings are important because they suggest certain consequences which may lead to a better understanding of the factor itself and may lead to a proper causal analysis.

One obvious consequence which would be expected in terms of the genetic contribution to the factor is that the same factor should appear in other cultures. Cross-cultural studies are notoriously difficult, but it is possible to translate questionnaires into different languages, administer them to large random samples of the population, and carry out correlational and factor analytic studies to demonstrate that the matrices of intercorrelations are sufficiently similar between cultures to give rise to similar or identical factors. This has been done on some 20 different cultures, ranging from European (England, Germany, France) to communist (Yugoslavia, Hungary, Bulgaria), Third World (India, Nigeria), to Asian (Japan, Hong Kong) countries. In all of these an essentially identical pattern of intercorrelations was observed, giving rise to identical or very similar factors to those observed in English populations (Eysenck and Eysenck, 1982). Thus factorially derived personality factors seem to be culture independent, which is an important addition to our knowledge.

The fact that a person's position on the dimension extraversion-introversion is determined largely by genetic factors, and the universality of the trait in many different cultures, suggests that it must have a firm biological basis and a physiological theory of extraversion-introversion has been put forward by the writer elsewhere (Eysenck, 1967, 1981). The theory links extraversion-introversion differences with differences in the habitual level of arousal in the cortex, mediated by the so-called reticular formation; it states, in particular, that people who have a habitually low level of cortical arousal will tend to develop extraverted patterns of behaviour, while people with a high level of cortical arousal would tend to develop introverted behavior patterns. People who are intermediate with respect to cortical arousal will also tend to be intermediate with respect to degree of extraversion-introversion. Such a theory can be tested along many different lines. One test would be using psychophysiological measures of various kinds (EEG, EKG GSR) on persons chosen on the basis of questionnaires to be high on extraversion or high on introversion; our knowledge of the physiology of arousal enables us to make quite specific predictions as to the differences to be observed (M.W. Eysenck, 1982). Stelmack (1981) has surveyed the evidence and finds that on the whole predictions tend to be borne out.

Psychological laboratory investigations can also be used to test the theory, because many of the behaviors they have studied (conditioning, learning, memory, etc.) are dependent on differences in arousal and hence testable predictions can be generated. Hundreds of such studies have indeed been carried out (Eysenck, 1976), and on the whole the evidence is strongly in favor of the theory (Eysenck, 1977, 1981).

A third line of investigation consists of looking at social behaviors (sexual, criminal, neurotic, etc.), where again predictions can be made from the theory and tested by observation and experiment. Many such studies have been done, including work on smoking, drinking, sexuality, and so on, and on the whole the results have again been favorable to the theory (Wilson, 1981). When seen in the context of such a nomological network, the "factor" of extraversion-introversion assumes a much more scientific and rational aspect than if it were simply the outcome of statistical manipulations of doubtful legitimacy, such as rotation of axes.

The same course of development that has characterized the concept of extraversion-introversion from theory from factor analysis to experimental study has characterized the development of other concepts, such as intelligence. Here, too, we start with a wide-ranging hypothesis, such as that put forward originally by Sir Francis Galton, go on to the construction of tests followed by correlational and factor analytic studies, and then determine genetic contributions leading to the hypothesis of underlying biological or physiological factors. These in turn are then hypothesized in a testable manner, and recent work using the evoked potential on the EEG has shown that very high correlations (up in the 80s) can be obtained between typical IQ tests such as the Wechsler or the

Ravens Matrices and the evoked potential (Eysenck, 1982a). Such findings make much more real and concrete the notion of "intelligence" as a heritable, biological foundation for "intelligent" behavior, problem solving, cognitive differentiation, and the like. In a very real sense it is only these last stages that properly legitimize the earlier ones; without these excursions into causality, critics like Guilford (1967) could still maintain that a fragmentation approach to intelligence, such as is embodied in his 120-dimensional "model of intellect," is equally feasible on a purely statistical, factor analytic basis, as is the hypothesis of a general factor of intelligence.

It will now be clear to the reader why I believe that factor analysis is an important and useful method, neither as conclusive in its results as its friends believe, nor as subjective and inconclusive as its critics think. For many purposes in personality research it represents an intermediate step between theory formation and measurement. The theorist conceives of a particular hypothetical trait, but this does not guarantee that such a trait "exists" in a meaningful sense or that the instruments of measurement available are capable of isolating it. If the correlations between different instruments or measurements define a factor reasonably close to the original hypothesis, we have clearly made an important advance, although alternative rotations of the factor structure might suggest alternative hypotheses. These can be ruled out only by going on to construct causal hypotheses which can be tested by means of univariate analyses the outcome and statistical significance of which can be assessed along orthodox lines. It is only when this process is complete that we can accept with real confidence the results of the factor analytic studies.

Thus the particular approach to factor analysis which I have always advocated stresses the importance of theory preceding the collection of data and the importance of causal analysis following the factor analytic work. Factor analysis undertaken merely to "see what happens" when a matrix of correlations between variables grouped together for no theoretical reasons is being analyzed, or reliance on the outcome of a particular rotation of factors without going on to make sure that this particular rotation was indeed psychologically meaningful, are both errors frequently found in the literature, tending to devalue factor analysis and make it unacceptable in the eyes of psychologists, who hold more rigorous views of acceptable evidence than do many psychometrists.

Indeed, it has to be admitted that many psychometrists are more interested in the statistical details of analysis than in the quality of the data that are being analyzed; it should be obvious that statistical refinements are appropriate only when the basic data are sufficiently accurate to make such refinements worthwhile. The same applies to the size of the population tested. Frequently, the number of subjects is too small to give factor loadings of any degree of reliability and replicability; in our work we found that groups of 500 are usually a minimum to guarantee reproducibility. A ratio of number of subjects to number of tests correlated of something like 5:1 seems to be a reasonable minimum to insist on for any scientifically valuable results to be achieved.

It is a sad reflection on the state of the art that while most textbooks of factor analysis will give detailed instructions on the statistical side, few if any deal with what one might call the logical basis of factor analysis and the rules that should be followed in applying it to theoretical and practical problems (Eysenck, 1953). Obviously, textbook authors feel safer when dealing with the reasonably objective methods of matrix algebra, but from the point of view of the reader who has to apply these methods to actual problems, there is a huge gap between the imposing structure of statistical sophistication and the abysmal reality of the psychological literature resulting from the application of the methods in question. It is always easier to learn the mechanics of scientific research than to discover the spirit, and factor analysis has been no exception. In equal measure, analysis of variance has frequently been abused by neglecting what in many investigations are the major factors in accounting for portions of the total variance, namely individual differences. Relegating these to the error term, which has been the almost unvarying practice of experimental, social, industrial, educational, and clinical psychologists, they have managed to reduce the portion of the total variance explained by the main factors and their interaction to a minute proportion of the total. Here again there has been a lack of theoretical consideration prior to experimentation, and the minute refinements of the statistical technique only served to obscure the absence of fundamental theoretical thought underlying the experiment. Truly, in psychology, the sophistication of the statistical methods we employ far exceeds the crude simplicity of the data for whose analysis they are used and the inadequate theorizing that usually precedes their application. In this respect, factor analysis is no better and no worse than any other statistical technique.

REFERENCES

Brengelmann, J. C. (1952). Kretschmers zyklothymer und schizothymer Typus im Bereich der normalen Persönlichkeit. *Psychol. Rundsh. 3*, 31–38.

Eaves, L., and H. J. Eysenck (1975). The nature of extraversion: a genetical analysis. *J. Pers. Soc. Psychol. 32*, 102–112.

Eysenck, H. J. (1952). Uses and abuses of factor analysis. *Appl. Statist.* *1*, 45–47.

Eysenck, H. J. (1953). The logical basis of factor analysis. *Am. Psychol.* *8*, 105–114.

Eysenck, H. J. (1956). The questionnaire measurement of neuroticism and extraversion. *Riv. Psicol.* *50*, 113–140.

Eysenck, H. J. (1958). A short questionnaire for the measurement of two dimensions of personality. *J. App. Psychol.* *42*, 14–17.

Eysenck, H. J. (1967). *The Biological Basis of Personality.* Thomas, Springfield, Ill.

Eysenck, H. J. (1976). *The Measurement of Personality.* M.T.P. Press, Lancaster, England.

Eysenck, H. J. (1979). *The Structure and Measurement of Intelligence.* Springer, New York.

Eysenck, H. J. (1981). *A Model for Personality.* Springer, New York.

Eysenck, H. J. (1982). *A Model for Intelligence.* Springer, New York.

Eysenck, H. J., and S. B. G. Eysenck (1969). *Personality Structure and Measurement.* Routledge & Kegan Paul, London.

Eysenck, H. J., and S. B. G. Eysenck (1982). Recent advances in the cross-cultural study of personality. In *Advances in Personality Assessment* (C. D. Spielberger and J. N. Butcher, eds.). Lawrence Erlbaum, Hillsdale, N.J.

Eysenck H. J., and D. Furneaux (1945). Primary and secondary suggestibility. An experimental and statistical study. *J. Exp. Psychol.* *35*, 485–503.

Eysenck, M. W. (1977). *Human Memory: Theory, Research and Individual Differences.* Pergamon Press, Oxford.

Eysenck, M. W. (1982). *Attention and Arousal.* Springer, New York.

Fulker, D. W. (1981). The genetic and environmental architecture of psychoticism, extraversion and neuroticism. In *A Model for Personality* (H. J. Eysenck ed.). Springer-Verlag, New York, pp. 88–122.

Guilford, J. P. (1967). *The Nature of Human Intelligence.* McGraw-Hill, New York.

Guilford, J. P. and R. B. Guilford (1936). Personality factors S, E and M, and their measurement. *J. Psychol.* *2*, 109–127.

Kretschmer, E. (1948). *Körperbau und Character.* Springer-Verlag, Berlin.

Royce, J. R. (1973). The conceptual framework for a multi-factor theory of individuality. In *Multivariate Analysis and Psychological Theory* (J. R. Royce, ed.). Academic Press, London.

Stelmack, R. M. (1981). The psychophysiology of extraversion and neuroticism. In *A Model for Personality* (H. J. Eysenck, ed.). Springer, New York.

Wilson, G. D. (1981). Personality and social behaviour. In *A Model for Personality*. (H. J. Eysenck, ed.). Springer, New York.

III
Hypothesis Testing

Are winters getting colder? Did different species of animals come into being by evolution? Is a particular medicine more effective than another? It would be marvelous if definitive answers could be given to such important questions, but unfortunately, these matters are far too complex for that. The widely used statistical technique of hypothesis testing, however, can throw light on even very difficult questions.

Despite the importance and popularity of statistical tests, most people initially find the concepts very confusing. One of the main difficulties is the jargon. A statistical test deals with the question of whether an observed event is a real or a chance effect; that is, it tells us how easy or difficult it is to explain a set of numbers on the basis of chance alone.

In Chapter 12, Jacqueline Dietz uses trends in weather data to introduce the ideas of statistical tests. At one time or another we all speculate or argue about the weather. Farmers discuss at length how much drier it is getting, while at the same time golfers complain about how much more rain there seems to be. Statistical tests provide a rationale for looking at sequences of weather patterns.

The next three chapters explore further the use of statistical tests. Bruce Weir delves into the fascinating world of genetics. As Dr. Weir points out, "Genetic material is now known to be DNA, which exists as long strands consisting of thousands of elements known as nucleotides in a row." The particular sequence of the nucleotides in the DNA in your body has not only decided that you are a human being but that you have a particular sex, eye color, and so on. Weir uses the renowned "chi-square" test to examine some of the futuristic aspects of modern genetics.

Michael Hendy and David Penny also considered nucleic acid sequences and show, in an ingenious way, how these could be used to test the theory of evolution.

During World War II, a statistician named Abraham Wald revolutionized procedures in the quality control of manufactured munitions by introducing the sequential probability ratio test. Howard Edwards introduces and uses Wald's test in the context of medical research.

12

Testing for Trends in the Weather

E. Jacquelin Dietz
North Carolina State University
Raleigh, North Carolina

INTRODUCTION

Does each winter seem colder than the preceding one where you live? Is each summer hotter than the one before? Do unusually cold and unusually hot days seem to come in "spells"? In this chapter, I discuss two procedures that can be used to test whether such conjectures about the weather are supported by actual weather data. I use these procedures to analyze two sets of temperature data from Raleigh, North Carolina. Detailed weather data for hundreds of U.S. cities are available in *Local Climatological Data*, a monthly publication of the National Oceanic and Atmospheric Administration of the U.S. Department of Commerce. Your local library may receive this publication for your city.

Table 1 gives average annual temperatures in Raleigh for the 40 years from 1942 through 1981. These data are plotted in a scatter diagram in Figure 1. Do you see a trend over time in these temperatures? The plotted data may suggest the presence of a cooling trend in Raleigh over the last 40 years. There is certainly considerable variability about any such cooling trend, however. In fact, any apparent trend may be due to chance rather than to any interesting meteorological phenomenon.

A second set of weather data is shown in Table 2. This table shows the departure from the normal daily temperature in Raleigh during the month of November 1981. Do warm days and cool days seem to come in

161

"spells"? Or do cool days seem to be followed by warm days, and vice versa? The data may suggest the presence of spells of relatively warm and cool days. Again, however, there is considerable variability in the data. Any apparent pattern may be due to chance.

Statistical methods can help us to decide whether the apparent trends in these two sets of data are "significant" or whether they are likely to have occurred simply by chance. The two statistical procedures I discuss, the Mann trend test and the runs test, are both nonparametric hypothesis tests. Nonparametric procedures are statistical methods that require relatively mild assumptions for their validity. For instance, such methods do not require the traditional assumption that data have been drawn from a population with a normal distribution—the familiar "bell-shaped curve." Nonparametric methods are often easy to understand and simple to apply. The two tests discussed here are intuitively appealing and require only simple arithmetic to perform.

THE MANN TREND TEST

Concordant and Discordant Pairs

There are many ways to measure and assess the significance of an apparent trend like that shown in Figure 1. To simplify the initial

TABLE 1   Average Annual Temperature in Raleigh, North Carolina, 1942–1981

Year	Temperature	Year	Temperature	Year	Temperature	Year	Temperature
1942	61.0	1952	60.3	1962	58.2	1972	58.5
1943	60.6	1953	61.3	1963	57.5	1973	59.8
1944	59.8	1954	60.1	1964	58.9	1974	59.7
1945	60.3	1955	59.6	1965	59.1	1975	59.6
1946	60.4	1956	59.9	1966	58.7	1976	58.7
1947	59.1	1957	60.1	1967	59.1	1977	58.6
1948	59.8	1958	57.4	1968	58.4	1978	59.1
1949	61.4	1959	59.9	1969	56.9	1979	58.6
1950	59.9	1960	57.9	1970	57.8	1980	59.3
1951	60.1	1961	58.7	1971	59.0	1981	58.3

Source: Local Climatological Data, Annual Summary 1981.

discussion of one method, consider the first four annual temperatures in Table 1.

Year	Temperature
1942	61.0
1943	60.6
1944	59.8
1945	60.3

Notice that 1942 was warmer than 1943, 1944, and 1945. Also, 1943 was warmer than 1944 and 1945. However, 1944 was colder than 1945. Thus, when the years are considered in pairs, in five out of the six pairs, the earlier year was warmer than the later year.

A pair of years such as 1944 and 1945, where the "larger" year is associated with the "larger" temperature, is referred to as a "concordant" pair. A pair of years such as 1942 and 1943, where the larger year is associated with the smaller temperature, is referred to as a "discordant" pair. Thus, for the years 1942–1945, there are five discordant pairs and one concordant pair.

The numbers of concordant and discordant pairs of years can be used to measure trend in the temperatures. A preponderance of concordant pairs suggests a warming trend over time; a preponderance of discordant pairs, a cooling trend. Nearly equal numbers of concordant and discordant pairs provide little evidence of either sort of trend.

In 1945 Henry Mann suggested that the number of concordant pairs minus the number of discordant pairs be used as a measure of trend. His

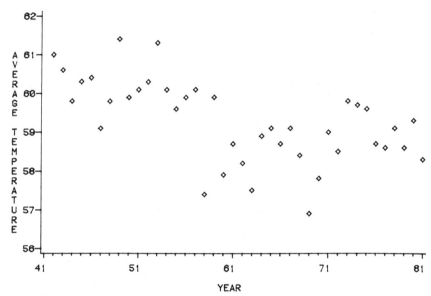

FIGURE 1   Average annual temperatures in Raleigh, North Carolina, 1942–1981.

test based on this measure (say, M) is a special case of a test for independence proposed by Sir Maurice Kendall (1975). Most textbooks on nonparametric statistics include a discussion of Kendall's test.

For the years 1942–1945, M is equal to $1 - 5 = -4$. Large negative values of M suggest a cooling trend, large positive values suggest a warming trend, and values near zero provide little evidence of trend. But what does an M value of -4 mean? Is this value of M sufficiently small to provide evidence of a cooling trend over the years 1942–1945?

Hypothesis Testing

To answer this question, we use the method of reasoning called a statistical hypothesis test. We begin by stating a so-called "null hypothesis," some conjecture about the true state of nature. Usually, the null hypothesis asserts that no interesting effect has occurred—in our case, that the weather shows no interesting trend or pattern. We then decide whether the observed data are consistent with the null hypothesis or whether, instead, the data contradict the hypothesis. If the data strongly contradict the null hypothesis, we conclude that the null hypothesis should be rejected in favor of an alternative hypothesis. This procedure is

TABLE 2   Departures from Normal of Daily Temperatures in Raleigh, North Carolina, November 1981

Date	Departure from normal	Date	Departure from normal	Date	Departure from normal
1	2	11	−2	21	−9
2	6	12	−4	22	−11
3	10	13	−6	23	−9
4	10	14	−1	24	−6
5	8	15	3	25	−3
6	6	16	3	26	−2
7	−4	17	3	27	14
8	−3	18	4	28	6
9	4	19	7	29	−3
10	1	20	9	30	−9

Source: Local Climatological Data, November 1981.

a conservative one in the sense that the null hypothesis is assumed to be true unless there is strong evidence to the contrary.

Our null hypothesis states that there was no real warming or cooling trend between 1942 and 1945; any apparent trend just occurred by chance. We need to decide whether an M value of −4 is consistent with this null hypothesis. Only if an M value of −4 strongly contradicts the null hypothesis will we take the somewhat drastic step of deciding that the weather in Raleigh is changing over time. But what do we mean by "contradict"? *We consider that our data contradict our null hypothesis if the chance of observing such data is very small, given that the null hypothesis is true.* The smaller this chance (known as the p value of the test), the less plausible the null hypothesis seems. Thus we need to evaluate the chance that we would observe an M value like −4 for four annual temperatures if there were really no temperature trend over those years.

Evaluating Chances

If there was no trend over the years 1942–1945, the ordering of the four temperatures 61.0, 60.6, 59.8, and 60.3 was just arbitrary—the temperatures could just as well have been 60.6, 61.0, 59.8, 60.3 or 60.3, 60.6, 59.8, 61.0. In fact, there are 24 possible orderings of the four temperatures. [The number $n(n − 1)(n − 2) \cdots (2)(1)$ (written $n!$ and read "n factorial") gives the number of possible orderings of n distinct objects. Thus, $4! = 4(3)(2)(1) = 24$ is the number of orderings of four distinct temperatures.]

TABLE 3   Possible Orderings of Four Temperatures and Associated M Values

Ordering	M	Ordering	M
59.8, 60.3, 60.6, 61.0	6	60.6, 59.8, 60.3, 61.0	2
59.8, 60.3, 61.0, 60.6	4	60.6, 59.8, 61.0, 60.3	0
59.8, 60.6, 60.3, 61.0	4	60.6, 60.3, 59.8, 61.0	0
59.8, 60.6, 61.0, 60.3	2	60.6, 60.3, 61.0, 59.8	−2
59.8, 61.0, 60.3, 60.6	2	60.6, 61.0, 59.8, 60.3	−2
59.8, 61.0, 60.6, 60.3	0	60.6, 61.0, 60.3, 59.8	−4
60.3, 59.8, 60.6, 61.0	4	61.0, 59.8, 60.3, 60.6	0
60.3, 59.8, 61.0, 60.6	2	61.0, 59.8, 60.6, 60.3	−2
60.3, 60.6, 59.8, 61.0	2	61.0, 60.3, 59.8, 60.6	−2
60.3, 60.6, 61.0, 59.8	0	61.0, 60.3, 60.6, 59.8	−4
60.3, 61.0, 59.8, 60.6	0	*61.0, 60.6 59.8, 60.3*	−4
60.3, 61.0, 60.6, 59.8	−2	61.0, 60.6, 60.3, 59.8	−6

If our null hypothesis is true, any of the 24 possible orderings of the temperatures would have been just as likely as the one that actually occurred. Each of these 24 orderings gives a value of M.

The 24 possible orderings of the temperatures and the corresponding values of M are listed in Table 3. The observed ordering is printed in italics. These results can be summarized in a "frequency distribution":

M	Frequency
−6	1
−4	3
−2	5
0	6
2	5
4	3
6	1

Note that M can range from −6 to +6 for four observations; the 24 values of M are symmetric about zero. Recall that large positive values of M suggest a warming trend; large negative values, a cooling trend; and values near zero, no trend. Of the 24 possible orderings, four give M values of −4 or −6. An additional four orderings give M values of 4 or 6. Each of these eight orderings gives as much evidence of a warming or cooling trend as does our observed ordering. That is, about eight times out of 24, we can expect to see as much evidence of a trend as occurred in

1942-1945, *just by chance.* The probability of observing an M value "as extreme" as -4 (i.e., -4, -6, 4, or 6) is $\frac{8}{24}$ or $\frac{1}{3}$ when the null hypothesis is true. Since this p value is fairly large, our M value of -4 is consistent with the null hypothesis. Therefore, the data from 1942 through 1945 provide no compelling evidence of a change in temperature in Raleigh.

Forty Years of Temperature Data

With just four years of data, it is actually impossible to obtain very strong evidence for a trend in temperature. Even the most extreme values of M, -6 and $+6$, occur about 2 times in 24 when the null hypothesis is true. Therefore, we now use the full sequence of 40 temperatures to test the null hypothesis of no temperature trend between 1942 and 1981. Computation of M is considerably more tedious for 40 years than for four years; the number of pairs of years is 780. [The number of pairs of n objects is given by $n(n-1)/2$. Thus there are $4(3)/2 = 6$ pairs for four years and $40(39)/2 = 780$ pairs for 40 years.] Of these 780 pairs, 215 are concordant and 544 are discordant. The remaining 21 pairs (e.g., 1945 and 1952) cannot be classified as concordant or discordant because the two years have the same annual temperature. The value of M for the observed ordering is $M = 215 - 544 = -329$.

The fact that M is negative is suggestive of a cooling trend. But is an M value of -329 small enough to provide compelling evidence for such a trend? How likely are we to observe a value of M this extreme, just by chance, if the null hypothesis is true?

Solution to Problem

If there was no warming or cooling trend or other temperature pattern during 1942-1981, the ordering of the 40 temperatures was just arbitrary. Any other ordering of the temperatures would have been just as likely as the observed ordering. There are 40! possible orderings, each with a corresponding value of M. Our job is to determine where our M value of -329 fits into the distribution of possible M values.

To attack this problem as before requires listing the 40! orderings and the corresponding M values. But 40! is bigger than 8 followed by 47 zeros! Fortunately, there are tables available giving the frequency distribution of M for different numbers of observations. Textbooks that include a discussion of Kendall's test usually contain such a table. These tables reveal that fewer than 0.001 of the 40! possible orderings of the temperatures give values of M less than or equal to -329. The distribution of M is always symmetric about zero when the null hypothesis is true; therefore, fewer than 0.001 of the possible orderings give values of M

greater than or equal to $+329$. Thus, if all 40! orderings were equally likely, the probability of observing an M value as extreme as -329, just by chance, is less than 0.002. The chance of observing data like ours is very small when the null hypothesis is true. Therefore, the data lead us to reject our null hypothesis; there is compelling evidence of a cooling trend in Raleigh during the years 1942–1981.

THE RUNS TEST

Counting Runs

Let's now explore the possibility of a pattern in the data of Table 2. Consider only the sign of each day's departure from normal, that is, whether it is positive or negative. Then we can represent the data in Table 3 as a sequence of 16 +'s and 14 −'s:

Sequence A: $+++++ +--++----++++++------++--$

We can gain insight into how to look at sequence A by considering certain extreme sequences. Suppose that we had observed the following sequence:

Sequence B: $+++++++++++++++++--------------$

Sequence B suggests strongly that warm and cool days come in spells. If instead, we had observed

Sequence C: $+-+-+-+-+-+-+-+-+-+-+-+-+-+-+++$

we would consider that cool days and warm days alternate. In both cases, we would conclude that the temperatures show a pattern. We can quantify the patterns in sequences B and C by considering the number of "runs" of +'s and −'s. Sequence B has two runs; sequence C, 29 runs. A small number of runs suggests clustering; a large number of runs, an alternating pattern. In sequence A, there are eight runs of +'s and −'s. Do eight runs provide evidence of a pattern in daily temperature?

The appropriate null hypothesis asserts that there was no pattern in the sequence of warm and cool days in Raleigh in November 1981. That is, the observed sequence of +'s and −'s is in a random order; any other ordering of 16 +'s and 14 −'s would have been just as likely.

TABLE 4   Possible Sequences of Three +'s and Three −'s

Sequence	Number of runs	Sequence	Number of runs
+ + + − − −	2	− − − + + +	2
+ + − + − −	4	− − + − + +	4
+ + − − + −	4	− − + + − +	4
+ + − − − +	3	− − + + + −	3
+ − + + − −	4	− + − − + +	4
+ − + − + −	6	− + − + − +	6
+ − + − − +	5	− + − + + −	5
+ − − + + −	4	− + + − − +	4
+ − − + − +	5	− + + − + −	5
+ − − − + +	3	− + + + − −	3

Example

To clarify our approach to this problem, consider a sequence of just three +'s and three −'s. There are 20 possible orderings of three +'s and three −'s, each with a corresponding number of runs. The 20 sequences and the number of runs for each are shown in Table 4. These results can be summarized in a frequency distribution:

Number of runs	Frequency
2	2
3	4
4	8
5	4
6	2

Thus when there are three +'s and three −'s, the number of runs ranges between two and six, with four the most frequent value.

Solution to Problem

In principle, we could use the same reasoning to obtain the distribution of the number of runs for 16 +'s and 14 −'s. If the null hypothesis is true, all possible orderings of 16 +'s and 14 −'s would have been equally likely. Each ordering has a corresponding number of runs. We want to know where the number eight fits into the distribution of possible numbers of runs.

Unfortunately, there are 145,422,675 sequences of 16 +'s and 14 −'s! [The number of sequences of n +'s and m −'s is given by $(m + n)!/m!n!$. In the previous example, $(3 + 3)!/3!3! = 20$; here, $(16 + 14)!/16!14! = 145,422,675$.] Although it is not feasible for us to list all the possible sequences of +'s and −'s, tables of the distribution of the number of runs appear in many textbooks on nonparametric statistics. For a sequence of 16 +'s and 14 −'s, the minimum and maximum number of runs possible are two and 29. (Recall sequences B and C.) The average number of runs for all the possible sequences is about 16. Fewer than 0.01 of the possible sequences have eight or fewer runs.

Thus, if 16 +'s and 14 −'s are in a random order, we expect to observe a sequence with as few as eight runs less than 0.01 of the time. Data like those in sequence A are quite unlikely when the null hypothesis is true. This suggests that we should reject our null hypothesis. There is evidence of a tendency toward clustering of cooler-than-normal and warmer-than-normal days in Raleigh in November 1981.

SUMMARY AND DISCUSSION

Hypothesis Testing

In this chapter some of the ideas of statistical hypothesis testing are introduced through the use of two nonparametric tests for trend. For each test, we begin by stating the null hypothesis that a sequence of temperature data (first, average annual temperatures, and second, signs of departures from normal daily temperatures) is randomly ordered. We then compute some measure of trend (either M, the number of concordant pairs minus the number of discordant pairs, or the number of runs). Such a measure, used in a statistical hypothesis test, is known as a "test statistic." Next, we evaluate the p value of the test—the probability of obtaining a value of the test statistic as extreme as the one observed, given that the null hypothesis is true. This is accomplished in each case by considering all possible orderings of the observations (annual temperatures or signs) and thereby obtaining the distribution of possible values of the test statistic. This method of enumerating equally likely arrangements of the data to find the distribution of a test statistic is used in many nonparametric hypothesis tests. If values of the test statistic like the one observed occur frequently when the null hypothesis is true, there is no reason to reject the hypothesis. But if the observed value of the test statistic is very unlikely given the null hypothesis, we reject the hypothesis. The smaller the p value of the test, the stronger the evidence in favor of rejecting the null hypothesis.

It is important to understand that this method of decision making is not infallible. In fact, there are two distinct types of errors we can make when using this reasoning. We may reject a null hypothesis that is actually true, if, just by chance, we observe an extreme value of the test statistic. Alternatively, we may fail to reject a null hypothesis that is actually false. This may occur for a variety of reasons, among them insufficient data, a poor choice of test statistic, or just bad luck—we may, just by chance, observe a value of the test statistic that is consistent with the null hypothesis. We can never completely eliminate the possibility of error in a hypothesis test. However, statisticians can often evaluate the probabilities of making these two types of errors and can devise hypothesis-testing procedures that minimize these probabilities.

The Weather (and More)

Will annual temperatures in Raleigh decline in the future? Have other cities in the United States or elsewhere shown a cooling trend over the last 40 years? Has there been clustering of relatively cool and relatively warm days in Raleigh in other months or years? Have other cities shown such clustering? We have not and cannot address such questions using the data given herein. Our conclusions are limited to Raleigh, North Carolina, during the time periods for which we have data. Moreover, the statistical significance of the temperature trends considered here tells us nothing about the meteorological events responsible for those trends. We conclude only that our observed patterns of data are unlikely under certain null hypotheses of randomness.

The Mann trend test and the runs test have been used here to test for temperature trends. Other questions we could ask about the weather include: Do rainy days and dry days come in spells? Is the total annual snowfall increasing (or decreasing) over time where you live? There is no need to restrict our questions to the weather. Is the annual energy consumption of your school or office decreasing? Have the winning scores in your favorite Olympic sport improved over time? Have there been runs of Democratic and Republican presidents of the United States? Counting concordant pairs and counting runs are two simple ways to explore apparent trends and patterns in the weather, as well as in sports, politics, and other aspects of daily life.

REFERENCES

Kendall, Maurice G. (1975). *Rank Correlation Methods.* Griffin, London.

Mann, Henry B. (1945). Nonparametric tests against trend. *Econometrica 13*, 245–259.

U.S. Department of Commerce, National Oceanic and Atmospheric Administration, Environmental Data and Information Service (1981). *Local Climatological Data.* National Climatic Center, Asheville, North Carolina.

13

How The Theory of Evolution Could Be Disproved, But Isn't

Michael D. Hendy and David Penny
Massey University
Palmerston North, New Zealand

HISTORICAL THEORIES

Some interesting problems in probability and statistics arise from the study of evolution. For example, how can "predictions" be made and tested about events that happened a long time ago? Can we make tests about the origin of the vertebrate animals that apparently arose more than 500 million years ago? It has been argued (Popper, 1972) that the theory of evolution is not a normal scientific theory because of the supposed difficulties of making predictions about past events. Most philosophers of science expect hypotheses (theories) to lead to predictions that can then be tested by observation or experiments. In this chapter we show that statistics can be used to make tests about unique events that occurred in the past.

It is important to divide the study of science into the two aspects of description and explanation. One of the best known examples comes from the history of science and involves Kepler's and Newton's laws. Kepler proposed that the planets moved in ellipses around the sun (this is the descriptive aspect), and Newton "explained" this motion with his model of gravitational force (and indeed predicted some deviations from elliptical orbits because of the influence of other planets). With evolution, the subject is again treated under the two areas of whether evolution has occured (description) and, if so, by what mechanism (explanation). We

consider mainly the descriptive aspect of whether evolution has occurred, and if so, by what pathways; this aspect was called the "theory of descent" by Darwin. He also proposed that "natural selection" of genetic variants in populations was the main mechanism that explained how evolution occurred.

In this chapter we use information from proteins to test some aspects of the theory. Proteins are formed as linear polymers (strings) of amino acids; there are 20 amino acids that can be included during protein synthesis, any one of the 20 can be included at each position in the polymer. The choice of amino acid at each position is controlled by the RNA sequence and the recognition enzymes. An example of a portion of human, rhesus monkey, and horse sequences of hemoglobin alpha are

Position in Sequence (Numerals)
Amino Acid Occurring (Three-letter Code)

4   5   6   7   8 15   16   17   18   19

Human   ...phe-ala-asp-lys-thr-....-gly-lys-val-gly-ala-...
Monkey   ...phe-ala-asp-lys-ser-....-gly-lys-val-gly-gly-...
Horse   ... ala -ala-asp-lys-thr-....-ser-lys-val-gly-gly-...

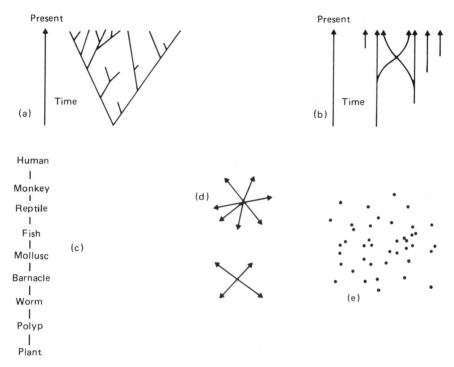

FIGURE 1   The five models that are considered for relationships between species. Darwin's theory of descent is a, Lamarck's suggestion is b, the "great chain of being" is c. Version d is an example of either a simple creationist model where an original protein has diverged since creation, or of an archetype (or master plan) where changes were made at the time of creation. Option e represents the case where there is no "relationship" between species. Additional details are given in the text.

The data are from Dayhoff (1972), a work which, together with its more recent supplements, contains an extensive collection of protein sequences. The next step is to examine theories as to how species may be related.

There are in principle many theories, but we will treat only the five examples shown in Figure 1. In each case we will give predictions that can be made about the similarities of amino acid sequences in proteins that come from species that are similar morphologically. The first example (Figure 1a) is based on the theory of Charles Darwin and A. R. Wallace and first appeared as a diagram in 1859 in Darwin's *On the Origin of Species*. There is a single origin of life (or possibly a few); living forms become modified (in Darwin's explanation) by the natural selection of "random" variants; there is splitting of lines of descent to form new species; other species become extinct. The important prediction is that similar species should have similar protein sequences because they have shared a common ancestor in the recent past.

The second model (Figure 1b) is an earlier theory of Lamarck (1802, 1809), who suggested that life could continually arise "spontaneously." Each newly developed species would then "improve" itself; if it was animal it may, given time, go through the series of worms, barnacles, mollusks, fish, reptiles, aquatic mammals, land mammals, and so on (Burkhardt, 1977). Lamarck believed that species did not become extinct because each species had the ability to adapt to new conditions. With simple versions of this model there is no need for species that are morphologically very similar (say, lions and tigers) to have similar protein sequences because the two species could have have arisen quite independently.

The third model (Figure 1c) represents the "great chain of being," which was a popular eighteenth-century idea that all living organisms could be arranged in one linear, ascending series with man (i.e., white males) at the top. Different people have interpreted this theory as either creationist (species being created in their present form) or as an evolutionary series. Evolution was usually supposed to go from simple to complex, but a few authors suggested that "lower" forms had "degenerated" from humans! On the model of the great chain of being there should be a continuous series of closely related species and the protein sequences should reflect this.

The last two models (Figure 1d and e) are versions of how a creationist model may appear when protein sequences are studied. The "star" (big-bang) tree (Figure 1d) represents a case where a single functional protein (say, hemoglobin beta) of each type was created and used for all species that possessed that protein. Since the supposed time of creation, there could have been changes in the sequences of individual species, but the changes would be independent of each other—by this is meant that a change in hemoglobin beta in penguins is independent of a change in lions—the penguins have no "knowledge" of what is happening to lion hemoglobin. On this simple model there is no reason for protein sequences to be more similar in morphologically similar species than in morphologically different species, nor for the trees derived from two different proteins to be similar.

The final model (Figure 1e) is a simple creationist model where each species is independently created to fit into their environment. A famous early anatomist (Cuvier) argued that similarities between species depended on functional constraints. By this was meant that an organism with a heart that pumped oxygenated blood would require either gills or lungs to oxygenate the blood, as well as a complex nervous system to control the heart and lungs. On this model there is no apparent reason why trees derived from different proteins should give similar trees.

TESTING THE THEORIES

One of the reasons people have been fascinated with theories of evolution is one of historical curiosity; however, the evolutionary process needs to be understood in order to predict outcomes of some gen*t*ic experiments and in order to assist in the classification of biological species. A natural question to ask is the relative relationships between different species such as whether humans are more closely related to the toad or to the snake. In our study we have concentrated on the relationships among a set of 11 mammalian species, represented by one species of each of monkey, sheep, horse, kangaroo, mouse, rabbit, dog, pig, human, cow, and ape. These were selected because of the availability of common data for these species.

The data that we use are from amino acid sequences of proteins, but the tree-building and comparison methods work equally well with nucleic acid sequences (RNA and DNA) or with information about the anatomy, morphology, physiology, or biochemistry of a species. The limitation is that data must have discrete states that can be expressed as integers (e.g., presence or absence of a character state, different nucleotide or amino acids present, etc.) It is not necessary to know which character state is the original ancestral state and which are derived (newer) character states.

If we assume that the observed similarities in the protein sequences have resulted from what Darwin called the "theory of descent," we infer that the sequences ought to be related in an "evolutionary treelike" manner. A common principle in choosing between two competing scientific models is to apply what is termed "Occam's razor" to select the simpler of the two models. The translation of this principle to constructing hypothetical evolutionary trees is the principle of "maximum parsimony" (i.e., "most economical"), where we select the tree that minimizes the differences (i.e., the number of mutations) between the observed sequences and the inferred sequences of their ancestors. We apply this principle to discover which tree, among all the mathematically constructible trees, represents the evolutionary scheme that would require the smallest number of changes (mutations) in the sequence. We call this number of changes the *length of the tree*. An analysis of this and other models, together with the difficulties involved in producing the trees, is given in Felsenstein (1982).

The major difficulty in applying any of the models is that a very large number of possible trees must be considered. For example, for a set of 11 species there are more than 282 million different ways of linking the species as a tree with each species assigned to a distinct end point. Many of these trees are not very discriminating in relating species to their hypothesized ancestors and from the analysis of protein sequences of reasonable length we usually find we obtain the most discriminating trees, the binary trees, in which each internal edge branches into two new edges of the tree. If we restrict our analysis to binary trees only, there are still more than 34 million possible trees to compare.

We developed an efficient searching program (Hendy and Penny, 1982) which enabled us to find all the possible trees of minimal and near-minimal length for sets of up to 15 sequences. We selected the set of 11 mammalian species listed above, for which the sequences of five different proteins were known. The underlying nucleotide sequences were derived for each sequence and then we determined the minimal and near-minimal-length trees spanning these 11 species for each of the five sequences in turn and also for the concatenated sequence obtained by linking each of the five sequences to create one large sequence. From this study we found there were 39 trees whose lengths were minimal or within 1.25% of the minimal length for the given protein. (These 39 trees are displayed in Figure 2.)

The natural question to be asked from examining such a display of trees is: Are they nearly alike or do they represent widely variant trees? We should not expect the trees derived from the different proteins to be identical as the mechanism of evolution as proposed by Darwin is a stochastic (i.e., with a random probabilistic component) rather than a deterministic (i.e., with no random component) process. However, there should be common patterns in the trees representing their supposed common evolutionary history. If the trees had little in common, our study would have revealed that the model was incorrect, or that the data did not reflect the evolutionary history of the species, or that Darwin's "theory of descent" hypothesis was invalid.

Robinson and Foulds (1981) described a method of assigning a distance measure between trees that was naturally associated with their evolutionary nature. Applying this measure to the $741 = 39 \times 38/2$ pairs of trees, we obtained tree difference values, ranging from 0 (1 pair) to 14 (8 pairs) with the majority (508) taking the values 6, 8, or 10. When comparing binary trees the differences must be 0 or positive even integers. The full range of values is given in Table 2.

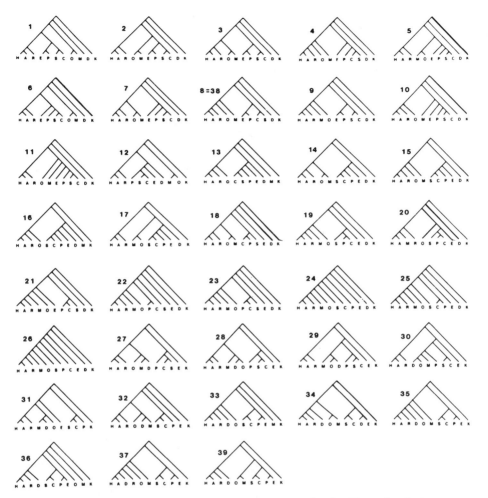

FIGURE 2   The 39 minimal and near-minimal trees for the 11 species that were studied. R, rhesus monkey; S, sheep; E, horse (*Equus*); K, kangaroo; M, mouse; O, rabbit (*Oryctolagus*); D, dog; P, pig; H, human; C, cow; A, ape.

TABLE 1   Numbers of Tree Pairs That Differed
by m in a Random Sampling of 1,100,000 Pairs,
Compared with the Theoretical
Numbers Expected

m	Monte Carlo	Theoretical
0	0	0
2	0	0
4	3	5
6	41	44
8	348	363
10	3,151	3,039
12	25,075	25,196
14	185,548	184,890
16	885,834	886,460

Finding that the value $d(T_8,T_{38})$, the distance between T_8 and T_8, was zero indicated that two of the trees were identical, an observation that was not made even after the trees had been drawn, as there are many different ways of drawing the same tree. What was now required was to analyze this observed distribution to determine whether the trees could be considered to be "nearly identical" or "widely scattered." At that stage no distribution of the tree difference values was known, so we estimated the frequency of the occurrence of these values using a *Monte Carlo simulation*, randomly sampling 1.1 million of the approximately 1000 (English) billion(!) pairs of possible binary trees spanning 11 species.

The Monte Carlo simulation method is a process often used to estimate statistical values of a very large data set which is too large to analyze completely. It is reminiscent of taking opinion polls of a random sample of people to estimate the feelings of the population. A subset of the data set is selected randomly, and the statistical values of the subset are used as an estimate of the corresponding values of the complete set. The statistics of the sample should approximate those of the complete data set with the accuracy of this approximation depending on the size of the sample.

The Monte Carlo simulation indicated that 81% of the randomly selected pairs differed by the maximum possible difference of 16, a further 17% differed by 14, while fewer than 0.33% pairs had a difference of 10 or less. Clearly, as long as the simulation gave an accurate indication of the distribution of these values, the trees we derived must be considered as being closely related.

TABLE 2 Numbers of Pairs of Trees, N of the 741 Pairs Which Differ by the Value m, Compared to the Expected Number of Pairs E, if the Trees Were Independent

m	0	2	4	6	8	10	12	14	16
N	1	53	87	163	200	145	84	8	0
E	0	0	0	0	0	2	17	125	597

Subsequent theoretical analysis of these distributions has confirmed that the distribution of these distances predicted by the simulation were fairly accurate. In Table 1 we contrast the number of times that each value of m occurred in the 1.1 million trials with the proportions of that total as calculated theoretically. In Table 2 we display the observed values from the 741 pairs in contrast to the expected values had the trees been unrelated. It is obvious from comparing the frequencies N and E in Table 2 that the observed differences between these trees is very different from that expected for independent trees. Hence the assumption that these trees are independent does not seem feasible. A more feasible assumption (hypothesis) is then that the trees are dependent or, in other words, related.

Many of these 741 pairs of trees are related because they are near minimal trees derived from the same protein, but if we consider pairs of trees from differing protein sequences only, we can estimate the probabilities that these diverse sequences *do not have* a common treelike relationship. We find that these probabilities range from 0.0011% for the fibrinopeptides A and B to 1.8% for the comparison between hemoglobin beta and fibrinopeptide B. A more detailed analysis of these distributions, together with the data sets, was reported in Penny et al. (1982).

As a conclusion we see that the probability that the five sets of protein sequences had been derived in a manner other than through the same "treelike" processes is very small. This observation is certainly consistent with Darwin's theory of descent and serves to reinforce our acceptance of this theory as being a useful explanation of the natural order of biology. The results also contradict Lamarck's theory, in that similar-looking species appear to have had common ancestors. The trees identified in our study are certainly not the type that could represent the "great chain of being" model (Figure 1c). The coincidences among the protein sequences seem to indicate that the models shown in Figure 1d and e are not in themselves sufficient to explain these. The trees are certainly not "starlike." A creationist model would need to be expanded to incorporate reasons for the consistent treelike relationships among the sequences that occur for each of the five proteins. A model that needs major modification to explain subsequent observed facts has less scientific merit than one whose predictions are subsequently verified.

Our studies have not proved the Darwinian theory of descent; indeed, no experiment can ever prove a scientific theory (Popper, 1972). However, they support our confidence in the theory. Neither do they disprove the creationist model, for that model itself is not a scientific theory, in the sense that it does not predict anything and therefore is logically irrefutable. But as that model makes no predictions, it is therefore of no scientific value.

REFERENCES

Burkhardt, R. W. (1977). *The Spirit of System*. Harvard University Press, Cambridge, Mass.

Dayhoff, M. O. (1972). *Atlas of Protein Sequence and Structure,* Vol. 5. National Biomedical Research Foundation, Silver Springs, Md.

Felsenstein, J. (1982). Numerical methods for inferring evolutionary trees. *Quart. Rev. Biol. 57*; 379–404.

Hendy, M. D., and D. Penny (1982). Branch and bound algorithms to determine minimal evolutionary trees. *Math. Biosci. 59*, 277–290.

Penny, D., L. R. Foulds, and M. D. Hendy (1982). Testing the theory of evolution by comparing phylogenetic trees constructed from five different protein sequences. *Nature 297*; 197–200.

Popper, K. R. (1972). *Objective Knowledge*. Oxford University Press, London.

Robinson, D. F., and L. R. Foulds (1981). Comparison of phylogenetic trees. *Math. Biosci. 53* 131–147.

14

Unraveling DNA Information

Bruce S. Weir
North Carolina State University
Raleigh, North Carolina

INTRODUCTION

The sciences of statistics and genetics have grown together since the 1890s when some scientists regarded the problems of animal evolution as being essentially statistical problems, and the foundations of regression theory were laid, in part, to account for observed patterns in inherited traits such as human heights. Mendel had showed, some 30 years previously, that individuals receive two copies of every gene, one from each parent, and pass on one of these two, chosen by chance, to each offspring. The regularity of this procedure has attracted mathematicians throughout this century, while the element of chance has required the tools of statisticians. With mathematical models based firmly on Mendel's laws, statisticians established methods for studying genetic data that generally consisted of observations on characters such as flower color in plants that could be considered to be controlled by one type of gene. The nature of such genes was unknown, and this whole theory of population genetics was derided by some people as being "beanbag genetics."

DNA SEQUENCE DATA

Since the 1950s, however, we have learned a great deal about the gene, and the nature of genetic data confronting statisticians has changed

considerably. In this chapter we give a very simplified picture of DNA sequence data, but complete accounts can be found in any modern genetics book, and a very good book is *Genes* by Lewin (1985). Genetic material is now known to be DNA, which exists as long strands consisting of thousands of elements known as nucleotides arranged in a row. There are only four kinds of nucleotides, depending on the base they carry and denoted by the letters A, G, C, or T. All of the genetic diversity that we recognize within and between species results from different sequences of these four types of nucleotides, so that we can regard genes as messages written in a language with a four-letter alphabet. The pathway from DNA sequence to the proper structure and functioning of living cells is also fairly well understood, and results from some DNA sequences coding for the production of enzymes which, in turn, direct various biochemical processes. Although humans receive about 3.5 billion nucleotides from each parent, it turns out that only a small fraction of them code for enzymes, and so can be regarded as functional genes, but modern methods can determine the nucleotide sequence of any portion of the DNA.

MITOCHONDRIAL DNA

The sheer magnitude of human DNA makes it unlikely that a complete sequence will ever be found for a single individual, and current studies focus on small regions. These regions may be of medical interest, such as

TABLE 1   Base Compositions for Heavy-Strand Mitochondrial DNA

	A	G	C	T	N
Human	0.247	0.302	0.139	0.312	14
Green monkey	0.259	0.293	0.132	0.316	4
House mouse	0.285	0.235	0.125	0.355	3

those coding for hemoglobin, but here we will consider mitochondrial DNA. Most human DNA is found in the nucleus of the cell, but about 1% is found in small bodies, mitochondria, outside the nucleus. These bodies are thought to control the respiration of the cell, and they contain a single circular DNA molecule that is about 16,500 nucleotides in length. A group of scientists determined the complete sequence of a mitochondrion for humans in 1981, and the sequence for mouse and cow has also been found.

Some properties of the mitochondrion can be found without complete sequencing. The DNA molecule is known to consist of two strands that can be characterized chemically. In Table 1 we show the relative proportions of the four base types in the mitochondrial "heavy" strand for humans, green monkeys, and house mice. The number of individuals supplying mitochondria for these studies are also shown, and the figures are taken from a paper published by W. M. Brown in 1981.

EVOLUTIONARY STUDIES

When mitochondrial sequences from different species are compared, a great deal of similarity is noticed, as indeed is the case when other sequences are compared between species. This, of course, is to be expected from evolutionary theory, which says that present species have diverged from a single ancestral species. Some of the differences between the sequences of present species are the results of mutations, or the substitutions of one type of base for another. It is estimated that about 1 base in 1 billion mutates in this way each year. There is still debate over the principal mechanisms of evolution. One school of thought holds that changes in DNA sequences will change the adaptation of individuals to the environment and so will be subject to natural selection. Most such changes will not survive in the population, but some occasional beneficial mutations will actually increase in frequency. The opposing school holds that the fate of new mutations is determined mainly by the chance mechanism that decides which genes are passed from parent to offspring. A comprehensive treatment of these issues is contained in a

recent book, *The Neutral Theory of Molecular Evolution* by M. Kimura
(1983).

Regardless of the forces acting on mutant sequences, it follows that
the more differences there are between the sequences of different species,
the longer it has been since they diverged, and a great deal of activity is
now under way to construct evolutionary trees on the basis of DNA
sequence data. Partly because of the technical difficulties in sequencing,
and partly because the regions being sequenced were thought not to be
different among members of the same species, evolutionary studies to
date have been based on a single determination of each sequence. Only
one individual has been used to represent a species. Although some
regions may indeed be uniform, the diversity we recognize among
members of our own species must be reflected in diversity of some DNA
sequences, and the within-species diversity must be taken account of
when between-species comparisons are made.

THE DATA

As a first step in measuring the amount of sequence variation in humans,
two American geneticists, C. F. Aquadro and B. D. Greenberg, in 1983
studied the sequences of a portion of the mitochondrial DNA from seven
individuals. Placental material was used to provide the mitochondria,
and a region of about 900 nucleotides was sequenced. These geneticists
found that 45 of the 900 sites showed variation among the seven people,
and the bases found at each of the 45 sites are shown in Table 2. The
position numbers shown there are the numbers used to describe human
mitochondrial DNA. The immediate conclusion is that there is indeed
substantial variation within a species. We would like to go further,
however, and see what the data can tell us about the mutation process
responsible for this variation.

The region studied by Aquadro and Greenberg is thought not to
encode for any enzymes, but to be involved in the replication process of
the DNA molecule. The question then becomes one of whether such
noncoding regions are subjected to any special rules. Since 90% of the
mitochondrial DNA does code for enzymes, we could suppose that the
base compositions in Table 1 are those appropriate for coding regions.
Does the composition differ at the variable sites in the noncoding region?
Does the mutation process in the noncoding region proceed "randomly"
in such a way that any base can be replaced by any other base, or must
the base composition still conform to that holding over the rest of the
molecule? We will consider how to answer such questions for each of the
45 variable sites. Although we will see that it is not particularly easy to

TABLE 2 Human Mitochondrial DNA Sequences

Position	Sequence 1	2	3	4	5	6	7
456	G	A	G	G	G	G	G
444	T	C	T	T	T	T	T
316	C	C	C	T	C	C	C
263	T	C	C	C	C	C	C
247	C	C	T	A	A	A	A
236	A	A	G	T	T	T	T
200	T	A	C	A	A	G	A
195	A	T	A	G	T	T	T
189	A	T	C	A	A	A	A
186	G	G	G	G	G	G	G
185	C	C	T	A	C	C	C
182	G	G	G	G	G	G	G
152	A	A	A	A	G	G	A
151	G	G	G	G	G	A	G
150	G	G	G	C	G	C	G
146	A	A	A	C	C	A	A
73	T	T	T	T	T	T	T
9	C	C	C	C	C	C	C
7	T	T	T	T	T	T	T
16519	A	A	A	G	A	A	A
16424	A	A	A	A	A	A	A
16362	A	A	G	A	A	A	A
16360	G	G	G	A	A	A	A

Position	Sequence 1	2	3	4	5	6	7
16356	A	A	A	A	A	G	A
16320	G	G	A	G	G	G	G
16311	A	A	G	G	G	A	A
16304	A	G	A	A	A	A	A
16294	G	G	T	G	G	G	G
16293	T	T	C	C	T	T	T
16280	T	T	C	T	T	T	T
16278	G	G	G	A	G	G	A
16243	A	A	A	A	A	A	A
16242	G	G	C	G	G	G	G
16230	T	T	T	T	T	T	T
16224	A	A	A	A	A	A	A
16223	G	G	G	G	G	G	G
16189	A	A	A	A	A	A	A
16188	G	G	G	G	G	G	G
16187	G	G	G	C	G	G	G
16172	A	T	A	C	A	A	A
16167	G	G	T	G	G	G	G
16163	T	G	A	G	T	T	T
16148	G	G	G	G	G	G	G
16134	G	G	G	G	G	A	G
16129	C	C	T	C	C	C	C

draw firm conclusions, the use of DNA sequences in evolutionary studies cannot proceed satisfactorily until such questions of within-species variation are answered.

PROBABILITY DISTRIBUTIONS

Before any statistical analyses can be undertaken, we need to take account of the way in which the data were generated. We refer to this as specifying the sampling model. The simplest assumptions here appear to be that every sequence is independent of the others, and that every sequence has the same chance of carrying a particular base at a nucleotide site. These assumptions lead to the multinomial sampling distribution, which tells us the likelihood of any particular arrangement of bases. If the probabilities that any particular position is occupied by bases A, G, C, or T are written as $Pr(A)$, $Pr(G)$, $Pr(C)$, or $Pr(T)$, then the probability of observing a, g, c, or t of the bases in a sample of N sequences is

$$\frac{N!}{a!g!c!t!}\,[\,Pr(A)]^{a}[\,Pr(G)]^{g}[\,Pr(C)]^{c}[\,Pr(T)]^{t}$$

To see if the data do support the idea that there is an equal chance of any base being at any position in the nucleotide sequence, so that mutation is "random," we set up a null hypothesis:

$$H:\ Pr(A) = Pr(G) = Pr(C) = Pr(T) = 0.25$$

If this hypothesis is true, we would expect to find $(0.25 \times 7) = 1.75$ of each type of base in each position among the seven sequences. Of course, we would expect some variation of 1.75. For one thing, the observed number must be a whole number. What we do expect, however, is a reasonably even spread without one type of base predominating. The role of the statistician is to give some meaning to "reasonably."

MULTINOMIAL TEST

The situation is rather like the situation described in Chapter 17. We first find a group of outcomes which are so different from what H predicts that they would arise with probability only 0.05 if H were true. Unfortunately, a sample of only seven is too small for the chi-square test to give the probability accurately, and we must use the multinomial distribution, which gives the exact probability of each arrangement. Those outcomes most different from what H predicts are those with just one or two base types appearing, and their probabilities are shown in Table 3 [$"Pr(H)"$].

TABLE 3   Rejection Region for Null Hypothesis of Equal Probabilities

n_A	n_G	n_C	n_T	Pr(H)	Pr(H')	n_A	n_G	n_C	n_T	Pr(H)	Pr(H')
7	0	0	0	0.0001	0.0001	2	0	5	0	0.0013	0.0001
0	7	0	0	0.0001	0.0002	2	0	0	5	0.0013	0.0038
0	0	7	0	0.0001	0.0000	0	5	2	0	0.0013	0.0010
0	0	0	7	0.0001	0.0003	0	5	0	2	0.0013	0.0051
6	1	0	0	0.0004	0.0005	0	2	5	0	0.0013	0.0001
6	0	1	0	0.0004	0.0002	0	2	0	5	0.0013	0.0057
6	0	0	1	0.0004	0.0005	0	0	5	2	0.0013	0.0001
1	6	0	0	0.0004	0.0013	0	0	2	5	0.0013	0.0012
1	0	6	0	0.0004	0.0000	4	3	0	0	0.0021	0.0036
1	0	0	6	0.0004	0.0016	4	0	3	0	0.0021	0.0003
0	6	1	0	0.0004	0.0007	4	0	0	3	0.0021	0.0040
0	6	0	1	0.0004	0.0017	3	4	0	0	0.0021	0.0044
0	1	6	0	0.0004	0.0000	3	0	4	0	0.0021	0.0002
0	1	0	6	0.0004	0.0019	3	0	0	4	0.0021	0.0050
0	0	6	1	0.0004	0.0000	0	4	3	0	0.0021	0.0008
0	0	1	6	0.0004	0.0009	0	4	0	3	0.0021	0.0088
5	2	0	0	0.0013	0.0018	0	3	4	0	0.0021	0.0004
5	0	2	0	0.0013	0.0004	0	3	0	4	0.0021	0.0091
5	0	0	2	0.0013	0.0019	0	0	4	3	0.0021	0.0004
2	5	0	0	0.0013	0.0032	0	0	3	4	0.0021	0.0009

The total probability of all of them is 0.046, conveniently close to 5%. We can therefore reject the hypothesis of equal base utilization in the noncoding region of human mitochondrial DNA if only two base types appear in the sample of seven. The value 5% is the type I error rate and is the probability H will be wrongly rejected. If we now look at the actual outcomes, we find that position 189 is the only one with more than two base types, and so is the only one for which H is not rejected. There is, therefore, some evidence of constraints on which bases are present even at the variable sites in the noncoding region.

POWER OF MULTINOMIAL TEST

The application of hypothesis tests carries the chance of making the mistake of rejecting a correct null hypothesis. Provided that we know the chance of this event, we need not be too worried. However, a full discussion of testing must include mention of the mistake of failing to reject an incorrect hypothesis. This second kind of mistake is called a type II error, and the chance of rejecting a false hypothesis is called the *power* of a test. Obviously, we would like our tests to have low significance levels and high powers.

To illustrate how power may be calculated, we return to the data in Table 1 that indicate an unequal distribution of base types in the heavy strand of human mitochondrial DNA. If these frequencies do apply to the variable sites in the noncoding region studies by Aquadro and Greenberg, how likely would we be to reject the null hypothesis of equal frequencies? The probabilities for the rejection region under the alternative hypothesis,

$$H': Pr(A) = 0.247, Pr(G) = 0.302, Pr(C) = 0.139, Pr(T) = 0.312$$

are also shown in Table 3 [“$Pr(H')$”], and they show that when H' is true, the chance of a sample falling into the rejection region we have established for H is 0.0722. In other words, the multinomial test has a desirably low significance level of 5%, but an undesirably low power of only 7%. This low power is a consequence of the very small sample of seven sequences, and we just cannot hope to distinguish between H and H' with such a sample.

DISCUSSION

Modern genetics is generating data that can be very extensive for any individual, but rather limited in terms of numbers of individuals. The first molecular geneticists to use sequence data to compare species did so on the basis of a single representative of each species, and the problem with that is illustrated by the fact that the variation within a species, as shown for the mitochondrial data in Table 2, is a substantial fraction of the variation between species, as shown in Table 1. The necessity for within-species sampling is now being recognized, but the currently small samples give low power to statistical tests. It is going to be difficult to make inferences about the base composition at any particular site if only a few independent copies of that site are available. Small numbers also require the use of fairly cumbersome procedures such as the multinomial test we have discussed, rather than the convenient and well-documented chi-square test.

A great deal of replication is provided by the many bases within each sequence, of course. With $N = 900$ bases, we would have no trouble in distinguishing between compositions for man and mouse in Table 1, but would still have trouble between man and monkey. The problem with bases within a sequence is that the multinomial distribution is no longer appropriate. It is known that adjacent nucleotides are not independent, since some types of pairs, such as a G followed by a C, can be much more common than others. Part of the dependence is simply a consequence of adjacent nucleotides being very likely to be transmitted together from parent to offspring, and we may find that for evolutionary studies, there is not a great deal more information in 100 consecutive bases than there is in just one. Analyses of DNA sequence data are going to require sequences from widely separated portions of the DNA or will require more sophisticated models. Other dependencies, such as those between sequences taken from related individuals, will also need to be taken into account.

The discussion in this chapter has also neglected other features of the bases. Their physical properties require that transitional changes, A for G or C for T, and vice versa, be treated differently from transversions, A for T, A for C, T for G, or G for C, and vice versa. If all 12 possible types of base substitutions were equally likely, transversions would be twice as frequent as transitions, but Aquadro and Greenberg (1983) point out that nearly all of their observed changes were transitions. Evidently, the mutation process is more complicated than considered here, but our statistical analysis of such data suggests areas in which geneticists need to seek biological reasons for patterns that cannot be attributed to pure chance.

More and more refinements of the underlying statistical and genetical models are going to lead more and more complex, although more realistic, analyses of DNA data. Although recombinant DNA technology has allowed geneticists to look directly at the gene, instead of inferring its existence from observations on the whole individual, the nature of DNA sequence data requires the continued collaboration of statisticians. The complete unraveling of all the information in a DNA sequence is providing one of the prime current examples of the fascination of statistics.

REFERENCES

Aquadro, C. F., and B. D. Greenberg (1983). Human mitochondrial DNA variation and evolution: analysis of nucleotide sequences from seven individuals. *Genetics 103*, 287–312.

Brown, W. M. (1981). Mechanisms of evolution in animal mitochondrial DNA. *Ann. N.Y. Acad. Sci. 361*, 119–134.

Kimura, M. (1983). *The Neutral Theory of Molecular Evolution.* Cambridge University Press, Cambridge.

Lewin, B. (1985). *Genes.* 2nd edition. Wiley, New York.

15

Sequential Experimentation, or Count Your Chickens as They Hatch

Howard P. Edwards
Massey University
Palmerston North, New Zealand

INTRODUCTION

One of the most important areas of statistics is the collection of data. The most common way of doing this is to take a sample of fixed size; for example, a public opinion pollster surveys the opinions of 500 citizens, or an agronomist measures the growth rate of 30 plants of a certain variety. Sometimes the size of the data set is affected by circumstances beyond the experimenter's control (several of the agronomist's plants could die of disease, for instance), but the design of the experiment is still that of a fixed number of measurements. An alternative sampling scheme is to sample sequentially; the data are collected in a series of groups, or one at a time. The most important difference between the two sampling schemes is that the size of a sequential sample is not fixed in advance; rather, it is determined by the data values as they are obtained. As will be shown below, there are many situations in which sequential experimentation is more efficient than fixed-size experimentation, and there are problems in which a sequential experiment is the only one possible.

AN EXAMPLE

Very small chips containing thousands of electronic circuits are manufactured on a large scale nowadays. Because of the critical nature of the manufacturing process, however, many of the chips produced are

defective. Therefore, a quality control inspector periodically takes a sample of 25 chips and tests them. If 5 or more of the chips are found to be defective, the manufacturing process is halted and extensive checks are made.

Now, no one would dream of sampling and testing 25 chips in this situation. Why? Well, suppose that the first 5 chips tested were all found to be defective. Clearly, there is no point in testing the remaining 20 chips, as we have already found out what we need to know; what is more, it has taken only 5 observations to do this as opposed to 25 in the fixed-sample-size case. A moment's reflection will show that 25 is, in fact, the maximum number of observations needed, and that the sequential sampling rule to use is as follows: continue testing until either 5 defective or 21 nondefective chips have been found.

This sequential experiment can be represented graphically. Assign each chip a score of 1 if it is nondefective or 0 if it is defective. If we keep a running total of the scores and call it Sn, where n is the number of chips tested so far, then Sn is just the number of nondefective chips out of a total of n tested (see Figure 1).

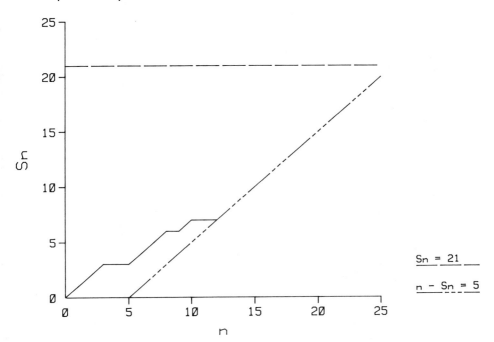

FIGURE 1   Number of nondefective chips.

Sampling stops if the number defective equals 5, in which case the process must be checked, or if the number not defective equals 21, in which case the process is acceptable. These numbers are $n - Sn$ and Sn, respectively, so the two straight lines on the graph represent the points where stopping occurs. The jagged line depicts one possible experimental outcome, namely the sequence *NNNDDNNNDNDD* (*N* = nondefective, *D* = defective). As soon as this line crosses one of the stopping boundaries, as they are called, the experiment is stopped and the appropriate decision is made. In this example the twelfth chip tested was the fifth defective to be found, and so the decision is made to check the process. (Incidentally, *Sn* is an example of what statisticians call a *random walk*, which has been used to model such diverse phenomena as the one-dimensional motion of a small particle, the contents of a water reservoir, and the capital of an insurance company.)

The efficiency of a sequential rule is measured by the number of observations required to reach a decision; since this number depends on the observations themselves, the *average sample number* (ASN) is used instead. In our example the ASN usually lies in the range 20 to 23 (the precise value of the ASN also depends on the true rate of defectives from the process, which is, of course, unknown). While this may seem a small reduction, the percentage savings (10 to 20%) are quite high, and this might be significant if the fixed sample size were larger or if the observations were time consuming or costly. Moreover, in the undesirable case where the true rate of defectives is not small, the sequential experiment will usually stop after a very small number of observations.

THE SEQUENTIAL PROBABILITY RATIO TEST

The sequential rule considered in the previous example is not truly sequential in the sense that the maximum sample size was fixed in advance. Under a truly sequential sampling scheme the sample size is completely unknown to the experimenter and is determined solely by the sample observations as they occur. Although this may not always be desirable, the following is an example in which it probably is.

A physician has available a new treatment (e.g., a drug) for a certain disease. It is claimed that this treatment is much superior to the existing treatment used for this disease; unlike the existing treatment, however, the success rate for the new treatment is not known. (The success rate of a treatment is defined as the proportion of all patients given it who make a complete recovery.) To decide whether the new treatment really is better, the physician carries out a clinical trial in which he administers the new treatment to a series of patients who have the disease, and observes the results. For simplicity, we shall assume that all the patients are alike with respect to other factors, such as age, sex, and race; this means that any differences observed may be attributed to the new treatment rather than to one or more of these factors.

Sequential experimentation is often used in medical trials such as this. One reason for this is that patients do not arrive at a hospital en masse; rather, they come as and when the disease is diagnosed or suspected. Another reason is the absence of restrictions on the sample size mentioned above. If the new treatment is in fact much superior (or much inferior) to the existing one, the trial should stop as soon as possible in order that only a few patients are given the inferior treatment; however, if the two treatments are of similar effect, the trial should be carried on for a longer period in order to be reasonably sure that the final decision made is the correct one.

Before proceeding any further, however, we need to know a little more about decision making in statistics. This problem (and many others) involves making a choice between two possible statements about the underlying phenomenon of interest. In this case the two statements (or *hypotheses*, as statisticians prefer to call them) are (1) the new treatment is better than the existing one, and (2) the new treatment is no better (and possibly worse) than the existing one. What the physician requires is a *test* for deciding on the basis of his trial results which of these two hypotheses is in fact true; the problem of hypothesis testing is to provide such a test.

The question that now arises is: What constitutes a "good" (or "best") test? The test used in the silicon chip problem seemed an intuitively reasonable one, but how does one go about finding and evaluating a test for any given problem?

Any procedure for deciding between two hypotheses can result in two types of error. In the clinical trial, one might decide that the new treatment is superior when in fact it is not, or one might decide that it is not superior when in fact it is. This all seems fairly obvious; however, it is the probabilities of making these two types of errors that statisticians use as the "goodness" criteria for a test procedure, be it sequential, fixed sample size, or whatever. A good procedure is one that has small error probabilities, that is, it is unlikely to result in either type of error. Just how small these error probabilities (call them a and b, respectively) should be depends on how serious the consequences of making these errors are, but values such as 0.01, 0.05, 0.10 (i.e., 1%, 5%, 10%) are fairly typical.

Unfortunately, it turns out that you cannot decrease one error probability without increasing the other error probability (this is the statistical analog of not being able to have your cake and eat it too), unless you increase the size of your sample (pay for some more cake). In other words, if you require a fixed-sample-size test for given values of a and b, your sample will have to be no smaller than a certain size. For simplicity, let us suppose that the existing treatment has a success rate of 75% and that the physician is interested in the new treatment only if it has a success rate of 90% or more; in other words, if the new treatment has a success rate between 75 and 90%, he is indifferent as to which treatment is declared better. Suppose also that the physician requires that the error probabilities a and b be no greater than 0.05 and 0.10, respectively. (One would expect a to be smaller than b because the physician would be more concerned about accepting the new treatment as superior when it is not, as its true success rate is unknown; the other error is given a larger probability because if he continues to use the existing treatment when the new treatment is in fact superior, he does at least know what success rate it will have.) It can then be shown that the fixed-sample-size test that satisfies these requirements is as follows. Give the new treatment to 54 patients, and if 46 or more make a complete recovery, declare the new treatment to be superior; otherwise, declare that it is not. Since the trial

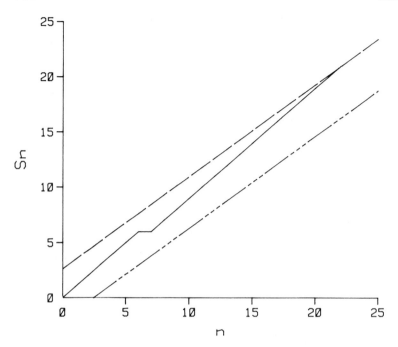

FIGURE 2   Outcome of an SPRT using a = 0.05 and b = 0.10.

proceeds sequentially we see that this test is of the same form as that in the chip example, that is, continue treating patients until either 46 make a complete recovery or until 9 do not; however, if new patients arrive more quickly than existing patients recover, it may still be necessary to treat all 54 patients.

For this hypothesis-testing situation there is a sequential test that is even more efficient than the sort of test used in the chip problem. It is known as the *sequential probability ratio test* (SPRT) and is described as follows. Two numbers A and B (called stopping boundaries) are chosen. As the sequence of observations is obtained, the probability or likelihood of obtaining that sequence under each hypothesis is calculated and then the ratio of the two probabilities is found. If the ratio exceeds B or drops below A, the corresponding hypothesis is accepted; otherwise, another observation is taken and the test is repeated. Figure 2 is a graphical representation of one possible outcome using the SPRT with the same error probabilities 0.05 and 0.10. In this case we see that only 22 patients were required before the trial ceased, and the new treatment is declared to be superior.

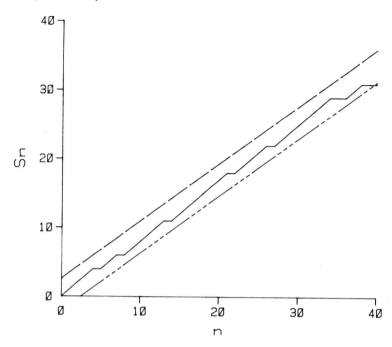

FIGURE 3   Outcome of an SPRT when the new treatment has success rate 80%.

Now comes the fascination of the SPRT. Clearly, the choice of boundaries A and B will depend on the desired error probabilities a and b. It turns out that these are the *only* quantities that A and B depend on; in particular, they do not depend on the probability distributions under the two hypotheses (in fact, these "distribution-free" boundaries are only approximations to the true values, but they are extremely good approximations). This is rarely the case for a fixed-sample-size test.

Second, the SPRT has a smaller ASN than the usual fixed-sample-size test under *either* hypothesis, regardless of the probability distributions. This in itself is a very powerful result, for it means that we can be sure that the SPRT will require fewer observations on average than the fixed-sample-size test even though we do not know which hypothesis is true. However, it turns out that the SPRT has a smaller ASN than *any* other test under either hypothesis. This is very surprising, because we might expect a test with a small ASN under one hypothesis to have a large ASN under the other hypothesis. The SPRT manages to minimize both simultaneously! Using the SPRT with the same error probabilities in the clinical trial, it turns out that the ASN's under the two hypotheses are 21.6 and 32.8, both of which are a lot smaller than the fixed sample size of 54. (Once again, the difference between the two ASNs is a consequence of the different values for a and b.)

To see what happens when the new treatment has a success rate close to that of the existing treatment, the outcome of the SPRT shown in Figure 3 was generated using a success rate of 80%. In this case 40 observations were required before the new treatment was declared to be no better than the existing one. Note that this number is significantly more than the ASNs under both hypotheses, as might have been expected, but it is interesting to note that it is also significantly smaller that the fixed sample size of 54.

PROBLEMS OF COMPARISON

Like many problems in statistics, the clinical trial involved a comparison—that between the new treatment and the existing treatment. In this case the success rate of the existing treatment was known, so it was not necessary to use it in the trial. However, it is often the case that a comparison is required between two (or more) treatments of unknown effect, so a decision must be made as to which treatment each patient is to be given. This poses an ethical dilemma for the physician that the statistician may not appreciate.

Suppose that a physician consults a statistician about the design of a clinical trial to decide which of two experimental treatments (call them A and B) has the larger success rate. The statistician elicits from the physician what error probabilities a and b he requires, performs a few calculations, and then announces that 64 patients will be required for the trial. "Of course," says the statistician, "you will also have to allocate your patients randomly as follows. For the first patient, flip a coin to decide whch of the two treatments he or she is to be given. Once this is decided, give the second patient the other treatment. Flip a coin again for the third patient, give the fourth patient the other treatment, and so on." The physician throws up his hands in horror. "Even if I had 64 patients," he says, "I could never agree to give 32 of them an inferior treatment. Suppose that of the first 10 patients, the 5 given treatment A recover while the 5 given treatment B do not. Do you expect me to tell the eleventh patient that she is to receive treatment B because I flipped a coin and it landed tails?" "Well," says the statistician, somewhat defensively, "we could run a sequential trial instead. If the situation you described occurs, a sequential experiment will probably stop at an early stage anyway." "But," says the physician, "does that still mean that half the patients in the trial will receive an inferior treatment?"

The answer to the physician's question is "yes" if either the sequential or the fixed-sample-size tests mentioned above are used. To this end, statisticians have tried to devise sequential tests that incorporate an experimental design; in other words, as experimentation proceeds, the choice of which treatment to administer next is made on the basis of the previous sequence of choices and outcomes. One such design is called the *play-the-winner* rule: if the treatment used on the most recent patient was successful, administer it to the next patient; if it was not, administer the other treatment next time. Another rule attempts to minimize the ASN allocated to the inferior treatment (of course, which of the treatments is inferior is not known). Note that the first rule tries to maximize the number of recoveries in the trial, while the second tries to maximize the number of times the superior treatment is administered during the second trial. The choice of stopping boundaries is more difficult for these sorts of rules, and in general they result in a larger total sample size than the usual "equal-allocation" rule (although the reverse holds as the number of treatments increases).

A problem similar to the above is known as the *two-armed-bandit problem*. Consider a gambler who has a choice of two slot machines (often called one-armed bandits) to play. Each time a machine is played the gambler either wins $1, say, or nothing. Given that she can play a fixed number of times, how should she play the machines so as to maximize her expected winnings? The essential difference between this and the clinical trial is that here the goal is not to decide which machine is better (a test of hypothesis); rather than designing an experiment as a means toward making a decision, the decision and the experimental design are one and the same problem here. Similar rules to those mentioned above have been proposed, and for certain variations of the problem best rules have been found; however, most are quite complex and nonintuitive, given the relative simplicity of the problem itself.

In concluding, then, it is probably fair to say that problems involving comparisons are extremely difficult to analyze when the sample size from *each* source of interest may be chosen sequentially. Given the difficulty and yet the potential savings in efficiency also, sequential allocation of experiments is one of the more unexplored areas of modern statistics.

REFERENCES

Hoel, D. G., M. Sobel, and G. H. Weiss (1975). A survey of adaptive sampling for clinical trials. In *Perspectives in Biometrics* (R. M. Elashoff, ed.). Academic Press, New York.

Wald, A. (1947). *Sequential Analysis*. Wiley, New York.

Wetherill, G. B. (1975). *Sequential Methods in Statistics*. Chapman & Hall, London.

IV
Estimation

The first lesson of formal logic is that one must not argue from the particular to the general. Unfortunately, generalizing from our personal experiences to the outside world is what we must do all the time. We have a gut feeling that the unseen will not be very different from the seen. Statisticians lay down very definite rules for making this process legitimate. The most important is that the sample observed must be selected randomly. When this is done, estimates can be calculated, together with a measure of their accuracy; but in some situations, selecting a random sample may be difficult.

Richard LeHeron's article describes the practical problems he faced when estimating the composition and motives of the crowd at a race meeting. At least he could see his crowd, even though it was not static, and a sample could be "captured" to respond to a questionnaire.

Animals in the wild are shy and not visible, so they are harder to count than human beings. Ken Pollock tackles this problem and describes how to estimate the number of raccoons in the wild using capture–recapture methods. The way he constructs a sample and calculates an estimate from it is an excellent example of the power of statistical procedures to estimate the unseen.

16

Estimating the Size of Wildlife Populations Using Capture Techniques

Kenneth H. Pollock
North Carolina State University
Raleigh, North Carolina

INTRODUCTION

How many trout are there in a stream? How many raccoons in a game reserve? In this chapter I discuss two approaches to addressing these types of difficult questions which are of such importance to fisheries and wildlife ecologists.

The first approach is the mark-and-recapture technique, which has been widely used for a variety of species. A sample of animals are captured and marked before being released back into the population. Later another sample is taken and the proportion of marked animals noted.

In some situations biologists find it preferable to use a second approach, which is based on permanent removal of the animals. This removal technique is also widely used, especially in fisheries, where the biologists may have to depend on samples collected by commercial fishermen.

To the statistician the problems posed involve *estimation*. The total population size (denoted by N) is an unknown *parameter*. Based on a statistical model, I derive an estimator of N using the data from the mark-and-recapture or removal experiments. Each model is discussed in detail with emphasis placed on the assumptions required to obtain a valid estimator. I also consider the important question of how many animals

should be sampled to obtain reliable estimators. Each technique is illustrated by a detailed practical example.

THE MARK-AND-RECAPTURE EXPERIMENT

Derivation of the Estimator

To begin with, I consider the simplest mark-and-recapture experiment. A sample of n_1 animals is taken from a population of size N animals. The animals are permanently marked in some manner, depending on the species involved. For example, many mammals are marked with metal ear tags, and birds with metal leg bands. There are, however, a myriad of other marking techniques. After marking, the animals are returned to the population. Later another sample of n_2 animals is taken. These animals are examined for marks and the number of marked animals (m_2) is recorded.

Intuitively, one might expect the ratio of marked to total animals in the second sample to reflect the same ratio in the whole population. This can be expressed mathematically by the equation

$$\frac{m_2}{n_2} \simeq \frac{n_1}{N} \qquad (1)$$

This equation can be solved for N to obtain

$$\hat{N} = \frac{n_1 n_2}{m_2} \qquad (2)$$

Let us illustrate these equations numerically. Suppose that the biologist marks 100 animals in the first sample and later captures 50, of which 5 are marked. The marked-to-total ratio in the sample is 0.1 (5:50). For the same ratio to apply to the population the population must consist of 1,000 animals (100/1000 = 0.1). Formally, using equation (2) we have

$$\hat{N} = \frac{n_1 n_2}{m_2}$$

$$= \frac{100 \times 50}{5}$$

$$= 1000$$

$\hat{N}$ is called a *point estimator* of N, the unknown population size. Statisticians are careful to emphasize the distinction between the *estimator* ($\hat{N}$), which can be enumerated from the data, and the *parameter* (N), which is the unknown population size. N cannot be found exactly without counting the whole population, which is impossible in practice.

One interesting mathematical property of the Petersen estimator needs mention. What happens if no marked animals are captured in the second sample ($m_2 = 0$)? We find that $\hat{N}$ is infinite. This is just telling us what we would expect intuitively. An experiment where no marked animals are captured gives us no useful information on population size.

Historical Perspective

The mark-and-recapture technique has a very long history, with the basic principle dating back at least to P. S. Laplace in the late eighteenth century. He used it to estimate the population size of France (Seber, 1982, p. 104). As his number "marked" (n_1) he used the size of a register of births for the whole country. His second sample consisted of a number of parishes of known total size (n_2). For these parishes he also obtained the total number of births (m_2). If one assumes that the ratio of births to total population size can be approximated by the ratio of births to total size in his known parishes, equation (2) can be used.

The use of the method on animal populations began much later. Carl Petersen, a Danish fisheries scientist, had realized the potential of the mark-and-recapture technique in a paper published in 1896. In 1930,

Frederick Lincoln, a U.S. Fish and Wildlife Service biologist, began to apply the method to waterfowl populations in North America. The estimator given in equation (2) is commonly called the Petersen estimator or Lincoln index after these two men. It should be emphasized, however, that "index" is a misnomer, because an index is usually defined to be an estimate of relative abundance. This estimator is for the total population size, which is absolute abundance.

In an interesting new monograph by White et al. (1982) a brief but very informative history of mark-and-recapture techniques is given. They have an interesting series of photographs of some of the important contributors to the statistical theory of these methods.

Model Assumptions

To obtain the estimator of N given in equation (2) we have built a statistical model of the mark-and-recapture procedure. This model depends on several critical assumptions which may not always be satisfied in real animal populations.

1. *Closure*: It is assumed that the population is closed during the period of the study. That is, there are no additions to the population due to birth or immigration. Also, there are no deletions from the population due to death or emigration.
2. *Equal catchability*: It is assumed that for each sample *every* animal has the same probability of being captured. However, the probability may differ for the two samples.
3. *Marks permanent*: It is assumed that marks are not lost before the second sample is taken.

The assumption of closure is often a reasonable one provided that one takes a short time period between the mark-and-recapture samples so that births and deaths can be ignored. Migration may be a problem in some studies where there is no well-defined natural boundary for the population.

The assumption of equal catchability is critical and is often violated in real populations. There are two different kind of departures from this assumption. There may be inherent *heterogeneity* in capture probabilities for animals due to many factors, such as age and sex. If the same animals tend to be easy to capture in the first and second samples, the number marked in the second sample (m_2) becomes too large and hence the population size estimator ($\hat{N}$) is too small. A statistician calls this a *negative bias* on the estimator. The second departure is that marking may affect an animal's capture probability in the second sample. For example, animals could become "trap shy" if their capture and marking was stressful, lowering their capture probability compared to unmarked animals. On the other hand, animals could become "trap happy" if their capture and marking was enjoyable. This often occurs if the traps contain food and results in marked animals having a higher probability of capture than unmarked animals. A "trap shy" response results in a positive bias on the population size estimates, while a "trap happy" response results in a negative bias. Use of different capture techniques in the two samples often greatly reduces problems with heterogeneity and trap response of the capture probabilities.

Mark loss can be a serious problem, especially in fisheries studies. It causes a positive bias on the population size estimator. Sometimes the biologist puts two marks on each animal and uses the information on how many animals retain one or two marks to estimate the rate of mark loss. The population size estimates can then be adjusted to eliminate the positive bias caused by the mark loss.

Practical Example

In a recent study on raccoons in northern Florida by Conner (1982), a sample of 48 animals was captured using cage-type live traps baited with fish heads. An unusual marking scheme was used on these animals. They were injected with a small amount of a radioactive isotope. The "recapture" sample involved an intensive search of the study area for scats (feces). Those feces which were "marked" (radioactive) were detected using a special scintillation counter.

Conner carried out his search for scats each week for 5 weeks. He decided to treat each week separately and obtain five different Petersen estimates so that he could consider the variation of the estimates. The data and resulting estimates for each week were as follows:

Week of collection	Number marked, n_1	Number of scats collected, n_2	Number of marked scats, m_2	Petersen estimate, $\hat{N}$
1	48	71	31	109.9
2	48	22	11	96.0
3	48	74	35	101.5
4	48	28	9	149.3
5	48	35	19	88.4

One approach is to assume that the five different estimates of N, the population size, are independent and have an approximate normal (bell-shaped) distribution. The mean of the five values in 109 animals, which is our "best" *point estimate* of the number of raccoons in the population. It is also informative to calculate a *95% confidence interval estimate* of *N* based on the mean of the five estimates. The theory is given in any good statistics text (e.g., Snedecor and Cochran, 1980, p. 56) and will not be presented here. For this example the confidence interval ranges from 79 to 139 animals. Let us summarize our results in a table:

Point estimate of population size	95% Confidence interval estimate of population size
109 animals	79–139 animals

The 95% confidence interval gives us an idea of the precision of our experiment. Here with approximately half of the animals in the population (48 of 109) marked and the scat collection lasting 5 weeks, the confidence interval had width 60 (139 - 79). To increase the precision of his study and reduce the confidence interval width, the biologist has two options. He can increase the number of marked animals in his population or he can continue the scat collection over more weeks. I calculated the approximate confidence interval width if he had the same number of marked animals but had carried out scat collection for 10 weeks at the same sampling intensity. I found that the confidence interval width would have been 34, which is around half the current width. Of course, this much more precise study would have cost a lot more to run. You do not get something for nothing in scientific work.

I believe that this is a sound study in terms of the assumptions behind the model. The biologist felt that the population was approximately closed over the period of the experiment. He also felt that there was no appreciable chance of not detecting radioactivity in the scats. The key assumption to consider here (as it is in most studies) is that of equal catchability of the animals. Here there is no danger of obtaining "trap shy" or "trap happy" animals because two completely different methods of sampling are used. These are trapping in the first sample and scat collection in the second. Also, because of the two different methods, I believe it is unlikely that animal capture probabilities in the traps are related to their scat collection probabilities. In other words, I do not believe that there will be any negative bias on the population size estimate caused by animals easy to capture in the traps also having their scats collected at a higher rate.

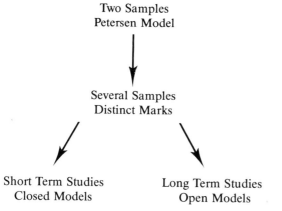

FIGURE 1   Extensions of mark-and-recapture models.

Extensions

I have barely scratched the surface of mark-and-recapture methodology. Often biologists find it necessary to mark and recapture animals over many samples. In these studies each animal has its own distinct mark (often a numbered tag) and a comprehensive capture history is built up for every animal captured at least once. The scope of mark-and-recapture metholodology can best be summarized in a diagram (Figure 1).

The closed population models can allow for unequal catchability of animals. The open population models also allow for the estimation of birth and death parameters, which are very important to biologists.

There is a very large literature on mark-and-recapture methods. The new monograph by White et al. (1982) is an excellent introduction, although it concentrates on closed population models. There are two good recent reviews written for biologists (Nichols et al., 1981; Pollock, 1981). An excellent, if rather technical book is that by Seber (1982).

THE REMOVAL EXPERIMENT

Derivation of the Estimator

Here I consider the simple two-sample removal experiment, which is closely related to the mark-and-recapture experiment considered in the preceding section. A sample of n_1 animals is captured from a population of size N animals. Now instead of being marked and returned, the animals are removed permanently. In fisheries they may sometimes be part of a commercial catch. For some species live traps may be used and the animals are taken to another site and released. Later another sample of n_2 animals is removed using the same amount of effort as for the first sample.

Given that the same amount of effort is expended for each sample, one might expect the proportion of the population removed to be constant for the two samples. This simple but powerful idea forms the basis for the estimator of N. Mathematically, the result is that n_1 removals out of a population of size N should be equivalent to n_2 removals out of the reduced population of size $(N - n_1)$. This can be expressed by the formula

$$\frac{n_1}{N} \simeq \frac{n_2}{N - n_1} \qquad (3)$$

Solving this equation gives us an estimator of N, which is

$$\hat{N} = \frac{n_1^2}{n_1 - n_2} \qquad (4)$$

Notice that this estimator has some interesting mathematical properties. If n_1 equals n_2, the estimate suggests to us that there are an infinite number of animals in our population. If n_1 is less than n_2, the estimate suggests to us that there are a negative number of animals in our population. If the amount of effort is low, then by chance n_1 could be less than or equal to n_2. These nonsensical results are really telling us that we do not have enough information to estimate the population size from such an experiment.

Model Assumptions

We have built a statistical model for the removal procedure which is very similar to the mark-and-recapture model. Again the model depends on several critical assumptions which may not be valid in real removal studies.

1. *Closure*: It is assumed that the population is closed during the period of the study.
2. *Equal catchability*: It is assumed that for each sample *every* animal has the same probability of being captured.

3. *Constant capture rate*: It is assumed that the amount of effort used for each sample is the same and that this means that all animals have the same probability of capture in sample 1 as in sample 2.

The assumption of closure may often be a reasonable one provided that there is a short period between the two samples to prevent migration. For example, if food is in short supply, then after animals are removed in the first sample, you may have animals moving in from surrounding areas.

Again the assumption of equal capture probabilities is crucial. If there is a tendency for the same animals to be difficult (or easy) to capture in both samples, the population size estimator $\hat{N}$ will have a negative bias.

The assumption that all animals have the same probability of capture in both samples is also crucial. The biologist can help ensure this by using equal effort in both samples, He or she should also attempt to conduct the two samples under identical weather conditions. Obviously, however, this may be impractical in some cases.

Practical Example

Sometimes fisheries biologists block off a small section of stream and carry out a removal experiment on the section. One capture technique, called electrofishing, is to pass an electric current through the water. Hand nets are then used to remove all the stunned fish that have come to the surface.

Seber and Le Cren (1967) have presented data for such an experiment, where 79 trout were removed in the first sample and 28 trout removed in the second sample. Use of equation (4) then gives us a point estimate of 122 trout for the population size. A 95% confidence interval of 104 to 140 trout is also given, but this theory will not be developed here.

Population size estimate:

$$\hat{N} = \frac{n_1^2}{n_1 - n_2}$$
$$= \frac{79^2}{79 - 28}$$
$$= 122 \text{ trout}$$

95% Confidence interval estimate:

$$122 \pm 18$$
$$104 \text{ to } 140 \text{ trout}$$

I believe that this is a valid estimate of the population size. Closure is guaranteed by there being a short period betwen samples and by having the stream blocked off at both ends to prevent fish escaping. The use of equal effort and the two periods close together suggests that the probability of capture should be constant over the two samples.

This is a very precise estimate for a biological study. The reason for this is obvious. The population size estimate is 122 trout and over both samples 107 trout (79 + 28) were actually removed. This degree of removal is often not feasible in practice.

Extensions

Often, one cannot hope for the high rate of removal of the electrofishing example. Sometimes, biologists extend the removals over several samples to compensate for this. It is possible to allow for some degree of heterogeneity of capture probabilities between animals in these multiple-sample-removal models.

Another extension relates especially to commercial fisheries. Often the effort in different samples is unequal but there is some measure of the relative effort. This leads to a whole class of what are called *catch per unit effort models*. There are versions for both closed and open populations. Again a summary diagram may be helpful (Figure 2).

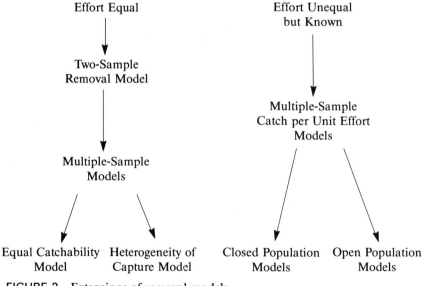

FIGURE 2   Extensions of removal models.

There is also a very large literature on removal and catch per unit effort models. An excellent reference for the removal models is the new monograph by White et al. (1982). Important references for the catch per unit effort models are Ricker (1975), which is written for fisheries biologists, and the more technical book by Seber (1982).

DISCUSSION

In this chapter four equations only have been presented. This was very deliberate on my part. I wanted to emphasize that the important concepts of statistics are easily understood because they involve intuition and common sense. Mathematics should be viewed as a tool the statistician uses. Unfortunately, all too often the layman becomes bogged down in the mathematical details of statistics and "cannot see the forest for the trees."

The aim of this chapter has been to show an interesting subset of statistical techniques that ecologists use in their work. For these methods to be used effectively it is necessary for the statistician and ecologist to work closely together. Therefore, I feel that it is essential for the statistician to have some knowledge of ecology and for the ecologist to have a sound training in the important concepts of statistics.

REFERENCES

Conner, M. C. (1982). Determination of bobcat (*Lynx rufus*) and raccoon (*Procyon lotor*) population abundance by radioisotope tagging. M.S. thesis, University of Florida.

Nichols, J. D., B. R. Noon, S. L. Stokes, and J. E. Hines (1981). Remarks on the use of mark-recapture methodology in estimating avian population size. In *Estimating Numbers of Terrestrial Birds*, Studies in Avian Biology, Vol. 6 (J. M. Scott and C. J. Ralph, eds.).

Pollock, K. H. (1981). Capture-recapture models: A review of current methods, assumptions and experimental design. In *Estimating Numbers of Terrestrial Birds*, Studies in Avian Biology, Vol. 6 (J. M. Scott and C. J. Ralph, eds.).

Ricker, W. E. (1975). *Computation and interpretation of biological statistics of fish populations*. Information Canada Bulletin 191, Ottawa, Canada.

Seber, G. A. F. (1982). *The Estimation of Animal Abundance and Related Parameters*, 2nd ed. Griffin, London.

Seber, G. A. F., and E. D. Le Cren (1967). Estimating population parameters from catches large relative to the population. *J. Anim. Ecol. 36*, 631–643.

Snedecor, G. W., and W. G. Cochran (1980). *Statistical Methods*, 7th ed. Iowa State University Press, Ames, Iowa.

White, G. C., D. R. Anderson, K. P. Burnham, and D. L. Otis (1982). *Capture-recapture and removal methods for sampling closed populations*. Los Alamos National Laboratory, Los Alamos, N. Mex.

17

Who Still Comes to the Races?

Richard B. Le Heron
Massey University
Palmerston North, New Zealand

INTRODUCTION

Social commentators during the 1950s and 1960s often acknowledged the liking of New Zealanders for rugby, racing, and beer. A national Totalisator Agency Board (TAB) established in 1950 did much to keep racing attractive and the industry prosperous. The organization took advantage of expanding disposable incomes and the virtual monopoly racing held at the time on gambling. In less than 20 years, however, the appeal of racing had waned and attendance at meetings had fallen away. The reasons were many for the reversal: strong competition for time and money from other leisure activities, the arrival of transistor radios (allowing race commentaries to be heard almost anywhere) and television coverage of premier races (creating "armchair racegoers"), a widely spread network of TAB agencies, and the availability of telephone betting. By the 1980s the percentage of real disposable income spent on betting was on a sharp decline, on-course betting turnover had flagged noticeably, and the rate of increase of off-course turnover had slowed. Racing officialdom, seeing the outlook as bleak, began to seek ways to arrest racing's decline in popularity.

One response by the New Zealand Racing Authority, the body empowered to oversee the development of racing and trotting in New Zealand, was to commission a study on racegoers and race attendance

patterns. The Authority was concerned about the erosion of racing club income, which comes in part from a percentage of on-course turnover—a figure closely associated with numbers at the races. The expectation was that information compiled would give a guide to what assumptions could be made about attendance patterns and racegoer preferences and enable some interpretation to be made about tendencies observed for racing as a whole.

This chapter is concerned with sample surveys of racegoers and examines the difficulties inherent in the design and application of sample survey methods. The surveys carried out enabled a description to be made of the composition of race-day crowds at eight race meetings held in the Manawatu area of New Zealand during the period March–June 1980. Interest is focused on sample characteristics and the representativeness of the samples. This is followed by an assessment of whether race-day crowds can be viewed as different from the regional population in terms of ethnic breakdown, proportions of males and females and age groupings, and whether the race-day crowds can, on a range of indicators, be regarded as constituting one population.

SAMPLE DESIGN

Two pragmatic concessions were negotiated at the outset of the investigation. Urgency dictated a compression of the inquiry into part of the racing season (spanning March–June in a September–July season), and practicality and cost restricted the investigation to eight meetings in the local Manawatu area. Two questions can fairly be asked about the wisdom of these concessions. Was the selection of one region sensible in relation to the organization of racing? Would the findings from the meetings studied be in any way representative of meetings held during the season?

Racing in New Zealand is organized to offer something approaching regional programs of racing. This administrative arrangement assumes that potential racegoers in different parts of the country are mainly interested in meetings close to where they live. Accordingly, it is realistic to think of a "regional system" of racing in the Manawatu area.

The conduct of meetings and the development of the activity has been influenced by government legislation and regulation by the Racing and Trotting Conferences and the New Zealand Racing Authority. This has imposed a stamp of functional similarity, although not necessarily uniformity, on any meeting and racecourse. Countering this, however, are many features making for a unique character to meetings or racecourses. Crowds at meetings can range from a few hundred to 30,000 plus. Meetings can last all day or be confined to afternoons or evenings. The days in the week for meetings are not constant. Clubs stage promotions and gala meetings at different times in the season. The starting time of the first race, the number of races, the duration of the meeting, and the intervals between races are decided by clubs. Stakes vary widely, affecting the class of horse entered and public interest in the meeting and particular races. Clubs may own or rent a racecourse, and some clubs hold meetings on different racecourses. Weather is always unpredictable. Club policy over complimentary tickets and advertising is often reviewed. The amenities at different levels of racecourses differ markedly. In the Manawatu area, one racecourse has regional status, in terms of the New Zealand Racing Authority's classification of racecourses, while five others are classified as subregional. Given such a variety there was no sensible criterion for selecting one meeting rather than another, so the dates of the study meetings were randomly chosen from the published Racing Calendar.

The unilateral rulings about the scope of the study reduced the design aspects to devising representative and effective ways to approach racegoers making up the target population. Decisions about aspects of the sample survey are now discussed.

Full or Sample Survey

As a race crowd is a very temporary gathering of people of unpredictable size, a survey of all attending was considered to be virtually impossible.

However, the adoption of a sample survey strategy, while manageable, introduced some major problems.

Making Contact

A race crowd consists of boisterous and preoccupied people, moving onto, around, and away from the racecourse. Arrival is concentrated in the early hours of the meeting, surging to peaks just before races and tapering off after four or five races. Critically, arriving patrons are intent on catching the next race. During the meeting a massive cirulation of patrons takes place between betting windows, stands, the birdcage, and refreshment and amusement stands. Departing racegoers exit in a pattern that mirrors arrival, but they are often hostile, subdued, or uncommunicative. Most racegoers come in cars, entering through members' or nonmembers' entrances, park their vehicles, and merge into anonymity in the enclosure and the stands. A few come by bus or on foot. In view of these matters a focus on the arrival stream was deemed desirable.

Unfortunately, club officials insisted that interviewing disrupt the flow of traffic and people as little as possible. This eliminated the tactic of stopping vehicles at gates. It was found, moreover, that patrons were most attentive just after they parked, and interviews were conducted at this time.

Target Population

In the absence of any reliable figures about either race attendance or patterns of arrival, gate monitoring was undertaken of all vehicles and people (including their sex) entering the racecourse. This enabled an accurate picture of patterns over the meeting to be established and was used to check the representativeness of the linked sample survey. A very important advantage of knowing the arrival distribution of racegoers was that the information approximated a sample frame, defining the population of racegoers from which a sample might theoretically be drawn. Although it was impossible to use the frame in the conventional way—its final form was only known at the conclusion of a meeting—it did provide, after several meetings, a working guide to the rate at which approaches should be made to racegoers during different stages of a meeting. Moreover, comprehensive monitoring widened the range of selection strategies open to the investigators. The target population, strictly speaking defined ex post, was thus *all racegoers* arriving in a vehicle or on foot at the race meeting.

Resources

Available funds allowed 15 to 20 interviewers in the field on any race day (with an equal number tied up with gate monitoring), and this severely limited the number of potential contacts. The primary objective was therefore reduced to constructing a picture of the race crowd at each meeting and not detailed information on groups making up the race-day crowd. The limited resources did impose some loss of precision. Under good conditions and at large race meetings, upward of 300 interviews were completed, but at small meetings and in adverse conditions only 100 to 150 interviews were achieved. This arose in part because arrivals were compressed into the first three hours of each meeting. Only by confining the inquiry to a simple description of the crowd was it possible to accept the sampling error associated with sample sizes on the order of 100 to 300 people.

Serious consideration was given to two alternative ways of selecting those for interview. First, efforts could be directed toward ensuring a sampling distribution that coincided with the arrival distribution established separately by monitoring at the entrances. For successful implementation of this strategy the pace of interviewing needed to be sped up or slowed down to reflect the pattern of arrivals—something that could be controlled by careful management. The significant advantage of this approach was the greater number of actual interviews that could be completed, remembering that most racegoers arrived over the space of an hour and increased sampling during that time would see a larger sample size. Second, a constant rate of sampling could have been adopted, paced by the arrival rates at the beginning and end of the meeting. The sample distribution could then be adjusted to reflect the arrival distibution, by giving more weight to the interviews taken at peak times and weighted estimates arrived at. In this strategy successful on-the-day operations could easily be achieved. A steady but unchanging rate of interviewing would not tax interviewers. However, only at very large meetings would the unweighted sample size be inflated by a wider time spread of racegoer arrivals. In the end it was decided that the first strategy kept the most options open—allowing at least some probing of diversity in composition of race-day crowds, rather than eliminating variation by tightly constraining the number of actual interviews.

Selection Process

The adopted selection process consisted of three parts: the time of arrival, the gate entered, and the mode of transport. Considerable reliance was placed on advice from club officials and car park attendants about the rhythm of arrival, the flow of vehicles once at the racecourse, and the proportions of the crowd who would come by car, bus, or on foot. This was of great assistance in resolving how to assign interviewers to different car parks and at pedestrian entrances. A simple weighting based on estimates of the capacity of car parks and of the proportion arriving by car was made in the assignment of interviewers.

Vehicles and pedestrians were chosen using a time-based selection interval. However, to avoid any bias because of periodicity in arrival patterns, the rate of interviewing over half-hour intervals (spanning successive races) was adjusted to reflect as closely as possible the monitored rate of arrival. In actuality there was a time lag between changes in the rate of arrival of patrons and the rate of interviewing. At this stage a random selection of a racegoer in each vehicle was made.

Summary

In spite of compromises in the selection process described above, it was felt that it would approximate an equal probability selection of racegoers and had the potential to yield an accurate sample. Certainly, it was judged to be satisfactory for the generation of broad estimates of direct use to clubs.

THE SAMPLE

Experience resulted in some refinement to the foregoing methods, but the on-the-day circumstances encountered often proved to be more or less extreme than was anticipated or were quite unexpected. In short, the procedures spelled out were adhered to but were subjected to considerable strain. It was important, therefore, to use a yardstick to tell whether or not the samples collected were unbiased. This was checked by examining representativeness on two points: arrival patterns and sex of racegoer.

A comparison of the sampling pattern over time with that of the arrival pattern amounts to a comparison between a sample distribution grouped into time periods and that of a population distribution over similar periods. A test appropriate for this purpose is the chi-square one-sample test, based on frequencies. In a similar way the sex of the sampled racegoers can be related to that of the target population compiled through gate monitoring. These tests of goodness of fit are especially useful in assessing *any* kind of difference between the sample and population

TABLE 1 Arrival Time Comparison of Sample with Gate (Race Meetings)[a]

Time period	1 O	1 E	2 O	2 E	3 O	3 E	4 O	4 E	5 O	5 E	6 O	6 E	7 O	7 E	8 O	8 E
1	18	29	31	26.5	61	107.5			51	55.7	26	73.6	47	62.5	88	96.4
2	121	116	93	109.6	82	52.9			34	25.0	66	27.6	70	42.6	70	64.3
3	74	75.7	61	54.0	36	27.4	46	53.8	10	16.1	16	11.7	13	24.9	28	18.3
4	41	38.5	47	36.7	30	21.0	61	58.8	9	7.2	9	6.1			6	13.2
5	42	36.8	16	21.2	12	10.0	17	11.4								
N	296		248		219		124		104		117		130		192	
χ^2	5.28		8.34		43.00		3.96		5.12		87.16		27.13		11.05	
p	0.3		0.08		<.001		0.1		0.2		<.001		<.001		<.002	
Decision	Retain H_o		Retain H_o		Reject H_o		Retain H_o		Retain H_o		Reject H_o		Reject H_o		Reject H_o	

[a]Observed values (O) are based on effective responses; expected values (E) are calculated from gate monitoring percentages applied to sample N; P, probability of a χ^2 value larger than the obtained value arising through random sampling effects.

distributions. The tests tell whether these differences are small enough to have arisen solely from the fluctuations in random sampling, or reflect particular influences.

Comparison of Arrival and Sampling Proportions

The sample and arrival distributions are displayed in Table 1. The observed values are effective response figures. Their use in preference to actual approach data imposes a slightly greater chance of rejecting the null hypothesis H_0 (the less desirable outcome here). A visual inspection of the table suggests that the samples at meetings 3, 6, and 7 were drastically astray. In contrast, the others appear to have broadly conformed with the arrival distribution. The chi-square test allows confirmation or rejection of this working hunch.

The differences are combined to give a single figure, the chi-square statistic:

$$\chi^2 = \sum_{i=1}^{k} \frac{(O_i - E_i)^2}{E_i}$$

where O_i = observed number of cases in ith category
E_i = expected number of cases in ith category under H_0

The details of key steps in the calculation are set out in the lower portions of Table 1. By treating the number of effective approaches as (O) and those from the gate monitoring as (E), a comparison can be made across the time periods (O − E). The calculated χ^2 value of 5.28 for race meeting 1 suggests that the differences between the sampling distribution and the gate-monitored arrival distribution could well be due simply to sampling fluctuations and therefore indicates a reasonable correspondence between the distributions. The table reveals both successes and failures in application of the selection procedure adopted. A recalculation using actual approach data "improves the fit" for all meetings and for meetings 7 and 8 results in retention of H_0. It should be noted that most refusals were experienced early in the day when patrons were impatient to get to feature races. There are reasonable ground, then, for concluding that the sample at most meetings was unbiased on this measure, and when it was not (meetings 3 and 6 especially) the magnitude of error was probably not great. Table 2 summarized the proportions of women observed at the gate with those in the samples. The most noticeable feature is the mismatch at meetings 2 and 4. The table shows proportions because they are easy to compare, particularly between different race meetings, but by multiplying the proportions by the sample size the observed and expected number of females can be found, and by subtraction the observed and expected

TABLE 2 Number of Females: Comparison of Sample and Gate

Race meeting	Female proportion		Sample size	χ^2	p	Decision
	In sample	In gate				
1	0.445	0.401	296	2.49	0.11	
2	0.330	0.423	248	8.67	0.003	Reject
3	0.374	0.334	219	1.61	0.20	
4	0.347	0.488	124	9.90	0.002	Reject
5	0.259	0.325	104	2.03	0.13	
6	0.410	0.413	117	0.004	0.95	
7	0.315	0.326	130	0.07	0.80	
8	0.375	0.426	192	2.04	0.15	

number of males. It is then possibe to carry out a χ^2 test in the same way as for the arrival distribution using just two categories, male and female. The test shows that several samples on this variable are biased. A check of the sex of those originally approached showed greater refusals from women, a tendency that helped inflate the male and decrease the female proportions at some meetings. Overall, though, the test results suggest that the sampling methods yielded reasonably representative samples as regards the sex of racegoers at most meetings.

RACE CROWDS AND THE REGIONAL POPULATION

Club officials often said that they needed to know the composition of crowds at their meetings, whether those going to the races were a cross section of the community, and whether they were justified in thinking of racegoers as a distinctive group. These questions deal with different points. Before presenting a profile of the raceday crowd, an attempt is made to answer the last two questions. Once an idea of the difference or similarity and the distinctiveness of the crowd is gained, attention can focus on the appropriate content and phrasing of a description of race crowds in the study area.

What is an appropriate regional population? Racegoers are reputed to travel considerable distances and each club and racecourse could have rather different patronage or catchment. One way around the definition of a regional population is simply to use parameters from a sizable area such as a statistical area or statistical division. The variation in parameter values is in fact small at this areal scale.

Table 3 summarizes the tests used to compare the sample parameters with regional population parameters. The strongest possible data are used for the female-male split, the proportions monitored at gates. Sample data are used for the age and ethnic composition comparisons. The results for Table 3 indicate that the female-male split in successive race-day crowds, with one exception, is quite different from the adult population living in the region. The ethnic composition does not deviate greatly, meeting by meeting, from the regional pattern. When it does (meetings 4 and 5) it is interpretable. One of these meetings was put on by the Otaki Maori Racing Club and the other was a meeting that "traditionally" attracts non-European support. Much speculation has surrounded the age structure of racegoers. Many officials have argued that gallops are supported by an older population. The findings suggest that racegoers in the area are weighted toward people over 45 years of age.

ARE RACE CROWDS SIMILAR?

Are there grounds for asserting that each sample is from the same population? The chi-square test can also be used to determine whether the observed differences between different race meetings can be attributed to sampling variations. Instead of basing the expected numbers on the regional poulation proportions, they are based on the proportion from the samples from all meetings combined. Results are shown in Table 3 for two variables, sex and ethnic groups. Not unexpectedly, the broad pattern for sex ratios is one of similarity rather than difference. In contrast, the ethnic compostion observed at the meetings should not be regarded as being drawn from a similar population. A large contribution to the deviations between observed and expected is confined to meetings 4 and 5, a reflection of the support base of particular clubs and racecourses. The pattern across racecourses, for age, show irregular deviations in the young and old groups. The small sample size at several meetings did create a problem in this part of the study. It precluded separate analysis of age patterns for men and women because cell totals were often too small for the chi-square test to be reliable. Data were also collected on occupation, reasons for coming, and means of entry (members, complimentary, or general public). All three showed differences between meetings that could not be explained as random sampling variabilitiy.

The tests suggest that it would be injudicious to argue that collectively, race crowds in the area make up a distinctive population. Every race meeting shows a lot of variety in race crowd profile.

TABLE 3 Comparison with Regional Population and Between Meetings, (Proportions)

	Regional population	Meeting 1	2	3	4	5	6	7	8	average	Between meetings
Ethnic composition											
Proportion non-European	0.09	0.077	0.097	0.119	0.193	0.173	0.094	0.130	0.114	0.113	
χ^2		1.53	0.07	1.87	14.8	7.8	0	2.16	1.18		22.0
p		0.2	0.8	0.2	<.001	0.005	1.0	0.1	0.3		0.01
Sex											
Proportion female	0.50	0.401	0.423	0.334	0.488	0.324	0.413	0.326	0.426	0.368	
χ^2		18.5	13.4	19.0	3.9	15.0	12.9	19.4	10.3		12.4
p		<.001	<.001	<.001	0.05	<.001	<.001	<.001	0.001		0.003
Age											
18–24	0.14	0.200	0.167	0.042	0.256	0.088	0.115	0.156	0.139		
25–34	0.25	0.175	0.158	0.198	0.155	0.186	0.105	0.260	0.161		
35–44	0.17	0.179	0.188	0.150	0.193	0.186	0.180	0.173	0.233		
45–54	0.16	0.164	0.197	0.217	0.193	0.127	0.228	0.148	0.183		
55–64	0.13	0.179	0.167	0.169	0.100	0.203	0.307	0.156	0.155		
65+	0.14	0.100	0.120	0.22	0.100	0.205	0.057	0.104	0.127		
χ^2		22.65	16.4	35.3	18.6	13.3	42.8	1.9	11.5		
p		<.001	0.006	<.001	0.002	0.02	<.001	0.9	0.04		

THE RACE CROWD

The facts collected and the analyses performed on them enable several critical points to be made about racegoers and race attendance patterns that are germane to club promotion of racing in the study area. These can be summarized as follows:

1. Race crowds should be recognized as being a mixture of different groups of people. There appears to be no stereotypical racegoer.
2. The number of males outweighs the number of females at meetings.
3. The ethnic origin of racegoers will vary from meeting to meeting depending very much on the ethnic makeup of the local population or local affiliations with clubs.
4. Racegoers are of all ages, with a bias toward older groups.

ESTIMATES AND EXPECTATIONS

The whole thrust of this chapter has been to examine the character of representative samples of racegoers gathered at a number of race meetings. Club officials and racing administrators were firmly convinced that race crowds had a fairly homogeneous character. The evidence from this study suggests that there are limited grounds for regarding crowds as similar across meetings, but that a reasonable case can be made viewing them as different from the population at large. Acceptance of these "regularities" should not, however, turn attention away from the considerable variety in race-day crowds. Some of this variety can be traced to the geographical setting of racecourses, particular race meetings, and the initiatives of clubs. The portrayal of racegoers by way of a profile depicting "typical" features might be a useful summary of information, but it may be hiding more than it reveals. One drawback to this approach is that while enabling a speedup of functional responses to the findings by individual clubs (in the form of advertising to target segments of the *population*, such as women or young people), it tends to mask the context and the reasons from which attendance at a race meeting emanated. It is for this reason that this study has answered "Who still comes to the races?" but has not penetrated the complexities surrounding "Who still goes ... ?"

V
Experimental

A feature of the scientific method is that the researcher usualy creates data so that questions can be answered in one way or another. The two common procedures for gathering data are experiments and surveys, and in both instances statistics has made a large impact in designing the procedures for collecting information. The two chapters in Part V are concerned with the design of experiments.

Experimental design has its roots in agricultural field research. Terms such as "plots" and "blocks," which were initially used in agriculture, have been carried across to other areas of application. It is not uncommon to find that, when the subject is taught in university courses, it tends to follow the historical development of agricultural experiments. It is therefore fitting that Chapters 18 and 19 have been written by statisticians with extensive practical experience in designing agricultural experiments, and furthermore, that both are concerned with historical aspects.

The article by Greg Arnold summarizes the main features of the extensive correspondence that took place in the 1930s between the late Dr. Hudson, a pioneer in agricultural research in New Zealand, and the famous British statistician W. S. Gosset, who published his work under the alias of "student." Over a distance of approxiamtley 12,000 miles in an age when it would take months for a letter to travel this distance, Gosset was the teacher and Hudson the student grappling with concepts of experimental design.

In Chapter 19, Arthur Rayner draws on a lifetime of experience in teaching experimental design and helping researchers to design their experiments. Again, much of the material in the article is derived from correspondence between the author and well-known statisticians. Readers who know something of the modern approach to this area of statistics will gain new insights into the way the subject has developed in this century.

The style in this part of the book is substantially different from other sections, mainly because of the historical theme. For those of you making a first acquaintance with experimental design, we suggest you read the chapter by Greg Arnold first, as it flows logically into the general picture of experimental design as described in the following chapter.

__ 18 __

Randomization: A Historic Controversy

Gregory C. Arnold
Massey University
Palmerston North, New Zealand

Sir Ernest Rutherford, the famous physicist, is alleged to have said that if you need statistics to analyze your results, you should have conducted a better experiment. In nineteenth-century physical science this might almost have been true. Those who have repeated nineteenth-century physics experiments in twentieth-century schools will well remember that a good experiment was one where straight-line graphs really were straight. However, Rutherford's statement would never have been true about biology. The inherent variability of plants and animals makes it necessary to distinguish between the differences caused by the experiments and the differences that would have been present anyway. Statisticians started arguing about this problem early in the twentieth century and have kept it up ever since.

In this chapter we try to show what the argument is about by quoting from a correspondence file belonging to A. W. Hudson, a crop experimentalist in New Zealand. The letters were written in the 1930s, and the other main correspondent was W. S. Gosset, a chemist turned statistician who worked in Ireland for Guinness Breweries. Gosset is better known by his pseudonym "Student" and the t test, which he developed. Hudson really had quite simple problems. He would have a few (often only one) new varieties of seed, and he would want to know how much better these were than the existing varieties. In the absence of any very well-developed

statistical procedure his practice followed pragmatic principles. To quote a letter he wrote to Gosset:

> I have been impressed with a few outstanding features of the layout of experiments in relation to their statistical examination. Chief amongst these are:

1. Frequency of replication is the most important factor for any given size of plot in determining the precision of the measurement of results
2. Close proximity of treatments being compared assists materially in reducing errors
3. The soil itself, apart from the influence of more or less uniform fertility slopes, provides randomization to treatments arranged in a deliberately regular order.

To understand what he meant, consider one of his experiments, which compared certified potato seed (C) with noncertified potato seed (NC). These two types of seed are the treatments mentioned in (2) above. The layout of the experiment is shown in Figure 1.

	Pair 1		Pair 2		Pair 3		Pair 4		Pair 5		Pair 6		Pair 7		Pair 8		Pair 9		Pair 10	
Seed type	NC	C	NC	C	NC	C	NC	C	NC	C	NC	C	NC	C	NC	C	NC	C	NC	C
Strip number	1	2	3	4	5	6	7	8	9	10	11	12	13	14	15	16	17	18	19	20

FIGURE 1 Layout of Hudson's potato experiment.

233

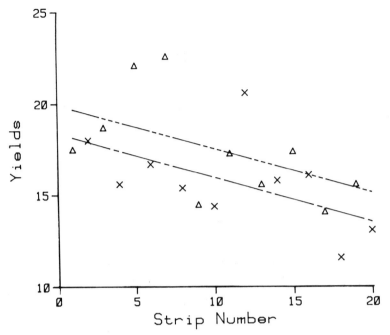

FIGURE 2 Potato yields by strip number. Δ, Certified seed;----,
certified seed trend line; X, noncertified seed;—-, noncertified seed trend.

There are a series of narrow strips of land, arranged in pairs, each
pair comprising a strip planted in certified seed and a strip planted in
noncertified seed. The results of this particular experiment showed that
the certified seed plots yielded 1.8 lb more potatoes on average than the
noncertified plots, and that this difference was so large that it was
unlikely to have arisen purely through the effect of biological
variability.

Now, what do Hudsons's principles mean? By "frequency of replication" he meant the number of pairs of strips; by "close proximity of
treatments" he meant that the strips within each pair, one of certified and
one of noncertified seed, were close together; and by "error" he meant
changes in yield caused by variation in soil fertility and other types of
biological variability. His idea was then that since the strips within each
pair are close together, the soil differences will be small (point 2) and
what differences are left will average out to zero (point 3) as long as there
are enough pairs (point 1). Perhaps this sounds quite sensible, but Gosset
was not entirely happy, and his assistant, Mr. Somerville, had this to
say:

There is an obvious fertility slope in this experiment and with such an arrangement of plots the results must give an exaggerated idea of the real difference. We can eliminate the effect of this fertility slope if we fit a straight line to the plots . . . .

The graph shown in Figure 2 is of yield against strip number and shows how the yield depended on the location of each strip. Although the main impression must be one of high variability, there is certainly a tendency for the strips with high numbers to have lower yields. The two straight lines, one for each seed type, show the average effect of this tendency, and show that on average there is a drop in yield of 0.24 lb from one strip to the next. They are constrained to have the same slope because the fertility changes they represent depend on the soil and not on the type of seed. Somerville's point was that because the certified seed was always on the left-hand strip, certified seed strips would be 0.24 lb better yielding than the noncertified seed even if there were no difference in yield between the seed itself. He went on to conclude that the amount by which the difference in mean yields (1.81 lb) was greater than 0.24 lb was within the bounds of what could plausibly be explained by biological variability. In short, the experiment had not convincingly demonstrated any difference between certified and noncertified seed. Indeed, the distance between the two lines, which represents the difference in yields betweeen the two seed types, is not all that large compared with the variation in distances between the pairs of points.

The point to take from this experiment is that, aside from biological variability, there were other possible causes for the differences between certified and noncertified strip yields, and that the conclusion from the experiment was rather different when one of these other causes was taken into account. Gosset's comment was:

It is better to rule out the simple fertility slope by arranging A B B A A B · · · B A and you can then fairly assume that the remaining soil differences are likely to be randomly distributed. Yet even then there is the possibility of a periodic variation due to previous cultivation, and that is why Dr Hilgendorf drill his half-drill strips diagonally to previous cultivations.

Hilgendorf, incidentally, was a New Zealand plant geneticist who now has a variety of wheat and a building named after him. With the regular arrangement suggested by Gosset and the strips drilled diagonally, is the problem solved? Unfortunately not. To see the next problem one has to understand how the seed potatoes were sown. A large seed box mounted across the back of a tractor had a number of tubes running across it leading to drills. The seed potatoes fell through these tubes into the soil. This seed box was divided in half and the middle drill blocked. This sowed the seed in two narrow strips side by side, certainly satisfying Hudson's principle (2). But suppose that one of the drills were a little bent or a tube slightly misshapen. Then the rate of seeding from that drill would always be different, and that variety could be penalized or advantaged. Hudson was aware of this problem, but regarded it as something the careful experimenter could avoid. Nevertheless, in a close examination of some experimental data some years later, Gosset discovered a regular fluctuation which he attributed to a partially blocked drill. This was not one of Hudson's experiments, but no doubt its supervisor regarded himself as careful. The difficulty is that however an experiment is laid out, a critical observer (and the only reason for carrying out experiments is to convince critical observers) can always point to some possible cause of the observed differences in any one experiment other than the treatment itself. You may feel that your opponent's alternative explanation is implausible, but if he or she thinks the same of yours, knowledge has not progressed very far.

There is a magical ingredient which overcomes the problem, and that is randomization. To quote Gosset:

> Nearly all the cases which illustrate the old crescendo "lies, damned lies and statistics" are due to lack of randomness in the data. Now in the case of agricultural experiments the only way in which you can be sure that your samples are perfectly random is to put the numbers of your plots into a hat and draw for position.

It is this ingredient, randomization, which statisticians argue about. Although Gosset was well aware of the difficulties of ensuring that the treatment differences could unambiguously be attributed to treatments, and even gave examples, he concluded:

> While I put these possibilities before you, you must not suppose I attach any very serious importance to them in the case of an investigator who keeps his eyes open, and in fact I have recently had a correspondence in "Nature" in which I upheld the view that such occurences are very unlikely. Nevertheless it behoves the investigator to keep his eyes open.

Hudson was dead against randomization and never lost an opportunity to express his disapproval. He believed himself to be (and undoubtably was) an investigator who kept his eyes open. Further, he believed that regular arrangements of treatments were less variable than random arrangements, and to prove it he made use of some data which

Regular	A	B	C	D	E	F	A	B	C	D	E	F
Random	D	C	A	E	B	F	F	A	C	B	D	E
Yield	653	692	649	540	489	544	560	526	512	550	521	537

Regular	F	E	D	C	B	A	F	E	D	C	B	A
Random	C	D	B	E	F	A	E	B	D	C	F	A
Yield	688	732	670	584	482	511	495	496	471	511	491	495

FIGURE 3 Layout of regular and random imaginary treatments.

TABLE 1   Means for Random and Regular Arrangements

Treatment	Random	Difference from mean	Regular	Difference from mean
A	545	(−13)	555	(−3)
B	551	(−7)	548	(−10)
C	601	(42)	564	(6)
D	594	(36)	558	(0)
E	539	(−19)	560	(1)
F	505	(−39)	566	(8)
Standard deviation (between means)	65		13	
Standard deviation (within means)	66		76	

had been collected from plots of ground before any experimental treatments had been applied. Such data are called *uniformity data* because they show how uniform a proposed experimental area is, and Hudson described what he did with it in a letter to Gosset.

I will tell you what I did, although I realise that it is all very elementary so far as you are concerned. I worked on data from uniformity trials of Mercer and Hall, Immen and Kulamkar in the following way:

1. Four, five or six imaginary treatments were allocated according to which was the most suitable to the full utilisation of the data
2. These were allocated to blocks in a regular balanced fashion, and then to the same blocks randomwise using various numbers or "units" per individual pot.
3. The amount by which the means of the plots of the different treatments differed from the mean of all plots was taken as an indication of the real error, the standard error being calculated by Fisher's randomized block method.

Elementary it may be, but it is nevertheless confusing. Figure 3 shows the layout of the plots that gave rise to one of these sets of data. This was another potato experiment, and the numbers are yields. Remember that there were no real treatments.

The differences between the yields of each plot represent no more than biological variability. Any differences between the mean treatment yields would be a consequence of biological variability combined with the way treatments were allocated to the plots. The top letters in Figure 3 show Hudson's regular arrangement of treatments and the lower letters his random arrangement. What Hudson hoped to show was that the differences between the treatment means were less when allocated by the regular arrangement than when allocated by the random arrangement. The results are shown in Table 1.

Clearly, the variation between the means is much greater when the treatments were allocated randomly. The standard deviation is a measure of this variation, but the difference is obvious without any technical measure. Hudson repeated this procedure for many arrangements, and the one shown here was the one showing the greatest advantage to the regular arrangement. The variation between the treatment means are what he calls "real errors." because the true difference between treatments was zero. He summarized his results as follows:

> When the number of replications of each treatment were as low as four, the real errors were considerably less under the regular arrangement in two cases, and about the same as random in the third. With from 8 to 32 replications the real errors were not markedly different, but in most cases were smaller under the regular arrangement.

For Hudson, this was sufficient. Regular arrangements gave more accurate results and had the additional merit of being easy for unskilled technicians to lay out in the field.

What did other people about that time think of randomization? A rather lukewarm comment was contained in a letter from I. J. Wishart, from the School of Agriculture, Cambridge:

> I am not sure that I am prepared to write an article advocating systematic as opposed to random arrangements, but the whole question certainly merits a closer examination than it has hitherto got, and if I may I will take a little time to think it over.

Sir Ronald Fisher, almost the inventor of randomization, was unequivocal.

> There is now no room for doubt that, as we put forward on theoretical grounds some years ago, systematic plot arrangements really preclude any genuine estimate of error (biological variability) at all, and therefore any valid test of significance.

In other words, without an estimate of biological variability there is no baseline with which to compare the differences between treatments.

We have so far ignored the need to estimate the biological variability (or "error" as it was usually called, rather misleadingly). In Table 1 the

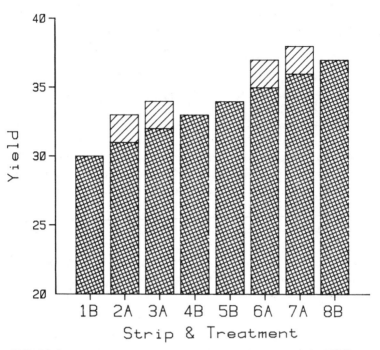

FIGURE 4   Natural yield and increase in yield for each strip. ▓▓▓ natural yields; ///, increase in yields.

standard deviation was used as a measure of the variability of the variety mean yields. A similar quantity can be calculated from individual yields within each variety. In Table 1 this was called the standard deviation within treatment means. This gives a separate measure of the variability of plots regardless of differences between varieties, and is therefore the all-important measure of biological variability. The standard statistical technique for assessing differences between treatments was (and still is) to compare the variability (standard deviation) between treatments with the biological variability. This comparison is the basis of the test of significiance Fisher mentioned, and he is claiming that if the biological variability is to be estimated correctly, the treatments must be applied randomly. Hudson noticed in his uniformity data that where the real errors (standard deviation between means) were large, the estimate of biological variability (standard deviation within means) was small. So we have a curious no-win situation. If an experimenter manages to arrange plots in a balanced way which eliminates some of the "error" so that the estimators are more accurate than would be obtained from a random arrangement, the estimate obtained of this "error" will be *increased*. Increasing the true precision of the treatment differences increases their apparent variability. To understand this rather paradoxical result, consider a very much simpliified example. Suppose that four pairs of strips are sown in a balanced pattern as advocated by Gosset, and suppose that the only biological variability is a uniform increase from one strip to the next which results in the right-hand strip always yielding 1 unit more than its left-hand neighbor. This is the natural yield shown by the crosshatched bars in Figure 4.

Now if treatment A increases the yield by 2 units, the yields would be as indicated by the total heights in the diagram. The differences between each pair would be 3, 1, 1, 3, respectively, and the estimate of the effect of treatment A would be the average of these four differences, or 2. The balanced experiment has given a precisely correct estimate. Now if the experiment had been randomized, there is a good chance that three pairs would have had treatment A on the right, and the corresponding differences would have been 3, 1, 3, 3. The estimate of the effect of treatment A would then have been 2.5, half a unit too high. But the apparent variability is less, since 3,1,3,3 is a less variable group of numbers than 3,1,1,3. Of course, a balanced arrangement could arise randomly, but checking all possibilities will show that the apparent variability cannot be greater than with the the balanced arrangement. Hudson's reaction to this was understandable: "if by randomization the real errors are increased, what use is a *valid* estimate of error"; and Gosset thought:

> The fact is of course that absolutely random samples are a luxury in statistical practice apart from these agricultural experiments they simply don't exist and though they add to one's peace of mind may be paid for too dearly if you are putting up your real error thereby—and so say all of us, eh!

Nevertheless, standard statistical techniques compare the treatment difference with the biological variability. A large estimate of biological variability means that a treatment difference has to be relatively large before it can be shown to be real. Hudson complained:

> I have never been satisfied with Fisher's plausible statement that any arrangement which reduced the real error "was at the expense of making them seem worse." They are certainly worse according to his statistics but I feel strongly that his analysis of variance is unsound. Is there any other statistical method which decreases the precision of the results when the dispersion of the individuals is reduced?

The analysis of variance is a statistical tool based on an algebraic equation, which says:

Total variability
= variability between treatments + variability within treatments

This equation predicts the effect that we have just described. The total variability is constant, so that if a regular arrangement lowers the variability between treatments, the variability within teaments (the estimate of biological variability) must rise to compensate.

The basic reason why randomization achieves all that is claimed for it is not too difficult to understand. Consider Hudson's potato experiment again. The difficulty was that any systematic arrangement could correspond to a naturally occurring effect which could cause a difference between the certified and noncertified seed plots, but which would not be a treatment effect. If a coin is tossed to decide how the certified and noncertifed seed should be applied, it is still possible that the certified seed would always fall to the left, but it is most unlikely. What is really important, though, is that we can calculate exactly how unlikely it is.

Its probability is equal to the probability of tossing 10 heads in a row (1/1024, in fact), and that is a figure on which everyone will agree. Randomization achieves even more than that. It does not matter what effects may be confusing the experiment; the probability of the certified seed falling on all the best plots is still 1/1024. Even if an experimenter were not being careful, as long as the randomization is correct, the probability of a misleading result being a consequence of the way the treatments are assigned to the plots can be calculated from the probability properties of tossed coins. The randomization process transfers a completely understood probability structure onto the experiment, and even hostile critics will agree about what happens when a coin is tossed.

Unfortunately, there is still a catch. Statisticians are in full agreement on the probabilities of any particular number of heads in several tosses of a coin, but they are not agreed on what "probability" itself means. We agree that as a coin is tossed more and more times, the proportion of heads will come closer and closer to half, and therefore that if an experimenter used a coin to allocate treatments, each treatment will come closer and closer to being equally favored by biological variability. But what happens in just one particular experiment? If Hudson has to decide which seed to recommend for a whole region on the basis of just one experiment, should he not use all his experience to allocate treatments in as fair a way as possible? Some would answer, yes. In practice, though, science is not like that. Once-and-for-all decisions are not made on the basis of one experiment. With randomized experiments a damaged drill on a plough, or a misconception about fertility trends, may cause rather variable results, but they will not cause consistently incorrect results. The position was summarized in a letter signed by K. Mather of the Galton Laboratory.

It seems to me that the real problem raised by the alternative of random or regular layouts is, what are the results wanted for? Hudson shows,and I think others, e.g., Tedin, have also shown, that with certain regular layouts the estimate of error is greater than the true one, or, to put it differently, the results are more precise than the data indicate. This deviation or bias may be small and insignificant for the experimenter under consideration. Furthermore the slight loss in apparent precision may be compensated for by the extra ease of agricultural operations involved. The real problem at issue however, in my opinion, are the results to be used for this experiment only? If so then everything is fine. The experimenter realises the limitations of the data and acts accordingly. But it may be that at some future date the experiment will be of value in some other connection either to the same or to a different worker. If this is the case a valid estimate of error is absolutely necessary. The bias of the regular layout will not be the same for two different effects that the experiment may bring out and, in any case, another worker would find it difficult to utilise experimental data which are known to be biased but to an unknown and incalculable degree. In other words the question to be decided is that of whether the results are to be simply of immediate importance or of real scientific value in the sense that they can be honestly utilised for the purposes other than those for which the experiment was originally designed by either the same, or , more specifically, other workers. It is very easy to say that the results are of no interest to other scientists or that only one experiment in twenty will be of wider interest. But how does one know which experiment is going to show some extra effect that may be of very wide interest either at present or in the future?

It would not be fair to nonrandomizers to conclude on that note. Remember Hudson's cry for another statistical method? Somerville used another statistical method when he analyzed Hudson's potato experiment. He fitted a line to the fertility trend and then looked at differences from that line. He was *modeling* the potato yield, and to do this he assumed that the yield was affected by a linear fertility trend, a difference between certified and noncertified seed and a purely random biological variability. That was his model of the way potatoes grew, and anyone satisfied with this model would accept his conclusions. Anyone with a different model could well arrive at different conclusions, and the experiment would not help decide who was correct.

Nowadays computers can fit very complex models. Many ideas can be tested, many possibilities examined, but they merely raise questions for testing in future experiments. Experimenters (like Rutherford) who believe that they fully understand the model underlying their experiment may abandon randomization and thereby gain precision. Unlike Hudson, they have a computer to fit almost any model they like. Paradoxically, though, the computer also gives much more scope to others wishing to criticize the data. The argument for randomization is therefore as strong as ever.

ACKNOWLEDGMENTS

Two officers of the New Zealand Ministry of Agriculture and Fisheries are to be thanked for the survival of Hudson's correspondence in usable form. Paul Lynch recognized its value when it was in danger of being thrown out, and Murray Jorgenson ensured that it was put in order and lodged with the New Zealand National Archives, where it now resides as documents Ag 51/1 and Ag 51/2.

Abe Hudson died in 1982. I thank his wife for permission to use his correspondence in this article. Also, I thank Professor Sir Kenneth Mather for permission to quote an extended portion of his letter. This letter is an imperfectly typed copy which omitted part of the first sentence as it appeared in the original.

19

Some Sidelights on Experimental Design

Arthur A. Rayner*
University of Natal
Pietermaritzburg, South Africa

R. A. FISHER AND ROTHAMSTED EXPERIMENTAL STATION

VJ Day, marking the end of World War II, occurred while I was at sea en route from New Zealand to the University of Edinburgh, where I spent two academic years studying under A. C. Aitken, duly producing an acceptable Ph.D. thesis entitled "The basic theorems of analysis of variance with special reference to field experiments."

I spent the 1946 and 1947 summer vacations, plus some extra months while awaiting passage back to New Zealand, working in the Statistics Department at Rothamsted Experimental Station with the lowly status of voluntary worker, later temporary worker (i.e., slightly paid). Rothamsted, founded in 1843 and situated at Harpenden in Hertfordshire, about 40 km north of London, owes its chief claim to fame (at least among statisticians) to the fact that here R. A. Fisher originated, and the major early developments were made in, the area of statistics known as the design and analysis of experiments. It seems to me fortunate that Fisher was concerned primarily with agricultural field experiments, where the difficulties are most acute; had the design and analysis of experiments originated in another area of application, such as industrial experiments, it is possible that a highly confused situation might have developed.

*Emeritus.

245

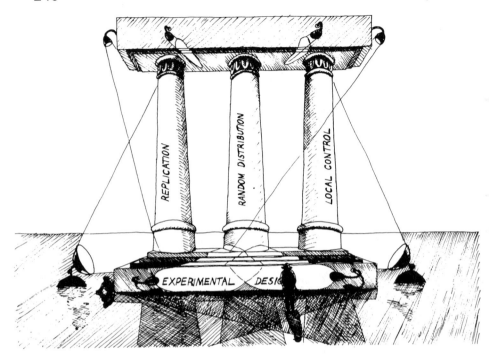

To my regret I never met R. A. Fisher. However, I did have as a colleague in Pietermaritzburg for several years Thomas Bogyo, who claimed that he saved Fisher's life at a genetics conference in Edinburgh, when during a visit to the zoo, Fisher was grabbed by an ape through the bars of its cage. This must have been the 1939 conference described by Box (1978, p. 371). I have yet to see an article: "The debt of statistics to T. P. Bogyo"! On a visit to Rothamsted in 1980 I observed a small weatherbeaten board bearing the legend "Fisher Laboratory" and remarked to John Nelder (the former head of Statistics at Rothamsted) that it seemed a rather humble memorial. Replied John: "What do you want? The Japanese expect a shrine!"

In embarking on these sidelights, I am very much aware that the recorded facts on the early history of experimental design are unavoidably somewhat sketchy. I am also aware that there are many living who came into the field much earlier than I did and who were more deeply involved. If I err on any point, I crave their indulgence.

THE RANDOMIZED BLOCKS AND LATIN SQUARE DESIGNS

Fisher came to Rothamsted in 1919. His immediate job was the extraction of information from old experimental data, and this soon led him to grapple with the problems of designing field experiments. In his famous book, *Statistical Methods for Research Workers* (1925), he devoted the last two sections to this problem, introducing the randomized blocks and Latin square designs. The following examples of field layouts are taken from the 1925–26 Rothamsted Report:*

Randomized Blocks

N	J	F	A	D	O	K	A
K	Q	O	D	L	B	F	N
B	C	M	L	H	P	G	E
H	E	P	G	M	Q	C	J
A	L	J	C	P	Q	B	E
K	B	G	O	C	H	J	O
E	F	Q	D	N	M	A	D
N	H	P	M	F	G	K	L

Treatments: All combinations of 0, 1, 2, and 4 cwt per acre of sulfate of
    potash with the same applications of sulfate of ammonia (i.e., a
    4 × 4 factorial arrangement)
"System of replication: Randomized blocks for all manurial combina-
    tions"
Crop: Potatoes

In this design there are four *replications*, that is, four repetitions of each of the 16 treatments (A, B, ..., Q), and the experimental area is divided into four blocks, each containing 16 plots, to which all 16 treatments have been allocated at random without repetition. Since the observed mean difference in yield between any two treatments is the mean of four within-block differences, the experimental error is effectively the variability of plots within blocks in the absence of treatment effects; this variability is likely to be less than the comparable variability of plots over the whole experimental area, which would be the appropriate experimental error if the treatments had been randomized over the whole experimental area

*The diagrams in this section and the following section are photocopies from the original publication.

without the restriction imposed by the blocks (the so-called completely random design). The use of blocks in this way is known as *local control*. The design also incorporates the property known as *orthogonality*; that is, each treatment appears once and once only in each block, which simplifies the statistical analysis of the plot data. However, Fisher soon realized that orthogonality was not necessary for the exercise of local control by blocking.

Latin Square

C	O	K	S	M
O	M	C	K	S
K	S	M	O	C
M	K	S	C	O
S	C	O	M	K

TREATMENTS:
S, sulfate of potash ⎫
M, muriate of potash ⎬ + basal
K, 30% potash salts ⎭
C, basal only (super S/A + N/S)
O, no manure
"SYSTEM OF REPLICATION: Latin square"
CROP: Sugar beet

In this design there is a double exercise of local control through two sets of blocks, called rows and columns, each treatment occurring once and once only in each row and in each column. The Latin square is likely to give a lower experimental error than a randomized blocks design, but has the disadvantage that the number of replications must be equal to the number of treatments, and this means that for a largish number of treatments the number of plots required may be considered excessive. The randomization of the Latin square was not described in detail until Fisher and Wishart (1930); it may be found, for example, in Cochran and Cox (1950, Sec. 4.33).

"The design of experiments is as old as the hills. It is not an area of thought that originated in the 1920's" (Kempthorne, 1975). Yates (1975) speculates that Archimedes, after his experience in the bath, immediately set about making experiments to confirm his principle. Maybe, but there is no evidence, and Kempthorne's hills get a bit younger when we advance to Renaissance times for the first recorded experiments by Gilbert and Galileo (see Rayner, 1969, Sec.2.10). However, the observations in these early experiments were of events unaffected by chance, and comparative experiments involving measurements had to wait a little longer. At least that is what has been traditionally viewed as the historical order of events. Yet Cochran (1976) in a detailed history of comparative experimentation referred to the discovery by S. M. Stigler, a noted modern statistical historian, that a certain Arab medical doctor recommended replication and the use of controls in clinical trials, and also warned against confounding variables, in the eleventh century! This does roll Kempthorne's hills back a little in time. Be this as it may, it is well known that experimental designs incorporating the principles of replication and blocking were used well before Fisher appeared on the scene (Yates, 1962).

What Fisher did between 1919 and 1924 was to meld these with the third essential component (randomization) into a coherent system, *pace* Kempthorne (1975), for the use of experimenters; more than this, he provided a method of statistical analysis of the resultant data (Fisher, 1925, pp. 224–232). The new designs and their analyses were tried out on Mercer and Hall's uniformity data (see Rothamsted Report, 1923-24), and their adoption resulted in a new-look Report for 1925-26. Thus were the foundations of the new science of the design and analysis of experiments well and truly laid.

The randomized blocks and Latin square designs are still—to borrow an expression from Stigler (1981)—the "automobiles" of experimental design. Together with the completely random design, they comprise the three basic experimental designs.

A DIAGRAM THAT DESERVES TO BE BETTER KNOWN

The diagram at the top of page 250 is due to Fisher (1931). There are a number of interesting points concerning this diagram.

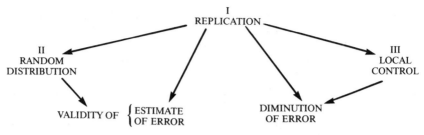

Principles of Field Experimentation

1. It may be labeled the forgotten diagram. The paper in which it appears is possibly unique in that in the five volumes of Fisher's collected papers (Bennett, 1971–1974) it is neither included as one of the reproduced papers, nor is it listed among the publications and other writings not included. Perhaps this is not surprising, seeing that Fisher himself never listed it in his bibliography in *Statistical Methods for Research Workers*.

2. Federer and Balaam (1972) list the paper, but curiously enough reproduce a diagram (p. 43) "similar to a diagram which is reported to have hung on the wall of R. A. Fisher's office at Rothamsted." Correspondence with John Nelder, however, has elicited the response that he has never seen it on display, and a statement from Frank Yates (who followed Fisher as the head of Statistics) that it was not on Fisher's wall in 1931 (also the year that Yates started work at Rothamsted).

3. The mystery thickens when one realizes that Fisher's diagram was hardly "state of the art" even for 1927, which may mean that it was around for several years prior to 1931. Federer and Balaam (p. 44) attempt an updated diagram which begins to look a little like one of those fierce mathematical models which have come upon the scene in recent years.

4. There is historical interest in that Box (1978, p. 153) reports the conjunction of two articles in the same journal (Russell, 1926; Fisher, 1926) expressing opposing views on experimental design, whereas by 1931 Russell, in a foreword to the Rothamsted Conferences publication just before Fisher's paper, showed that he had come round to Fisher's ideas, quoting at some length directly from the 1925–26 and 1927–28 Rothamsted Reports. [Sir John Russell was the Director of Rothamsted Experimental Station. In fact, in 1926 Fisher repudiated his boss, which seems deserved, since "the new method of field experiments was introduced ... in 1925, and has been used exclusively in all the new field experiments both at Rothamsted and at Woburn" (Rothamsted Report, 1927–28).]

The diagram attempts to encapsulate the basic principles of experimental design and their purpose (not confined to field experiments), although it makes no more than an oblique reference to analysis. In the following sections I shall attempt to examine how well the diagram does this, and give some indications of how it has become outdated by subsequent developments, these often due to Fisher himself and the Rothamsted school.

REPLICATION → LOCAL CONTROL → DIMINUTION OF ERROR (PATH 1)

The name "local control" does not seem to be used much outside the United Kingdom, but it was used in the 1923-24 Rothamsted Report, which, depending on publication dates relative to Fisher (1925), possibly contains the first account of the new experimental methods, with mention of the Latin square design, but not randomized blocks. It seems reasonable to assume that the section concerned, "Statistical control of the field and laboratory observations," was written by Fisher himself. The principle was spelled out as follows:

> The lowering of the experimental error may be achieved to a greater extent than has hitherto been attempted by the systematic adoption of the principle of local control, by which plots to be compared are set out on land of comparatively similar quality, without vitiating the estimate of the experimental error calculated from the totality of the results.

The origin of the name seems to be in the phrase "local differences between blocks" (Fisher, 1925, p. 227).

Fisher (1925, pp. 224-230) also explained how in the *analysis of variance* these differences between blocks could, in line with the purpose of the design, be eliminated from the estimate of experimental error by calculating a sum of squares for blocks as well as for treatments. Subtraction of these from the total sum of squares of the plot yields then lays bare the residual random (or error) variation (i.e., the error sum of squares). Division of the latter by the corresponding number of degrees of freedom in the analysis of variance table gives the error mean square as an unbiased estimator of the hypothetical true error variance (σ^2). For the Latin square two such sums of squares (for rows and columns) must be calculated to remove differences due to these sources of variation. Practitioners who used "erroneous methods of estimating the error, which failed to eliminate in the arithmetical procedure elements of variation which had in fact been eliminated from the real errors by the arrangement in the field" were duly castigated (Fisher, 1931).

Like a good salesman, Fisher probably felt compelled to emphasize the diminution of error due to local control and did not present examples where no diminution occurred. However, as indicated by the name "local control," the real objective is to *control* error (i.e., keep it within bounds) and there is no reason to grieve if in a particular experiment the blocks mean square is not larger than the error mean square (Rayner, 1969, Sec. 13.10).

The association of replication with local control in the diagram and the association of blocks and replications in the randomized blocks design are unfortunately likely to lead a student to think that replication as such reduces the error variance (σ^2). Worse still, a student is likely to be misled into taking out a sum of squares for "replications" in a completely random design. As I say: "Replications aren't necessarily blocks, and blocks aren't necessarily replications."

Blocks are not replications in confounded factorial designs, for example, which were described at an incredibly early date (Fisher, 1926). The simplest illustration of confounding is when an even number of treatments is divided into two equal groups and each replication is divided into two equal blocks, with the treatments in the two groups always occupying separate blocks within which they are separately randomized. This halves the block size as compared with a randomized blocks design. However, the treatment comparison represented by (group 1 - group 2) is identified with the difference between the two blocks in every replication and is therefore subject to the block variability, not the plot variability or error variance (σ^2). It is said to be *completely confounded* with blocks. This loss of information on the confounded comparison can be accepted if the experimenter is not interested in this particular comparison, since the halving of the block size may well mean a lower σ^2 and since all comparisons uncorrelated with (orthogonal to) the confounded one are unaffected by the confounding. Also, the loss of information can be spread by means of *partial confounding* (i.e., varying the confounded comparison from replication to replication). With factorial treatment arrangements (there is an example with two factors at four levels each in the randomized blocks experiment in the second section of this chapter) it is possible to identify treatment effects [e.g., interactions (especially high-order interactions)] and interaction components, which are suitable for confounding. Moreover, when the number of the levels of the factors is suitable, there are commonly used designs with the block size $\frac{1}{3}$, $\frac{1}{4}$, $\frac{1}{8}$, or $\frac{1}{9}$ of what it would be if each block contained a complete replication. Hence confounded factorial designs are an example of *incomplete block designs*. Other types of incomplete block designs intended for experiments with a large number of varieties (for example) were still ahead (Yates, 1936b;c).

REPLICATION → DIMINUTION OF ERROR (PATH 2)

The direct path from "replication" to the same terminus as path 1 can even more actively mislead a student into believing that replication per se reduces σ^2. This is an important reason why I have never used the diagram as a teaching tool, but only in an examination question such as "Use this diagram as a basis for the discussion of the basic principles of experimental design."

It was not until *The Design of Experiments* that Fisher (1935, pp. 66–67) presented an explanation clearly involving the formula σ^2/r for the variance of a treatment mean based on r replications. Compare, for example, Fisher (1933): "whereas the object of diminishing the experimental error is much aided by replication, replication is only one of the many methods of furthering this aim." (Fisher was not using "replication" to imply blocking, which was discussed later in his article.) Bearing in mind that 8 years earlier he labeled the error line of the analysis of variance as "experimental error," one can only regard this as misleading.

However, Fisher (1931), in explaining the diminution of error by *increased* replication (without any formula, though), linked paths 1 and 2 in countering the argument that such diminution will not be fulfilled if greater soil heterogeneity is encountered because of the larger experimental area required:

> However many blocks may be used in such an experiment the error of our comparisons will be due wholly to soil heterogeneity within blocks and this element of the heterogeneity has no tendency to increase as the number of blocks is made greater. The increased heterogeneity of the whole area is in fact wholly accounted for by the increasing disparity in yield between different blocks. . . . It is thus seen that . . . the diminution of error may, if a sufficient number of plots can be used, be carried to any required degree of precision, at least if the primary principle of replication is supplemented by the Principle of Local Control.

A modern researcher would doubtless be concerned that under these circumstances two assumptions usually made in analyzing data from a randomized blocks design, namely that

1. the yield differences of any pair of treatments are constant from block to block apart from experimental error, and
2. the error variance is constant over all blocks

may not be justified. When the former assumption is untrue, there is said to be an interaction of blocks and treatments.

BLOCK SIZE IN FIELD EXPERIMENTS

This topic nominally relates to path 1, but is intermingled here with path 2 and is therefore treated separately. By block size is meant the number of plots per block, rather than physical size.

According to Boyd (1973):

> The text books recommend small compact blocks and an important factor in the new style of experiments has been the realisation that so long as reasonable care is taken in choosing experimental sites, the loss of accuracy with large blocks is likely to be trivial. Indeed when in 1962 most [British soil scientists] increased the number of plots per block in their fertilizer experiments with potatoes from the conventional nine or ten to from thirty to forty, plot errors were unchanged; since then some hundreds of fertilizer experiments have been done in large blocks. Previous work on the effect of block sizes was suspect because it assumed that the errors of large-block experiments could be estimated from those of small-block experiments merely by ignoring the block correction; the possibility that, in choosing sites for large blocks, some small-block sites would have been rejected altogether was overlooked.... Experience shows that any small loss of accuracy from using large blocks is far outweighed by their greater flexibility, simplifying design and analysis by minimizing the need for confounding.

Of course, this seemingly strikes right at the heart of the motivation for confounded and incomplete block designs. Intrigued, I entered into a correspondence with Boyd which was terminated by his retirement and recent death. Let us examine the evidence.

Yates (1933a) in the pioneer paper on confounding certainly did not pay much attention to motivation, contenting himself with a statement that without confounding "the magnitude of soil differences within the same block may ... become unduly great." Yates (1935) admitted that he did not know of any extensive investigation into the increase in experimental error due to increased block size, but he did present figures based on Rothamsted experience of the gains in efficiency due to confounding. In summary: "In less than 10 per cent of the [66] experiments has the efficiency been more than doubled ... but in over half of the experiments it has been raised by more than 25 per cent. In about a third of the experiments the gain was either trivial or nonexistent." In a more extensive investigation based on 454 cereal trials Patterson and Ross (1963) found that despite major changes in field technique since 1935, the gain in efficiency due to the use of small blocks was still about the same as Yates had found it. They also derived an empirical relationship between error variance and block size in the form $\sigma^2 \propto n^{1/4}$. According to this formula, the average gains in efficiency for the estimation of unconfounded treatment effects are: $\frac{1}{2}$ block size, 19%; $\frac{1}{3}$ block size, 32%; $\frac{1}{4}$ block size, 41%.

From Rothamsted Reports between 1925 and 1936 I found that block sizes in 304 experiments conducted under Fisher himself and in the initial years of Yates's headship had hardly varied over these years from a mean of 8.3 plots per block, despite the increased use of confounding. Only 19 experiments (6%) had block sizes greater than 12, and only 7 (2%) had block sizes greater than 16, bearing out my statement that 16 is a reasonable maximum (Rayner, 1969, Sec. 22.1). However, I could see no evidence that large blocks had been tried and discarded.

Boyd denied my suggestion that 1962 might have been a favorable year for low experimental errors, stating that the conclusion was not based on a single year. He also did not agree that it was unlikely that experimenters would overlook the need for as uniform a site as possible (e.g., to avoid block × treatment interaction) (Rayner, 1969, Sec. 13.5.4): "Most [British] experimenters are convinced they need only achieve uniformity within a block." As regards the work of Patterson and Ross (1963), he wrote: "I remember after Patterson's paper FY [Yates] remarked to me that he wondered why Patterson was making such a song and dance about small blocks when the gain in efficiency was so small anyway."

It seems to me that the position is similar to blocks versus no blocks. You may much of the time gain little by small blocks, but they are a safeguard against a very high error which might have arisen without them. In other words, one cannot work on averages. Also, if blocks of 40, why any blocks at all?

The last word belongs to G. J. S. Ross (1983, personal communication), who expressed the view that there are two situations:

1. When experiments involve many factors at many levels and when there is access to an almost unlimited choice of land, so that uniform sites may be chosen (Boyd's case), then large blocks may be supported.

2. Workers who have to use locally available land and who have different objectives find smaller blocks advantageous and necessary.

REPLICATION → ESTIMATE OF ERROR (PATH 3)

It is not at all clear why path 3 should exist separately from path 4. However, at one stage Fisher seems to have considered replication to be the primary principle (or sine qua non) of experimental design—cf. the quotation from Fisher (1931) on page 253 and the heading "Replicated experiments" (as opposed to the old type) in the Rothamsted Reports from 1925–26 to 1932. Fisher (1925, p. 224) wrote: "The first requirement which governs all well-planned experiments is that the experiment should yield not only a comparison of different . . . treatments . . . but also a means of testing the significance of such differences as are observed. Consequently all treatments must at least be duplicated, and preferably further replicated. . . . " After only one more sentence, however, he continues: "For our test of significance to be valid . . . ," and we are onto path 4. Indeed, this pairing of replication and randomization in order to achieve a valid estimation of error is a constant theme of Fisher's writings. The main exception is Fisher (1931), where randomization is barely mentioned. In a section entitled "Validity of the estimation of error," Fisher (1935, p. 69) states that the main purpose of replication, "which there is no alternative method of achieving, is to supply an estimate of error . . . ," but randomization is quickly brought into the picture to complete the section.

I have been unable to find any reference to the discovery, or rather realization, that in factorial experiments high-order interactions could be used to estimate error, and that therefore absolute replication could be dispensed with. According to Yates (1975), the motivation came from "the need of experimenters for a design with a modest number of plots which would give information on the responses to the three standard plant nutrients, N, P, K and their interactions," which led to the confounded single-replicate 3^3 design with error estimated from the unconfounded three-factor interaction components and the nonlinear components of the two-factor interactions. Certainly, an experiment with this design was conducted at Rothamsted in 1933 (in which year the heading "Replicated experiments" was dropped from the Rothamsted Report!). Yates's account is fully verified by the first formal presentations (Fisher, 1935, p. 111; Yates, 1935); the contrast with the quotation from Fisher (1935) just above will be noted.

However, this history is rather surprising in view of the computing tools available at that time. One would have thought that pooling high-order interactions with error in replicated factorial experiments—the first explicit mention of which in the literature may be that by Yates (1937, p.

65)—would have been a way of life as a time-saver, but computational convenience was secondary in Yates's motivation.

As Yates (1975) says, the extension to fractional replication was logical, except perhaps to those who were still struggling to understand confounding! At Rothamsted in 1946 I had just got used to single-replicate designs when M. H. Quenouille, who was acting as my mentor, asked if I would like to try the analysis of a fractional-replicate design. My reply was the 1946 equivalent of "You've got to be kidding!" In the first paper on fractional replication Finney (1945) thanks Yates for suggesting the possibility of such designs. In 1982, David Finney visited South Africa and I was privileged to hear his account of how this arose; subsequently (Finney, 1983, personal communication), he gave me the full details. In the United Kingdom during World War II much land under grass became required for arable cropping, but wireworm buildup under grass could be dangerous to the crops in the first one or two years after ploughing the old grass. However, as Finney (1941) discovered in a survey of sugar beet experiments, loss of seedlings due to wireworms did not necessarily mean a proportional drop in yield, because of decreased competition. Yates suggested that it would be helpful to study experimentally the effects of random removal of seedlings to simulate wireworm attacks, and in order to make use of existing experiments suggested that it ought to be possible to incorporate an additional four-level factor (proportion of plants removed) into existing 2^n fertilizer experiments in a balanced way; in order to do minimal harm to the existing experiments, only a small area of each existing plot should be utilized to test the new factor. Thus Finney developed the theory for the partial conversion in 1940 of a single-replicate 2^5 factorial into a quarter-replicate of a 4×2^5 design, and in 1941 of a single-replicate 2^6 factorial into a quarter-replicate of a 4×2^6 design, both in blocks of eight plots. These experiments are reported on pages P/8 to P/14 of Vol. II of *Results of the Field Experiments 1939–1947,* Rothamsted Experimental Station, and are also mentioned in a footnote on p. 41 of *Wireworms and Food Production,* Bulletin No. 128, Ministry of Agriculture and Fisheries (1944).

REPLICATION → RANDOM DISTRIBUTION → VALIDITY OF ESTIMATE OF ERROR (PATH 4)

The reader might be surprised to notice the use of "random distribution" instead of the more familiar "randomization," which dates from Fisher

(1926), where, however, it seems to occur incidentally rather than as the deliberate naming of a new principle. The phrase favored then, and also in Fisher (1925), was "random arrangement." Soon, however, "randomization" was elevated to "the primary requirement" in the distribution of plots to treatments (Fisher and Wishart, 1930), and finally came of age in *Design of Experiments* (Fisher, 1935), appearing four times in the list of contents and being discussed in five substantial sections.

Fisher's (1926) justification of path 4 is particularly interesting:

> The estimate of error is valid, because, if we imagine a large number of different results obtained by different random arrangements, the ratio of the real to the estimated error [in a completely random design], calculated afresh for each of these arrangements, will be actually distributed in the theoretical distribution by which the significance of the results is tested.

Box (1978, p. 147) concludes that "Fisher here claims that the error ratio will, over all randomizations, have the distribution appropriate when observations have been drawn independently from a normally distributed population." This is typical of the sort of extrapolation needed to interpret Fisher's writings, but I have difficulty in equating "a large number" to "all." In other words, I do not think that this is randomization theory yet.

At this stage it is necessary to take note of the competition, in the form of so-called systematic designs—really patterned designs based on the idea of balancing—which had the aim of ensuring that all treatments shared as equally as possible in the fertility of the experimental area. From 1925 to 1935 Fisher condemned systematic designs:

> With such an arrangement . . . we have no guarantee that an estimate of the standard error derived from the discrepancies between parallel plots is really representative of the differences produced between the different treatments, consequently no such estimate . . . can be trusted, and no test of significance is possible. (Fisher, 1925, p. 228)

> Consequently the skill and judgement devoted to obtaining plot arrangements which involve errors between the comparisons less than those of random arrangements may indeed make the experiment really better, but at the expense of making it seem worse. (Fisher and Wishart, 1930)

On the contrary, Fisher (1926) argued that "an estimate of error (derived from differences of plots treated alike) will only be valid . . . if we make sure that, in the plot arrangement, pairs of plots treated alike are not nearer together or further apart than, or in any other relevant way, distinguishable from pairs of plots treated differently." Systematic designs introduced "a flagrant violation" of these conditions, whereas one way to ensure compliance was "to arrange the plots deliberately at random" (i.e., the completely random design). Moreover, Fisher maintained that the use of local control, especially the Latin square design, in conjunction with randomization "will give actual errors as small as even the most ingenious of systematic arrangements" (1926).

The justification of randomization in terms of what is now known as randomization theory came in Fisher (1935, pp. 50-54): "It seems to have escaped recognition that the physical act of randomization ... affords the means, in respect of any particular body of data, of examining the wider hypothesis [i.e., that two groups of measurements are drawn from the same population] in which no normality of distribution is implied."

Statistical opinion seems to be that randomization is one of Fisher's original and most important contributions to statistical science: "The next and equally fundamental step was the introduction of randomization. This idea, I believe, was entirely due to Fisher ... " (Yates, 1962). However, on the point of originality, Stigler (1980), as reported by Bancroft (1982), has again thrown a spanner into the works by noting that Peirce and Jastrow (1885) conducted an experiment not too dissimilar from the lady tasting tea (Fisher, 1935), but involving the detection of the heavier of two nearly equal weights from differential pressure on a subject's finger. The order of presentation of the weights was randomized by shuffling playing cards ! (This briefly summarizes a complicated arrangement.) However, it seems improbable at any significance level that Fisher was aware of this development.

Acceptance of randomization by experimenters on the one hand and by statisticians on the other was by no means immediate. As Hotelling (1951) wrote about *Statistical Methods for Research Workers*: "Biologists have often had great difficulty in trying to understand the book, but not so much difficulty as mathematicians. . . . " The latter, of course, felt the need for proofs. The experimenters were undoubtedly won over by the growing pre-eminence of Rothamsted in the area of experimental design, but controversy resurfaced in the exchange between "Student" and Fisher in 1936–1939 (see Chapter 18). In New Zealand in 1944 I found that randomization had been routinely adopted by the Crop Experimentalist's Section of the Department of Agriculture, except for the half-drill strip, which I worked to eliminate. In the South African Department of Agriculture in 1949 I found that , probably due to the pioneer work of A. R. Saunders, the Rothamsted methods had been universally adopted. The same was not true, however, for the rest of Africa, and I can remember receiving for analysis data from the most elaborately constructed systematic designs which would have been excellent as patterns for tiling or brick driveways. I understand that there are some countries where systematic designs are still used.

Fisher (1925) did not deal with the mechanics of randomization, but the 1925–26 Rothamsted Report did, in an article possibly written by Fisher, although he is referred to in the third person. For the randomized blocks design the recommended method was "writing the possible arrangements on cards, shuffling them, and drawing one out." For the Latin square a similar procedure is required, but there is a wry comment that "a surprisingly large number of arrangements are possible." Fisher himself (1926) confirmed that one of *all possible* arrangements is sought. After the appearance of random number tables (Tippett, 1927), Fisher and Wishart (1930) described as an alternative to drawing numbered cards the now-familiar process of drawing pairs of random digits and obtaining remainders from division by the number of treatments.

Randomization used to worry me considerably. In discussing the drawing of cards or marbles by shuffling or shaking the receptable thoroughly, Tippett (1927) added in brackets: "if indeed that be possible." I felt the same way. The Crop Experimentalist's section in New Zealand in 1948 had some numbered marbles, but even when a special apparatus was constructed so that the draws could be made untouched by human hands, the sort of thing that, after randomizing three blocks of a design, a certain treatment would be in the first or second place, or a particular pair of treatments alongside, in all three blocks, would occur with sickening frequency. A visit was even paid to an office in Featherston Street, Wellington, to see the "Art Union" (lottery) drawn, but this was the ultimate disillusionment. As regards random numbers, after reading how these were laundered to avoid ontoward occurrences, I never knew whether to be concerned about this or to welcome the protection. In any case the results were often similar to those with the marbles.

Yates (1975) says that Fisher "ducked [the question of] what to do when a 'systematic' arrangement is obtained." In one sense it is clear that he did not, since the early evidence is that he would have been unmoved by such an occurrence. Thus he (1926) described how an experimenter might be shocked after a randomization to find how unequally the treatments were distributed, and warned against rejecting or "cooking" a seemingly bad arrangement. I put this question to Professor Aitken in 1946 or 1947, and after a pause he replied: "Put the numbers back in the hat." It is interesting to read that Savage et al. (1962), as reported by Holschub (1980), in a 1952 conversation put this very question to Fisher himself, who replied that he thought he would draw again, but also that "ideally, a theory explicitly excluding regular squares should be developed." According to Yates (1970), in the 1930s the idea of deliberately rejecting a randomization would have been regarded as "heretical," but prior to Savage's question Yates had suggested the concept of restricted randomization in connection with certain confounded factorial designs in quasi-Latin squares, in which an unfortunate arrangement could otherwise arise too frequently for comfort (Grundy and Healy, 1950). In any case there had already been some erosion of the all-possible-arrangements prescription by Yates (1933b), who for Latin squares of order 7 to 12 presented only one square each, suitable randomization of which "will give sets of squares amply large enough to serve all agricultural purposes" (words which have never been explained). Also, Cochran and Cox (1950) seemed to shake the establishment slightly by producing tables of random permutations, seemingly inviting randomizers to draw for ever from these limited sets, and Fisher and Yates (1963) later followed suit with an even smaller selection.

Yates was not perhaps above a little ducking on his own account. At Rothamsted in 1947 I experimented by randomizing a design using the last two figures on the dates of British pennies, mixing up the coins in my trousers pocket before each draw. As it happened, this operation was interrupted by tea, and already in the first few replications some disconcerting arrangement had turned up. I mentioned this to Yates, explaining my method in case he might be able to fault it, and expressing the fear that the trend might continue in the remaining replications. Yates pooh-poohed this possibility and asked me if I would make a small bet on it. Of course, I had to put my money where my mouth was, even though the odds were against me, and I lost the bet.

I plead guilty, under provocation, both to cooking and re-randomizing! A particular occasion I recall was when my architect, having designed a ceiling for my lounge divided into 35 squares in a 7×5 rectangular layout, invited me to suggest how, in the then contemporary style, these might be colored. Preferring seven colors to five, I selected a Youden square after the fashion of the dust-jacket of *Design of Experiments*. Alas, certain predominant colors seemed to group themselves in my randomization in a manner less than aesthetically pleasing, and *for this reason alone* a re-randomization was performed. Today, with the change of fashion, the ceiling reflects a uniformity trial.

A decade or so earlier than the quasi-Latin square problem Yates was called upon to do some more serious firefighting when the pavement of path 4 seemed to be cracking badly. From the quotations *ex* Fisher given earlier in this section, it was not surprising that there was a general belief that, as the diagram shows, any design incorporating replication and randomization would give a valid estimate of error, presumably in the sense that on the null hypothesis of no treatment effects, the expectations of the treatments and error mean squares in the appropriate analysis of variance would be equal, the expectations being based on the normal theory assumptions necessary for the F test. Examination of the quotations in their context, however, shows that they refer to the three basic designs, the only possible exception being that from Fisher (1925, p. 51) containing the words "in respect of any particular body of data" (see p. 259), which are capable of a wider interpretation than that probably intended.

Looking back on this, Yates (1975) says:

I have been asked at this Symposium how Fisher hit on randomization. I have little doubt that originally he considered it as a device for

(a) ensuring that no treatment was unduly favoured,
(b) giving substance to the basic assumption of Gaussian least squares that the errors are uncorrelated

That the estimate of error is unbiased over all random patterns of the simpler designs is obvious, and was regarded by Fisher as a criterion for a good design—I would say rather a necessary but not sufficient condition.

This is an intensely interesting remark which gives rise to a number of questions or comments:

1. Why did not someone (e.g., Yates himself) put this question to Fisher while he was alive? [Yates (1983, personal communication) confirms that he cannot recall any explicit discussion with Fisher on this point.]
2. I believe that Yates has probably given the correct answer (see Fisher, 1933), but there is no mention of unbiased treatment comparisons in the diagram and the only words I have come across in Fisher's writings relevant to this aspect are when he instructs an experimenter to carry out "a physical experimental process of randomization, using means which shall ensure that each variety has an equal chance of being tested on any particular plot of ground" (Fisher, 1935, p. 56). Even this passage does not specifically mention unbiased treatment comparisons, which are at least as important as an unbiased estimate of error, and in the early days there were doubts

about the fairness of randomization as opposed to systematic designs.

3. If randomization was to ensure uncorrelated errors in a least squares analysis, why must unbiasedness of the estimate of experimental error be considered over all possible randomizations instead of in terms of the least squares analysis? In such an analysis the deviations mean square (= error mean square of the analysis of variance) always provides an unbiased estimator of the random variance (Kempthorne, 1955).

Yates (1933b) had indeed given a proof of the unbiasedness (based on randomization expectation) of the estimate of error in a Latin square design, and this stood him in good stead in pronouncing the semi-Latin square to be biased (Yates, 1935; 1936a). From the published discussion on the 1935 paper it is apparent that this was news to Wishart, and possibly Fisher. In the 1936 paper Yates established Latin squares with one row, column, or treatment omitted (or missing) as "valid experimental arrangements" in the foregoing sense. Commenting on this Kempthorne (1975) said:

It is most curious that one should consider distribution over the induced randomization population for validity and then, having determined that a design is valid in this sense, one could revert to normal law theory. I am grateful that Frank [Yates] agreed in conversation that there is a bit of a mystery to be explained, and he assured me that he would give it his attention.

Yates and Mather (1963) appear to state, or to give Fisher's view, that randomization is a necessary and sufficient condition for the estimate of error to be unbiased: "The principles of randomization were first expounded in *Statistical Methods for Research Workers* (1925). Points emphasized were ... that only randomization can provide valid tests of significance, because then, and only then, is the expectation of the treatments mean square equal to the expectation of the error mean square when both are averaged over all possible random patterns. ..." However, the seeming contradiction to what Kempthorne (1983) has termed "Yatesian unbiasedness" disappears on realization that the words "only randomization" must refer to the completely random design in the context of comparisons with alternatives such as systematic designs.

Quite apart from the need to fill in the gaps arising from Fisher's "intuitive leap" (Box, 1978, p. 147) (i.e., randomization), there has continued to be a series of papers reexamining this principle. In one of the most recent of these, Kempthorne (1977), who gives a fairly complete list of these papers, seems to express a feeling of unease on his own account. Perhaps one might paraphrase Winston Churchill and say that randomized designs are the worst type of experimental design except for all those other types which have been tried from time to time!

ACKNOWLEDGMENTS

I am grateful to Dr. J. A. Nelder, Dr. F. Yates, and Dr. G. J. S. Ross of Rothamsted Experimental Station for their assistance in clearing up certain points of information, and to Drs. Yates and Ross for permission to quote, or give the gist of, opinions expressed in letters to me. I also thank Ms. Julie Drop for typing the manuscript.

REFERENCES

Bancroft, T. A. (1982). Review of *American Contributions to Mathematical Statistics in the Nineteenth Century*, Vols. 1 and 2, (S. M. Stigler, ed.). Arno Press, New York. *J. Am. Stat. Assoc. 77*, 212.

Bennett, J. H. (1971-1974). *Collected papers of R. A. Fisher*, Vols. 1-5. University of Adelaide, Adelaide, Australia.

Box, J. F. (1978). *R. A. Fisher, the Life of a Scientist*, Wiley, New York.

Boyd, D. A. (1973). Developments in field experimentation with fertilizers. *Phosphorus Agric. 61*, 7-17.

Cochran, W. G. (1976). Early developments of techniques in comparative experimentation. In *On the History of Statistics and Probability* (D. B. Owen, ed.). Marcel Dekker, New York, pp. 3-25.

Cochran, W. G., and G. M. Cox (1950). *Experimental Designs*. Wiley, New York.

Federer, W. T., and L. N. Balaam (1972). *Bibliography of Experiments and Treatment Design Pre-1968*. Oliver & Boyd, Edinburgh.

Finney, D. J. (1941). The relationship of plant number and yield in sugar-beet and mangolds. *Emp. J. Exp. Agric. 9*, 57-64.

Finney, D. J. (1945). The fractional replication of factorial arrangements. *Ann. Eugen. 12*, 291-301.

Fisher, R. A. (1925). *Statistical Methods for Research Workers*. Oliver & Boyd, Edinburgh.

Fisher, R. A. (1926). The arrangement of field experiments. *J. Minist. Agric. (GB) 33*, 503-513.

Fisher, R. A. (1931) Principles of plot experimentation in relation to the statistical interpretation of the results. Rothamsted Conferences XIII. The technique of field experiments, 11–13.

Fisher, R. A. (1933). *The contributions of Rothamsted to the development of the science of Statistics.* Report for 1933, Rothamsted Experimental Station, Harpenden, England, 43–50.

Fisher, R. A. (1935). *The Design of Experiments.* Oliver & Boyd, Edinburgh.

Fisher, R. A., and J. Wishart (1930). *The arrangement of field experiments and the statistical reduction of the results.* Technical Communication No. 10, Imperial Bureau of Soil Science, Rothamsted Experimental Station, Harpenden, England.

Fisher, R. A., and F. Yates (1963). *Statistical Tables for Biological, Agricultural and Medical research*, 6th ed. Oliver & Boyd, Edinburgh.

Grundy, P. M., and M. J. R. Healy (1950). Restricted randomization and quasi-Latin squares. *J. R. Stat. Soc. B 12*, 286–291.

Holschuh, N. (1980). Randomization and design: I. In *R. A. Fisher: An Appreciation*, Lecture Notes in Statistics, Vol. 1. (S. E. Fienberg and D. V. Hinkley, eds.). Springer-Verlag, New York.

Hotelling, H. (1951). The impact of R. A. Fisher on statistics. *Biometrics 46*, 35–46.

Kempthorne, O. (1955). The randomization theory of statistical inference. *J. Am. Stat. Assoc. 50*, 946–967.

Kempthorne, O. (1975). Inference from experiments and randomization. In *A Survey of Statistical Design and Linear Models* (J. N. Srivastava, ed.). North-Holland, Amsterdam, pp. 303–331.

Kempthorne, O. (1977). Why randomize? *J. Stat. Plann. Inference 1*, 1–25.

Kempthorne, O. (1983). Contribution to the discussion of the paper by Wilkinson et al. *J. R. Stat. Soc. B 45*, 199–200.

Patterson, H. D., and G. J. S. Ross (1963). The effect of block size on the errors of modern cereal experiments. *J. Agric. Sci.* (Cambridge) *60*, 275–278.

Peirce, S. C., and J. Jastrow (1885). On small differences of sensation. *Memoirs of the National Academy of Sciences for 1884*, 75–83.

Rayner, A. A. (1969). *A First Course in Biometry for Agricultural Students.* University of Natal Press, Pietermaritzburg, South Africa.

Rothamsted Experimental Station. Report 1923–24. Harpenden, England.

Rothamsted Experimental Station. Report 1925–26. Harpenden, England.

Rothamsted Experimental Station. Report 1927–28. Harpenden, England.

Russell, E. J. (1926). Field experiments: how they are made and what they are. *J. Minist. Agric. (G B) 32*, 989–1001.

Savage, L. J., (1970). *The Foundations of Statistical Inference* (2nd impr). Methuen, London.

Savage, L. J. (1976). On rereading R. A. Fisher. *Ann. Stat. 4*, 441–500.

Stigler, S. M. (1980). Mathematical statistics in the early United States. In *American Contributions to Mathematical Statistics in the Nineteenth Century*, Vol. 1 (S. M. Stigler, ed.). Arno Press, New York.

Stigler, S. M. (1981). Gauss and the invention of the least squares. *Ann. Stat. 9*, 465–474.

Tippett, L. H. C. (1927). *Random Sampling Numbers*. Tracts for computers No. 15, Cambridge University Press, Cambridge.

Yates, F. (1933a). The principles of orthogonality and confounding in replicated experiments. *J. Agric. Sci.* (Cambridge) *23*, 108–145.

Yates, F. (1933b). The formation of Latin squares for use in field experiments. *Emp. J. Exp. Agric. 1*, 235–244.

Yates, F. (1935). Complex experiments (with discussion). *Suppl. J. R. Stat. Soc. 2*, 181–247.

Yates, F. (1936a). Incomplete Latin squares. *J. Agric. Sci.* (Cambridge) *26*, 301–315.

Yates, F. (1936b). Incomplete randomized blocks. *Ann. Eugen. 7*, 121–140.

Yates, F. (1936c). A new method of arranging variety trials involving a large number of varieties. *J. Agric. Sci.* (Cambridge) *26*, 424–455.

Yates, F. (1937). *The design and analysis of factorial experiments*. Technical Communication No. 35, Imperial Bureau of Soil Science, Harpenden, England.

Yates, F. (1962). Appreciation of R. A. Fisher. *Biometrics 18*, 442–447.

Yates, F. (1970). Author's note on The formation of Latin squares for field experiments (1933). *Experimental Design: Selected Papers of Frank Yates, C.B.E., F.R.S.*, Griffin, London.

Yates, F. (1975). The early history of experimental design. In *A Survey of Statistical Design and Linear Models* (J. N. Srivastava, ed.). North-Holland, Amsterdam, pp.581–592.

Yates, F., and K. Mather (1963). Ronald Aylmer Fisher 1890–1962. *Biogr. Mem. Fellows R. Soc. London 9*, 91–120. Reprinted in *Collected Papers of R. A. Fisher*, Vol. 1 (J. H. Bennett, ed.). University of Adelaide, 1971, pp. 23–52. Adelaide, Australia.

VI
Prediction

At times, many of us wish that we could predict the future and see what will happen in a day or two or perhaps a year or two. Some people are so intrigued with this idea that they spend considerable time and energy studying their astrological stars, tarot cards, or crystal balls. The media bombard us with predictions on such matters as economic trends, the weather, and patterns of crime. When we stop to think of it, we are continually using our past experiences as a basis for consciously planning for the future, or at least unconsciously making assumptions about what will happen.

In the five chapters that follow, statistical methods have been applied to widely varying situations involving some predictions of future events. Each chapter includes collecting data and deciding on certain assumptions that lead to the selection of a statistical model. These ingredients are also contained in part VII and the line between these two parts is very indistinct. Most of the chapters in Part VI, however, include the additional idea of predicting a future event.

In Chapter 20, Hugh Morton looks at the intriguing question of how fast athletes can run. There certainly must be some limits to the time taken by runners to cover a given distance; yet one by one such supposed barriers as the four-minute mile have been shattered. In this analysis, only one variable is considered, namely, the time itself.

David Rhoades considers earthquake prediction. As both the author and the editors live in an earthquake-prone area of the world, we find this a topic of considerable relevance. It is known that some animals seem to be able to sense the approach of earthquakes; but even though scientists make use of very sophisticated equipment, the forecasting of disastrous events is very difficult and mathematical solutions are very elusive.

Alan Lee and Alastair Scott consider data obtained from amazing ultrasound techniques that produce measurements on unborn babies. Before these babies are born, these methods can identify those who may be at risk from certain medical conditions.

In Chapters 20 and 23 only a few variables are considered, but the remaining chapters, by comparison, involve a large number of variables. Keith Petrie and Kerry Chamberlain consider statistical methods of deciding whether people who have attempted suicide will try it again in the near future. If this is likely, they should be given intensive care and supervision. To involve staff and resources on everyone who has attempted suicide would be inordinately expensive, so it is essential to isolate those at risk of making further attempts.

Improving the quality of livestock is an important ongoing problem. It requires the careful collection of masses of data and then the even more difficult task of processing this information. Robert Anderson explains how statistical procedures can be employed to predict the performance of dairy cattle to build up genetically superior herds.

In each situation, the question must arise as to how good the prediction is likely to be. Steady improvement in the areas of athletics and herd improvement can be noted over the years, providing a check on the efficacy of the model used. Other areas–suicides and earthquakes, in particular—lead to more dramatic and cataclysmic events. If the researchers in these areas only succeed in reducing the risks by a small amount, however, the prediction could be seen as a very worthwhile activity.

20

You Can't Catch Me:
I'm the Gingerbread Man

R. Hugh Morton
Massey University
Palmerston North, New Zealand

We know that the gingerbread man ran faster than a little boy, and an old man and an old woman, and three farmers, and a bear, and a wolf. For all we know, he might have been able to run faster than the fox, too, had not the wily creature outsmarted him. Notwithstanding the children's story, one wonders just how fast the gingerbread man, or any real man or woman, could run, or be expected to run.

An ultimate limit of human endeavor in any of the widest range of activities, if it even exists, is surely a topic guaranteed to raise animated discussion in any gathering. Athletics, and in particular running races, is certainly one of the more frequently discussed such activities, and the 1500-m "metric" mile is one of the glamor events. With such thoughts in mind, let us critically examine the trend in the world records for this event to see what indications there may be.

The data go back to the late nineteenth century, to the first of the modern-day Olympics. Not all the records were ratified by the International Amateur Athletic Federation, but they do seem to have been recognized as the world's best performance at the time. Some of the earlier records may not be very reliable, and the timekeeping equipment may not have been very accurate. Furthermore, there may be some debate as to whether the very first record in such a series is indeed representative of the true world's best performance. However, there is little that can be

done about it, for such matters are historical and, in any case, the more recent records are more relevant in helping us reach some answer to our question. For the 1500 m, since Australian Edwin Flack's 1896 record of 4-33.2, the world's best time has been equaled or bettered 40 times up to now (October 1983), culminating in Britain's track star Steve Ovett's current record of 3-30.78 set in September 1983. This progressive list appears as Table 1. A graphical presentation will give a good idea of the progression, and this is presented as Figure 1.

The pronounced downward trend over the years is very evident, of course, but there are two other indications also. They are first, that the initial observation indeed appears questionable, for it seems out of alignment with all the others even allowing for the evident fluctuations. Second, there is a slight tendency toward curvature becoming apparent, even if the first observation is disregarded, for it contributes much toward this tendency. This suggests that a horizontal asymptote, that is, an ultimate limit, might be indicated by the data. However, we shall not jump to conclusions, but subject the data to some statistical analyses.

It is frequently both helpful and useful to "model" a relationship, such as is depicted in Figure 1, by a mathematical expression, that is, to hypothesize that some curve fits the data. The simplest such "curve" is a straight line, and if you imagine a straight line, sloping downward to the

TABLE 1   World Best Times for the 1500 m

Runner	Time	Date
Flack (Australia)	4-33.2	30 March 1896
Lermusiaux (France)	4-10.4	28 June 1896
Bennett (U.K.)	4-06.2	15 July 1900
Lightbody (U.S.)	4-05.4	3 September 1904
Wilson (U.K.)	3-59.8	30 May 1908
Kiviat (U.S.)	3-59.2	16 May 1912
Kiviat (U.S)	3-55.8	8 June 1912
Zander (Sweden)	3-54.7	5 August 1917
Nurmi (Finland)	3-53.0	23 August 1923
Nurmi (Finland)	3-52.6	19 June 1924
Peltzer (Germany)	3-51.0	11 September 1926
Ladoumege (France)	3-49.2	5 October 1930
Beccali (Italy)	3-49.2	9 September 1933
Beccali (Italy)	3-49.0	17 September 1933
Bonthron (U.S.)	3-48.8	30 June 1934
Lovelock (New Zealand)	3-47.8	6 August 1936
Hägg (Sweden)	3-47.6	10 August 1941
Hägg (Sweden)	3-45.8	17 July 1942
Anderssen (Sweden)	3-45.0	1 July 1943
Hägg (Sweden)	3-43.0	7 July 1944
Strand (Sweden)	3-43.0	15 July 1947
Lueg (Germany)	3-43.0	29 June 1952
Bannister (U.K.)	3-43.0	6 May 1954
Santee (U.S.)	3-42.8	4 June 1954
Landy (Australia)	3-41.8	21 June 1954
Iharos (Hungary)	3-40.8	28 July 1955
Tabori (Hungary)	3-40.8	6 September 1955
Nielson (Sweden)	3-40.8	6 September 1955
Rozsavolgyi (Hungary)	3-40.5	3 August 1956
Salonen (Finland)	3-40.2	11 July 1957
Salsola (Finland)	3-40.2	11 July 1957
Jungwirth (Czechoslovakia)	3-38.1	12 July 1957
Elliott (Australia)	3-36.0	28 August 1958
Elliott (Australia)	3-35.6	6 September 1960
Ryun (U.S.)	3-33.1	8 July 1967
Bayi (Tanzania)	3-32.2	2 February 1974
Coe (U.K.)	3-32.1	6 July 1979
Ovett (U.K.)	3-32.1	17 August 1979
Ovett (U.K.)	3-31.36	27 August 1980
Maree (U.S.)	3-31.23	26 August 1983
Ovett (U.K.)	3-30.78	4 September 1983

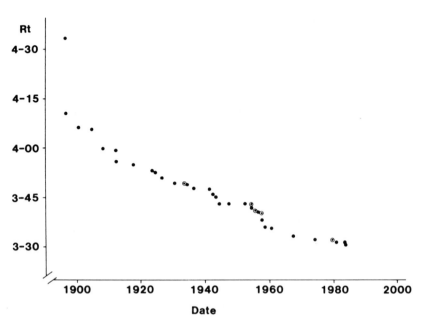

FIGURE 1   World best times for the 1500 m.

right, almost all of the points do cluster quite closely around it. Statistical techniques known as linear regression are a well-recognized tool of analysis for such fitting and are widely available for computers and calculators. I have utilized such a technique to obtain the following straight line as a fit to the data:

$$R_t = 4.123 - 0.0073t \qquad (1)$$

where R_t represents the record at date t. (The seconds are expressed as decimals of a minute, with the days and months expressed as decimals of a year, the first record being adjusted to having taken place at $t = 0$.)

The correlation coefficient r measures the "goodness" of the fit, and for this fit it is given by -0.783, the negative sign indicating a downward slope. A value of -1 would be perfect but unobtainable in practice, so the fit is quite reasonable . Alternatively, r^2 gives the proportion of variability in R_t explained by the linear fit, and this is 0.612, or 61.2%, which although acceptable, still leaves room for improvement.

The regression analysis should include checking the residuals, which are the differences by which each data point lies above or below the fitted straight line. These are commonly plotted and examined for the presence

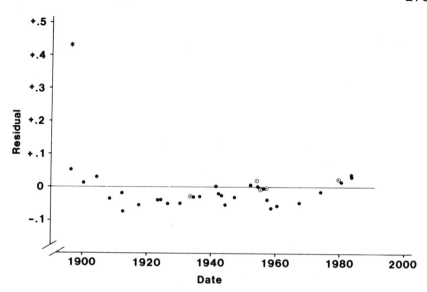

FIGURE 2   Residuals from straight-line fit.

or absence of certain types of behavior. Figure 2 is a plot of the residuals resulting from fitting equation (1), the plot being in (date) order of occurrence. This plot shows two things. The most obvious is the anomalous behavior of the first record. Its residual lies almost off the top of the graph and is nearly six times as large in magnitude as the second largest residual. This confirms our earlier suspicions. The second, more subtle characteristic of the plot is the fact that residuals at the start and at the end of the series are positive, while those in the center are for the most part negative. This is typically the case when a straight line has been fitted to data that conforms better to a curved line, as we suspected earlier also. Since the first observation contributes to this curvature, we shall not discard it yet, but rather investigate a plausible alternative to the straight line.

Intuition, of course, strongly suggests that some nonzero ultimate limit to the series must exist, and that the line must curve gently toward the horizontal, flattening out very near this limit. The theory of relativity at the very least, could be invoked to justify its existence. There probably is not one among us who would accept that a human could ever generate enough power to run faster even the speed of sound, let alone approach the speed of light. Notwithstanding, from the statistical point of view we need only assume that some limit exists, nonzero or otherwise. Techniques exist for fitting curved lines with asymptotes to data, and then we could test the fitted asymptote to see if it could reasonably be regarded as zero or not.

What we require as a model, therefore, is a smooth ever-decreasing curve, with a horizontal asymptote to the right. The most appealing and obvious such model is the exponential decay. It has been found to approximate closely a wide variety of naturally occurring phenomena, such as the decay of radioactive material. It has the additional feature of exhibiting a constant percentage decline over time toward the ultimate limit. There are, of course, other possible models, but few have the sorts of characteristics that make the exponential my first choice. Mathematically, it is expressed as follows:

$$R_t = L + ae^{-bt} \qquad (2)$$

where L, a, and b are the parameters that characterize the decay curve and its limit.

More specialist techniques are required to fit a nonlinear equation of this form to data. A single computer program, denoted P3R, from the Biomedical Computer Programs (BMDP-82) package provides the necessary procedure. Other than converting the times and dates to decimals, the shift of the date scale zero to March 30, 1896 was intended to make the parameter a in equation (2) of manageable size (there is no effect on L or b). The result of this fit to the full data set yielded the equation

$$R_t = 2.986 + 1.157e^{-0.010t} \qquad (3)$$

This represents a projected ultimate limit of 2-59.2, with a standard error of ±2.14 seconds and a percentage rate of decay of about 1% per annum toward this limit. The coefficient of nonlinear correlation is now -0.849, revealing a good fit, certainly an improvement on the value of -0.783 obtained for the straight line.

However, before accepting the exponential model above, and the projected limit, we should examine the residuals, in particular the first one. As might have been expected, it is not as bad as previously, although it is still nearly four times larger than the next biggest. Such magnitude remains very suspect and consideration should be given to its omission. The procedure is to omit the first point and examine the new set of residuals from the thereby obtained new fit, in comparison to the previous residuals. In so doing, the sum of squared residuals was reduced by an amount which can be shown to be statistically significant. Furthermore, there are no other aberrant observations, and a plot of the new set of residuals shows no discernible patterns. It is justified therefore

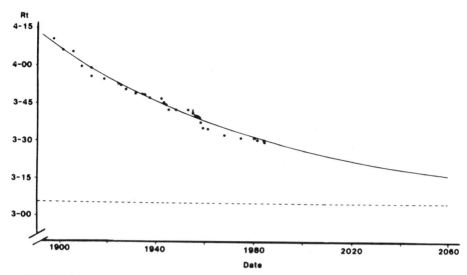

FIGURE 3   Amended world times and fitted line.

to omit this first value from the set of observations. It would appear that Edwin Flack was lucky enough to have turned on a good performance in the right place at the right time!

Proceeding in the same way, without the first data point, yields the following estimated equation:

$$R_t = 3.095 + 1.053 \, e^{-0.0102t} \qquad (4)$$

This represents a projected ultimate limit of 3-5.7, with a standard error of ±1.0 second, with again about a 1% annual decay toward the limit. This limit is quite clearly nonzero! The correlation coefficient has increased to −0.963, revealing the significant improvement expected. You will note how the standard error has reduced also, revealing increased precision of estimation. Figure 3 presents these amended data together with equation (4) and displays the ultimate limit clearly.

The position in the late-1983 is that we are still well short of the ultimate record, by 25.1 seconds, or 12%. The current record lies just below the curve, so it would be a little unexpected for the record to be broken again in the near future. In an earlier analysis, a year ago, the record of that time lay above the curve, suggesting some expectation of an impending improvement. This did indeed occur. Equation (4) can be used to make various deductions; for example, we might expect a 3½-minute 1500 m late in 1989, while in another hundred years the projected record is 3-15.0, closer but still appreciably short of the ultimate. Our

TABLE 2

Distance (m)	Limit	Current shortfall (%)
100	9.15	7.6
200	18.15	8.3
400	39.33	10.2
800	1-33.0	8.6
3000	6-16.9	16.6
5000	11-19.3	12.9

readers may care to make some deductions of their own from Figure 3 and equation (4). It is probably also of interest to our readers that I have repeated the foregoing type of analysis on several other races, and Table 2 represents a selection of results.

In conclusion, within the limitations of the assumptions I have made, it seems that trends which had previously been believed to be not yet evident are now becoming so. The selection of one feasible curvilinear model has revealed good fits in several cases and an estimable nonzero limit is available. It must be conceded, of course, that should there occur a sudden rapid series of improvements in times, the appearance of this tendency could be spoiled.

REFERENCES

This chapter is based on a paper published in the *Australian Journal of Sport Sciences*, referenced below. In addition I have included some other references that interested readers may wish to consult.

Craig, A. B. (1963). Evaluation and predictions of world running and swimming records. *J. Sports Med. Phys. Fitness 3*, 14–21.

Morton, R. H. (1983). The supreme runner: what evidence now? *Aust. J. Sport Sci. 3*(2), 7–10.

Quercetani, H. (1964). *A Modern History of Track and Field Athletics*. Oxford University Press, London.

Ryder, H. W., H. J. Carr, and P. Herget (1976). Future performance in footracing. *Sci. Am. 234*, 108–119.

21

Ultrasound in Ante-Natal Diagnosis

Alan J. Lee and Alastair J. Scott
University of Auckland
Auckland, New Zealand

BACKGROUND

As a pregnancy progresses, an important part of the obstetrician's task is to monitor the growth of the unborn child and ensure that the fetus is growing in a normal manner. Early detection of any departure from the normal pattern of growth is obviously important for the efficient diagnosis and treatment of problems arising during pregnancy. Accordingly, an important problem in obstetrics is to establish the normal pattern of growth and determine the relationship between the age of the unborn infant (the gestational age) and the size of the infant. Put another way, it is desirable to establish the normal range of fetal size for any given gestational age.

One difficulty is that both of these quantities were, until recently, difficult to measure. The precise time of conception is often difficult for the obstetrician to measure due to such factors as faulty recall on the part of the mother or variations in the menstrual cycle. The size of the fetus was even more difficult to determine precisely; palpating the abdomen and various methods based on biochemical measurements were notoriously inaccurate. However, in the last few years great advances have been made with the employment of ultrasound techniques, using high-frequency sound waves to produce an electronic picture of the infant in the womb. The principle is similar to that employed in radar except that

sound waves are substituted for electromagnetic waves. High-frequency sound waves are emitted from a transducer which is passed over the mother's abdomen. These waves pass harmlessly through the mother and infant, being reflected or penetrating according to the type of tissue they encounter, and the reflected waves are in turn detected by a receiver in the transducer and processed electronically by a small computer connected to the receiver. The different rates of reflection of the sound waves, after processing, give an astonishingly detailed picture of the fetus, which is displayed on a television (TV) monitor attached to the computer. The mother-to-be is thus confronted with a live-action TV image of her child, which may be viewed from different perspectives as the transducer moves over her abdomen. Measurements of fetal size are made directly by positioning two "cursors" or marks on the TV screen by means of controls similar to those employed in video games, and automatically recording the distance between the cursors at the touch of a key on the computer console.

Two measurements that are easy to make are the width of the skull [the biparietal diameter (BPD)] and the width of the trunk [trunk width (TW)]. The former is made by positioning the cursors on opposite sides of the image of the baby's skull at the ends of a line at right angles to the line

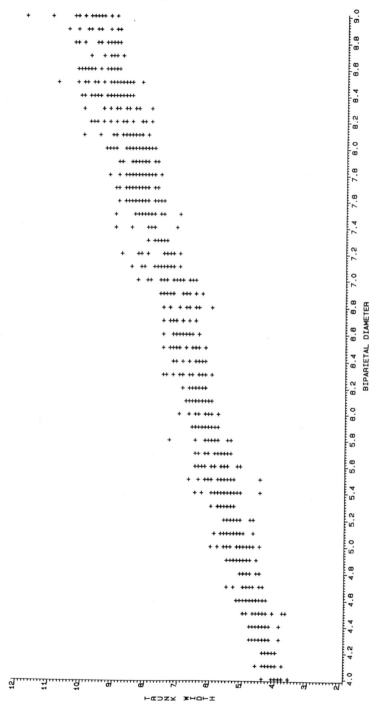

FIGURE 1   TW versus BPD for single births.

dividing the hemispheres of the cranium. Being the only straight line in the body, this line is an easy landmark to recognize in the ultrasound image. Similarly, the cross section through the trunk represents the only ellipsoidal shape in the body and hence is also easily recognised, enabling measurements of trunk width to be made.

The biparietal diameter is very closely related to the gestational age of the fetus; it is thought that the head draws nourishment preferentially compared to the rest of the body and so responds much more slowly to growth retardation than does the trunk width. (Some idea of the relationship between BPD and gestational age for one particular data set can be obtained by looking at Figure 9.) In what follows, BPD is taken as a measure of gestational age and trunk width is taken as the fundamental measure of fetal size. In medical science as in other disciplines, we are often forced to use an easily measured variable related to the real quantity of interest as a *proxy* for the quantity itself, which may be difficult to measure. For the reasons cited above, gestational age is difficult to measure exactly, and in any event is not always available, whereas BPD is easily and accurately measured via ultrasound, so that BPD is particularly suitable proxy.

Thus by comparing the two ultrasound measurements, the progress of the unborn infant may be checked and any necessary medical action taken in good time. More information about the use of ultrasound in obstetrics can be found in *Diagnostic Ultrasound in Obstetrics,* edited by J. C. Hobbins (1979).

THE RELATIONSHIP BETWEEN FETAL SIZE AND GESTATIONAL AGE

If we want to pick out abnormal growth patterns, we first need to know something about the normal relationship between size and age. In the light of the discussion in the preceding section, we interpret this to mean the relationship between trunk width and biparietal diameter.

To establish this relationship, the ultrasound measurements of TW and BPD made on 868 separate occasions on infants in an Auckland maternity hospital were subjected to a statistical analysis. The first step was to plot one variable against the other; the resulting scatter diagram is depicted in Figure 1, and it is immediately apparent that:

1. The relationship is approximately linear (with a slight curvature).
2. The cases with high BPD have more variable TWs than do cases with low BPD. This phenomenon, known as a "funnel effect," is quite common.

FIGURE 2 TW/BPD versus BPD for single births.

Standard statistical methods exist for establishing relationships such as those above; one variable (in this case, TW) is taken as a "response" that is to be explained in terms of an "explanatory variable" (taken here to be BPD).

Statisticians often assume that their data have been generated according to some idealized mechanism or "model." One such, often used in the situation above, is to imagine that each response is obtained by taking some function of the explanatory variable and adding to it a random perturbation (or "error"), so that we have

$$y = f(x) + e$$

where x denotes the value of the explanatory variable, e the error, and y the response. We assume that the error is selected at random by nature from some population of errors; we can imagine nature as having a bag of "errors" into which it dips and selects one at random. The bag is characterized by the error standard deviation, which we shall denote by σ. The larger the standard deviation, the more variable are the errors, and the more "fuzzy" is the relationship between the explanatory variable and the response.

The function f may be thought of as describing an idealized relationship between the response and the explanatory variable, which in practice is corrupted by the errors to an extent measured by the error standard deviation. If σ is large, large deviations between the actual response and that predicted by the idealized relationship are quite likely to occur. On the other hand, if σ is small, the large deviations are rare. Thus the model consists of two parts:

1. An idealized relationship f
2. A measure σ of the size of the discrepancy between the idealized relationship and the data actually observed in practice

Implicit in the formulation above is the idea that the errors do not depend on the value of the explanatory variable. This does not seem a reasonable assumption for the data plotted in Figure 1 in light of the clear funnel effect. The standard statistical methods work best when the errors do not depend on the explanatory variable, so a natural next step is to transform our ultrasound data to make them more nearly conform to the model.

Specifically, we consider the variables

$$y = \frac{TW}{BPD}$$

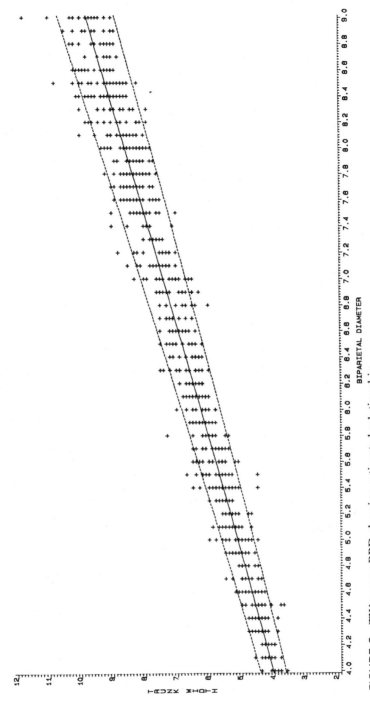

FIGURE 3   TW versus BPD showing estimated relationship.

and

$$x = \text{BPD}$$

and plot y against x. A glance at the resulting plot (Figure 2) indicates that the relationship between the variables is reasonably linear and that the funnel affect has disappeared. (This is by no means the only way to achieve such a result; plotting log TW against BPD, for example, has very nearly the same effect.) It seems reasonable to assume that the "idealized relationship" f is, in fact, a straight line, so we take

$$f(x) = a + bx$$

where a and b are constants, and our model for the data is

$$y = a + bx + e$$

Note that this is equivalent, in the original scale, to

$$\text{TW} = a\text{BPD} + b(\text{BPD})^2 + e\text{BPD} \qquad (1)$$

where, of course, e does not depend on BPD.

We could get informal estimates of a and b by fitting a straight line by eye to the points in Figure 2. More formally, we can obtain estimates of a and b using the method of least squares in which we choose values of a and b to minimize the sum of the squares of the deviations between the observed values, y, and the predicted values, $a + bx$. The method also enables us to calculate an estimate of σ, the error standard deviation. Denoting these estimates by $\hat{a}$, $\hat{b}$, and $\hat{\sigma}$, respectively, we find that $\hat{a} = 0.951$, $\hat{b} = 0.015$, $\hat{\sigma} = 0.061$. The estimated relationship curve,

$$\text{TW} = 0.951\text{BPD} + 0.015(\text{BPD})^2$$

is shown as the solid line in Figure 3, together with the individual data points.

If we want to use this curve as a diagnostic tool, we need some extra information on how much variation about the curve we might expect in normal cases. If we assume that the errors have a (bell-shaped) normal distribution, about 90% of the errors will be less than 1.65σ in magnitude. (See Chapter 2, for more details about the normal distribution.) This means that about 90% of the points should fall between the curves

$$\text{TW} = 0.951\text{BPD} + 0.015(\text{BPD})^2 \pm 1.65\hat{\sigma}\text{BPD}$$

which are shown as dashed lines in Figure 3. Points outside the dashed lines represent cases for which the trunk width is abnormally large or abnormally small for the measured BPD, and would merit closer investigation by the obstetrician. A close examination of Figure 3

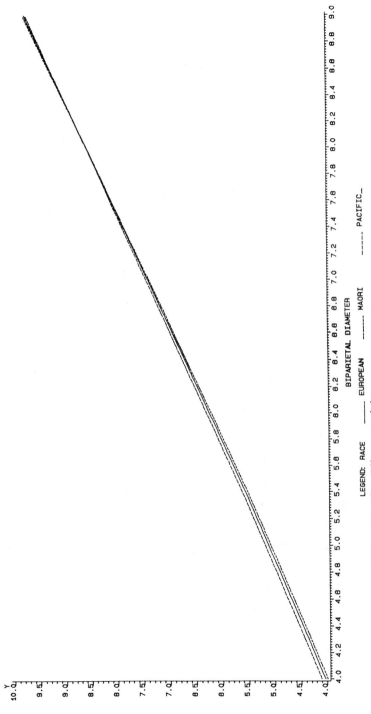

FIGURE 4 Estimated relationships for different racial groups.

indicates that our model is certainly not perfect. For example, there are more points above the dashed lines than there are below, and for some purposes it might be advantageous to fit a more elaborate model.

Of course, we should not take the assumption that the errors have a normal distribution, nor indeed any of the other assumptions of the model, on faith. Our basic tool for validating a model is an analysis of the *residuals* (i.e., the differences between the observed trunk widths and the values predicted by the model postulated). In practice, a large part of any statistical analysis is taken up by plotting the residuals in a variety of ways in order to show up any deficiencies in the data or the model. (A number of transcription errors in this data set were picked up in this way.) A good account of these procedures is given in *Interactive Data Analysis* by D. R. McNeil (1977). In this case the assumption of normality seems very reasonable.

We see that the investigation has had two facets:

1. The establishment of an "idealized relationship" between BPD and TW represented by the solid line in Figure 3
2. The establishment of a "normal variation" about this relationship, represented by the dashed lines in Figure 3

RACIAL DIFFERENCES IN THE RELATIONSHIPS

In the preceding section we have analyzed the data in Figure 1 as if they were drawn from a single homogeneous population, but in fact they can be classified into three groups according to whether the mothers were European, Maori, or Pacific Islanders. This raises the question of whether or not we should postulate the same idealized relationship for all points; growth patterns may well be different for the three groups. Figure 4 depicts relationships estimated separately for the three groups, and we see that slight differences do exist, although the three curves are very similar.

Suppose that we took more measurements from mothers in these three ethnic groups. Then the three curves estimated from the new data would differ slightly from those in Figure 4 since they are computed from different numbers. If we took still another set of data, we would get still another set of estimated relationships, and so on. It is thus apparent that there is natural variation in the estimated relationship curves. This poses an obvious question: Are the observed differences in racial groups shown in Figure 4 real or are they simply due to this natural variation?

To answer this question, the statistician draws on the theory of probability. Assuming that the mechanism generating the data is that

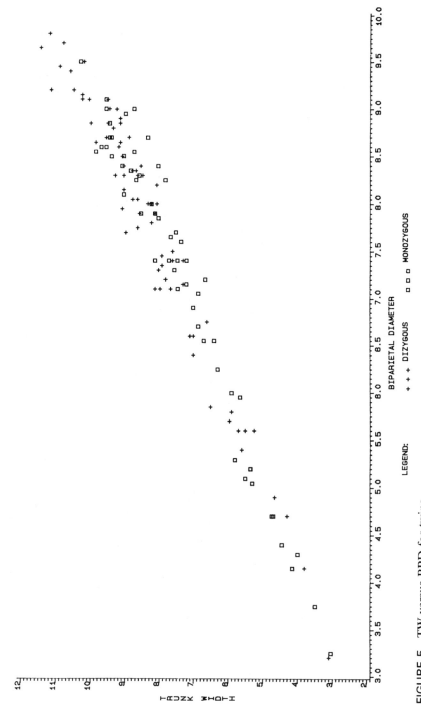

FIGURE 5 TW versus BPD for twins.

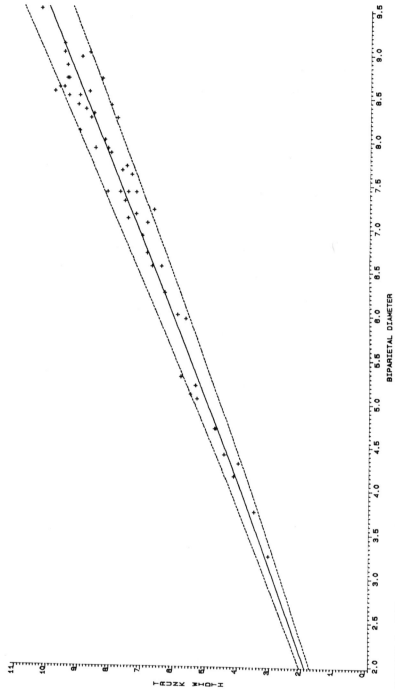

FIGURE 6 TW versus BPD for monozygous twins.

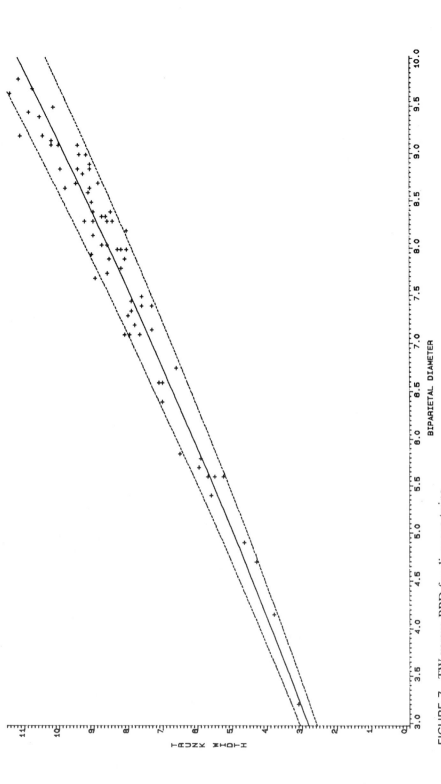

FIGURE 7 TW versus BPD for dizygous twins.

given by equation (1), it is possible to calculate how large the observed differences are likely to be when they are merely a consequence of natural variation in the data.

In fact, it turns out that differences equal to or greater than those observed in the present case are quite likely to occur purely as a consequence of natural variation. There is thus no reason to believe, at least on the basis of these data, that the idealized relationships for the three racial groups are different. Thus treating all the observations as a single group and using Figure 3 as a screening device for all mothers, regardless of their ethnic origin, seems a reasonable procedure in this hospital at least.

TWINS

The relationship between BPD and trunk width discussed above was based on the analysis of single births only. There is considerable medical interest in the question of whether the relationship for twins differs from that for single births. In fact, it is possible that the relationship is different for monozygous (i.e., identical) and dizygous twins. (In general, monozygous twins share the same placenta, while dizygous twins have separate placentas.) To investigate this question, ultrasound measurements made on 131 separate occasions on mothers bearing twins were available for analysis. For each ultrasound session, measurements of BPD and TW were made for both fetuses.

Now one feature of the analysis employed so far that we have not mentioned is that the estimate of σ (from which we obtain our bounds on normal variation) in valid only when the errors are independent. However, it seems very likely that the errors for twins in the same womb are strongly related, so that observations on twins are not independent.

An easy way to get round this problem is to average the BPD measurements for each twin, and similarly for the TW measurements. It is now quite reasonable to assume that measurements on different wombs are independent, so the averages should more nearly follow the assumptions made in the model. From now on BPD means "average BPD" and the same for TW.

Figure 5 shows a plot of TW versus BPD for the twins. Monozygous and dizygous twins are plotted with separate symbols. Once again a slightly curved relationship and a definite funnel effect are apparent. Again we can get rid of the funnel effect by considering the variables $y = $ TW/BPD and BPD.

To investigate the possibility that the two types of twins have different relationships between TW and BPD, we considered the models

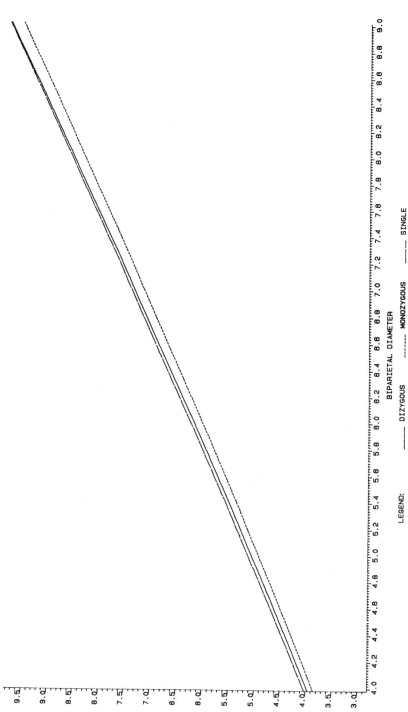

FIGURE 8   Estimated relationships for different twin types.

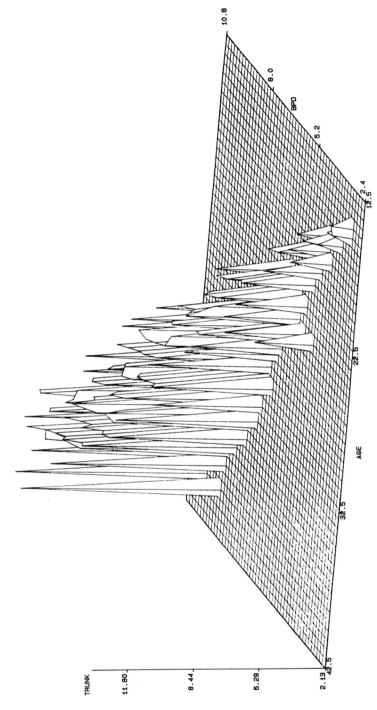

FIGURE 9 TW by age and BPD.

$$y = a_M + b_M \text{BPD} + e$$

for monozygous twins, and

$$y = a_D + b_D \text{BPD} + e$$

for dizygous twins.

Once again the coefficients were estimated by least squares, and the estimated relationships (transformed back to the original variables) are $TW = 0.936\text{BPD} + 0.012(\text{BPD})^2$ for monozygous twins, and $TW = 0.874\text{BPD} + 0.024(\text{BPD})^2$ for dizygous twins. The relationships, together with the original points, are shown in Figures 6 and 7.

In Figure 8 we show the three relationships for singles, monozygous twins, and dizygous twins. As with the three ethnic groups the curves look very similar, but it turns out in this case that the differences are *not* likely to have come about purely through natural variation. In statistical jargon, we say that the differences are (statistically) significant. We need to be very careful in our interpretation of statistical significance; although we conclude that the observed differences probably reflect real differences in the underlying relationships, these real differences may well be too small to be of any *medical* significance. This, of course, is a medical judgment which cannot be made by the statistician, although the estimated coefficients for the three groups, together with the standard errors of these coefficients, provide useful information for making this judgment.

For this particular set of data we also happen to have reasonably reliable information on the gestational age of the twins. The joint relationship between the three variables (gestational age, trunk width, and BPD) is conveniently displayed in a three-dimensional scatter diagram as in Figure 9. Each data point is represented by a pyramid whose height is equal to the corresponding trunk width. It is clear that all three variables are closely related. In particular, the plot gives reasonable support to the practice of using BPD as a proxy for gestational age.

ACKNOWLEDGMENTS

We are very grateful to Dr. J. D. Mora and Ms. J. Mitchell and the staff of the Radiology Unit at St. Helens Hospital for introducing us to the statistical problems in using ultrasound measurements, and for providing all the data used in this chapter.

REFERENCES

Hobbins, J. C., ed. (1979). *Diagnostic Ultrasound in Obstetrics.* Churchill Livingstone, New York.

McNeil, D. R. (1977). *Interactive Data Analysis.* Wiley, New York.

22

Predicting Suicide: Statistical and Clinical Strategies

Keith Petrie
Waikato Hospital
Hamilton, New Zealand

Kerry Chamberlain
Massey University
Palmerston North, New Zealand

THE CLINICAL PROBLEM: PREDICTING SUICIDAL BEHAVIOR

Two cases of attempted suicide had arrived at the hospital's emergency department. Julie, a 20-year-old student, had swallowed four handfuls of tablets from the medicine cabinet. She had been arguing bitterly with her parents. George, a middle-aged man, was found hanging in the garage. His wife, who found him by accident, acted quickly and cut him down. He was still breathing.

Two days later the condition of both patients had stabilized. A decision must now be made regarding the likelihood of Julie or George reattempting or commiting suicide, and their future contact with limited psychiatric services. This is a clinical decision with major consequences and brings the clinician into the area of assignment of suicidal risk.

The range of suicidal behavior may be seen as falling into distinctive, yet overlapping populations. These broad clusters are ideators, threateners, attempters, and completers. Attempted suicide patients such as Julie or George have a high risk of reattempting or commiting suicide. About 1% of attempted suicides will commit suicide in the 12 months following their attempt. Ten percent will end up killing themselves (Stengel, 1972). Approximately 18% of attempted suicide patients will reattempt in the following year (Bancroft and Marsack, 1977).

295

The dilemma for the clinician assigning suicidal risk is how to identify the subgroup of attempted suicides in danger of further suicidal behavior. The clinician assigning suicidal risk may choose to consider the *base rates* of suicidal behavior within the general population. Base rates are the proportion of people in a given population that engage in suicidal behavior. The overall rate of suicide in the population is low—around 12 per 100,000 persons in the United States will kill themselves in a year. However, there are certain subpopulations that commit suicide more frequently than others. We know that the following groups commit suicide disproportionately to the overall rate in the general population; males (generally in the ratio of 2 or 3 males to 1 female), the divorced or widowed, those that have been admitted to psychiatric hospitals (particularly for depression), older people, those that live alone, alcohol and drug abusers, and individuals suffering from a physical illness (Brown and Sheran, 1972). Several tests or checklists have been developed to aid risk assignment through the use of base rates. An example of such a test is one developed by Tuckman and Youngman (1968) shown in Table 1.

Unfortunately, the accurate prediction of suicidal behavior is not as easy as categorizing patients as high risks from their score on a suicide prediction scale. With a low-frequency behavior such as suicide, even a very accurate test will result in a large number of *false positives*. There are patients that are identified as having high suicidal risk but never actually commit suicide. Prediction by psychological tests of such a low-frequency behavior is very difficult, and even a test with good predictive power will result in a prohibitive number of false positives (Rosen, 1954). Take, for example, a test that can accurately identify 80% of suicides and nonsuicides. In a population of 2000 psychiatric patients there may be 20 suicides a year. By giving the test we could expect to classify correctly 16 of the 20 patients who commit suicide, but also the test would misclassify as suicidal another 92 of the 2000 patients. Because of the consequences of classifying persons of high suicidal risk such as close surveillance, restriction of freedom, and special medical care, coping with a large number of false positives is generally impractical. Raising the cutoff score to reduce the number of false positives would result in fewer suicides being identified.

USING STATISTICS TO IMPROVE THE PREDICTION

As with most human actions, we find there are a great many factors that can influence suicidal behavior. We know, for instance, that many separate variables, such as age, sex, being depressed, or social isolation, are each associated with suicide. Each of these variables makes some

TABLE 1   Scale for Assessing Suicide Risk in Attempted Suicides

Factor	High risk	Low risk
Age	Over 45 years	Under 45
Sex	Male	Female
Race	White	Nonwhite
Marital status	Separated, divorced, widowed	Single, married
Living arrangements	Alone	With others
Employment status	Unemployed, retired	Employed
Physical health	Poor	Good
Mental condition	Nervous or mental disorder	Normal
Medical care (within 6 months)	Yes	No
Method	Hanging, firearms, jumping, drowning	Cutting or piercing, gas or carbon monoxide, poison, combination, other
Season	Warm months	Cold months
Time of day	6.00 a.m.–5.59 p.m.	6.00 p.m.–5.59 a.m.
Where the attempt was made	Own or other's home	Other premises, out of doors
Time interval between attempt and discovery	Almost immediate	Later
Self-reported intent to kill	No	Yes
Suicide note	Yes	No
Previous attempt or threat	Yes	No

Source: Tuckman and Youngman (1968).

contribution toward the behavior, and each relates to a different aspect of the behavior. If we could assess how these variables interact and operate in combination, we should be able to improve our prediction of the behavior considerably.

For example, knowing that John is divorced does not in itself give him a high probability of attempting suicide. Knowing also that he is an older male increases the probability, and knowing further that he experiences feelings of hopelessness and that he lives alone may place him at high risk of attempting suicide.

Putting this in statistical terms, we would say that variation in the dependent variable (suicidal behavior) is a function of concomitant variation in a number of explanatory variables (age, sex, severity of feelings of hopelessness, number of people in the household, etc.) all acting simultaneously. The statistical technique we use to examine this is multiple regression. This technique allows us to predict the score on the dependent variable from our knowledge of the scores on the set of explanatory variables in combination better than we could by predicting from each of the explanatory variables alone.

Let us say, for example, that if we knew a person's score on a hopelessness measure, we could account for 30% of the variance in suicidal behavior, and if we knew the number of people in the household, we could account for 15% of the variance in suicidal behavior. By incorporating both of these scores in a multiple regression equation, we would be able to improve on the amount of variance in suicidal behavior that we can account for. Although it is unlikely that this would increase to 45%, we should be able to account for something between 30 and 45% by using both variables together.

The reason that we are unlikely to reach 45% is because the two explanatory variables, hopelessness and number of people in the household, are themselves associated or correlated. The strength of this association is taken into account by the procedure in determining the joint relationship of these variables to suicidal behavior.

If we introduce more variables, these will probably themselves be intercorrelated, and they may or may not improve our prediction. The multiple regression technique can help us by locating an optimal set of variables from those available which in combination provide the best possible prediction of the dependent variable. In practice, the improvement in our prediction that we gain from using several variables is not likely to be great after the first five or six variables, provided that these are ones which relate in important ways to the dependent variable. The gain from adding more and more variables is quite limited. Table 2 gives an example of the outcome from conducting a multiple regression analysis relating five variables to suicidal behavior.

The first point to notice is that the technique gives us a multiple correlation (multiple R). This is simply the correlation between the dependent variable and the set of explanatory variables, and it can be interpreted in the same way as a standard correlation coefficient. That is, the coefficient squared gives us the amount of variation in the dependent variable which is accounted for by the set of independent variables in

TABLE 2   Multiple Regression Summary Table

Dependent variable: suicidal behavior
  Multiple R   0.69945
  R-square   0.48923

Independent variable	B	Beta	Beta^2
Age	0.121	0.263	0.069
Household size	−0.239	−0.286	0.081
Anxiety	0.022	0.035	0.001
Hopelessness	0.367	0.392	0.153
Depression	0.175	0.316	0.099
(Constant)	(−2.749)		

combination. In our example, the R value is 0.69, indicating that the relationship between the set of variables and suicidal behavior is positive and that together the variable set accounts for about 49% of the variance in suicidal behavior (as R^2 = 0.49). Although this may not appear to be a large amount, it is more than we would be able to account for from any of the variables alone.

The main point of conducting the multiple regression in our situation is to help us predict who will attempt suicide. From the analysis we obtain a straightforward linear equation. We can use this to calculate a score for suicidal behavior for any person whose scores we have on the set of independent variables.

This equation has the general form

$$Y = B1X1 + B2X2 + B3X3 + B4X4 + B5X5 + C$$

Here Y stands for the dependent variable that we are trying to predict, and $X1$ to $X5$ are the explanatory variables from which we are trying to predict Y. C is just a constant, and the values $B1$ to $B5$ are numbers, called weights, which are used to multiply the respective variables. The multiple regression technique is concerned with finding the values for these weights so that the explanatory variables in combination give the best possible prediction.

From our example, if we know Julie's age, the number of people in her household, and her scores on hopelessness, anxiety, and depression, we can enter those into the equation and obtain her predicted score on suicidal behavior. This we can interpret as a measure of how likely she is to attempt suicide.

From Table 2 we can determine that the equation will be

Suicide score = 0.121(age) − 0.239(household size) + 0.022(anxiety)
                + 0.367(hopelessness) + 0.175(depression) − 2.749

In other words, if we enter Julie's scores for the variables, multiplied by the appropriate weights and add the constant term, we will obtain her predicted score for suicidal behavior. If this is very high, she will be at high risk of engaging in suicidal behavior.

The next question of concern is to ask which variables are the most important in predicting suicidal behavior. Here the multiple regression cannot help us. Knowledge of which variables are involved in suicide must come from our knowledge of prior research in this area and from our theoretical understanding of the nature of suicide. This theoretical and empirical knowledge will, of course, be informative in helping us to decide which variables to include in our studies of suicidal prediction. In the light of recent research, for example, we would be remiss if we did not include hopelessness as one of the variables (Beck et al., 1975; Kovacs et al., 1975).

However, once we have made the decisions on which variables to include, multiple regression can help us decide the relative importance of these variables in making the prediction. The weights, $B1$ to $B5$, in the regression equation appear to determine the relative contribution of each variable. If one of these weights is large (close to 1), that variable contributes a large amount to the equation. This approach has one major problem—the variables are all measured in different units and so are not comparable. Age is measured in years, hopelessness is measured as a score out of 40, and anxiety is measured on a 9-point scale. If we were to standardize all these variables (transform them to standardized scores so that they each have a mean of 0 and a standard deviation of 1), they would be comparable since they would now all be in the same units. Table 2 presents another column of weights beside the Bs, labeled betas. These are, in fact, the weights that would be used in the regression equation if it was calculated using standard scores for the explanatory variables. Since they are measured in the same units, the betas do give us a good indication of the relative importance of the variables that we have included. Thus hopelessness turns out to be the most important variable (beta = 0.392) and anxiety turns out to be the least important variable (beta = 0.035).

We can take this interpretation of the beta weights even further, as they are a form of partial correlation coefficient. The betas squared can be interpreted as giving the proportion of variance in the dependent variable accounted for by each independent variable. In our example, we can say that hopelessness accounts for 15% of the *explained* variance in suicidal behavior, while anxiety accounts for less than 1%. The total variance accounted for under this interpretation is less than the *R* squared because the variables in our equation are intercorrelated, so the interpretation needs to be made cautiously.

Several further points should be noted about the multiple regression technique. First, it assumes that the relationship between the dependent variable and the explanatory variable set is linear, and it operates to calculate the optimal linear combination of independent variables to maximize the multiple correlation. If relationships between the dependent variable and the explanatory variables are not linear, as may happen in suicide research (Diggory, 1974), steps such as transforming variables or the use of alternative procedures must be undertaken.

We have already mentioned that the gain in prediction from including more than a limited set of variables (about five or six, usually) is small. Further, it should be noted that we obtain the optimal prediction when the explanatory variables are related strongly to the dependent variable but are related weakly to each other. Finally, the prediction equation that we obtain will vary somewhat from sample to sample, depending on the nature of the particular sample that we have obtained. In practice, we should obtain the best sample possible, and we assume that the relationships found are good estimates of the actual relationships existing in the overall population.

GOING IT ALONE: AN ALTERNATIVE PREDICTION PROCESS

The clinician assigning suicidal risk may opt to make a decision subjectively. That is reaching a decision by combining intuitively or at "a gut level" what he or she has learned from this patient with the experience gained from similiar patients in the past. This intuitive process is called *clinical prediction* and contrasts with a *statistical* (also called *actuarial*) prediction process like multiple regression.

The two prediction processes exemplify extreme positions. They differ on what information is included and how that information is combined. In statistical prediction there is always the same set of predictive variables entered into the predictive equation regardless of the patient. With clinical prediction the number and type of variables may alter with each case. Statistical prediction requires the variables to be combined in a consistent manner, whereas the clinical process may give weightings to differing pieces of information depending on what the clinician feels is relevant to a particular case.

For example, consider the two patients we began with. In Julie's case the clinician may consider her a low suicide risk because she is female, shows no psychiatric symptoms and has many supportive friends. However, for George, the fact that he had a well-developed suicide plan, combined with the lethal nature of his attempt and his depressed condition, may sway the clinician to classify George as high risk. These decisions show the clinical prediction process to be a volatile one—the number and importance of predictive variables change depending on the characteristics of each case. The nature of the clinical approach means that it is often difficult to deduce how a clinician arrived at a prediction decision.

Although the statistical method is precise, many clinicians prefer to predict suicidal behavior using a clinical prediction process. This is due to the strong influence of the individual case and the lack of a strong theory that explains suicidal base rates.

When assessing an attempted suicide patient the clinician may feel that the statistical prediction process lacks individuality. A multiple regression procedure may not be seen as capturing the uniqueness of an individual. Base rates are, of course, based on the likelihood of suicide over a population, and when confronted by an individual the clinician may ask whether the particular trend will hold true in this particular case. In fact, there are common predictive biases that may influence the clinician's judgment here. First, research has shown that judges assigning risk or the probability of an event occurring are insensitive to base rates even when these are well known. Judges prefer to base their decision on the perceived similarity of the individual's characteristics to their stereotype of the person being predicted (Tversky and Kahneman, 1974; Kahneman and Tversky, 1973). Generally, human decision makers are poor at assimilating and integrating a wide variety of (often contradictory) information. A study of suicide lethality judgments suggests that even judges trained to use a wide variety of factors in their assessment of suicide risk based their decision primarily on only three: suicide plan, age, and prior suicidal behavior (Brown, 1970).

Although the statistical approach incorporates who we know to be at risk of suicide, the demographics of suicidal behavior in fact present a superficial view of what is actually causing the self-destructive behavior. For example, demographic statistics do not tell us why older males have a high suicide rate, or why younger females are more frequent attempters. For the clinician it is the psychological causes behind these figures that are crucial for assigning a risk level in an individual case. Factors such as hopelessness, depression, or the lack of social support in the patient's environment need to be combined rationally with demographic variables to assess risk in the individual case. It may be that one day we will know enough about suicidal behavior that these psychological predictive factors will overtake the use of demographics.

John Greist and his coworkers (Greist et al., 1973, 1980) have presented an innovative example of the systematic combination of psychological and demographic variables in suicide prediction. They used a computer to interview attempted suicide patients and assign a suicidal risk. The computer branches the patient through a clinical interview eliciting psychological information such as the level of depression and frequency of suicidal thoughts as well as demographic data such as age, sex, and marital status. The computer then processes the information statistically, assigning a probability of reattempting suicide for each patient. Research using this computer program shows the computer to be significantly more accurate than clinicians in predicting which patients reattempt suicide. In a follow-up study, the computer assigned a high probability of reattempting to 91% of those patients who actually reattempted, whereas clinicians assigned a high level of probability to only 16% of reattempters. In a retrospective study, using files of known reattempters and nonreattempters, the computer assigned 60% of reattempters a high probability compared to 40% for clinicians.

Research of this sort suggests that clinicians could make a better prediction if they incorporated a statistical approach in their decision making. Although clinical experience is useful in the assignment of suicidal risk, it is likely that the optimal prediction will be gained by using clinical judgment in combination with a good statistical procedure. Current work on statistical models of suicide prediction is likely to be of great help in future identification of the suicidal person.

CONCLUSION

Clinicians assigning suicidal risk face a real problem, and their decisions carry serious consequences. Their predictive task is made harder by the fact that suicidal behavior occurs infrequently. Some approaches to prediction have attempted to utilize the base rates for behavior in subpopulations. Certain groups, such as males, the socially isolated, and alcohol and drug abusers, engage in suicidal behavior more frequently than the general population. In trying to incorporate base rates into the predictive process, the clinician is faced with the problem of a large number of false positives. As well, the clinical approach is liable to judgment biases that undermine the consistent use of base rates, and affect decisions. For a more comprehensive understanding of suicidal behavior, psychological variables need to be considered as well as demographics. To improve prediction we need a systematic method to combine the many variables that influence suicidal behavior, and the statistical technique of multiple regression gives the clinician a procedure for doing this. Multiple regression allows us to use a set of selected variables acting in combination to predict suicidal behavior. The approach provides the clinician with a useful tool which helps eliminate judgment biases, and provides a systematic procedure for assessing the effects of the many variables, both demographic and psychological, that are operating. Used in conjunction with clinical judgment, the statistical approach has great value in improving the overall quality of suicidal behavior prediction.

REFERENCES

Bancroft, J., and P. Marsack (1977). The repetitiveness of self poisoning and self injury. *Br. J. Psychiatry 131*, 394–399.

Beck, A. T., M. Kovacs, and A. Weissman (1975). Hopelessness and suicidal behavior: An overview. *J. Am. Med. Assoc. 234*, 1146–1149.

Brown, T. R. (1970). The judgment of suicide lethality: A comparison of judgmental models obtained under contrived versus natural conditions. Unpublished doctoral dissertation, University of Oregon.

Brown, T. R, and T. J. Sheran (1972). Suicide prediction: A review. *Life Threat. Behav. 2*, 67–98.

Diggory, J. C. (1974). Predicting suicide: Will-o-the wisp or reasonable challenge. In *The Prediction of Suicide* (A. T. Beck, H. L. P. Resnik, and D. J. Lettieri, eds.). Charles Press, Maryland.

Greist, J. H., D. H. Gustafson, F. F. Strauss, G. L. Rowse, T. P. Laughren, and J. A. Chiles (1973). A computer interview for suicide-risk prediction. *Am. J. Psychiatry 130*, 1327–1332.

Greist, J. H., D. H. Gustafson, H. P. Erdman, J. E. Taves, M. H. Klein, and S. D. Spiedel (1980). Suicide risk prediction by computer interview: A prospective study. Unpublished manuscript.

Kahneman, D., and A. Tversky (1973). On the psychology of prediction. *Psychol. Rev. 80*, 237–251.

Kovacs, M., A. T. Beck, and A. Weissman (1975). Hopelessness: An indicator of suicidal risk. *Suicide 5*, 98–103.

Rosen, D. H. (1954). Detection of suicidal patients: An example of some of the limitations in the prediction of infrequent events. *J. Consult. Psychol. 18*, 397–403.

Stengel, E. (1972). A survey of follow-up examinations of attempted suicides. In *Suicide and Attempted Suicide* (J. Waldenstrom, T. Barson, and N. Ljungstedt, eds.). Nordiska Bokhandelns Forlag, Stockholm.

Tuckman, J., and W. F. Youngman (1968). A scale for assessing suicide risk of attempted suicides. *J. Clin. Psychol. 24*, 17–19.

Tversky, A., and D. Kahneman (1974). Judgement under uncertainty: Heuristics and biases. *Science 185*, 1124–1131.

23

Predicting Earthquakes

David A. Rhoades
Department of Scientific and Industrial Research
Wellington, New Zealand

INTRODUCTION

Scientists in earthquake-prone countries have recently been devoting a lot of effort to the problem of how to predict damaging earthquakes. The forces and rock structures in the crust of the earth that give rise to earthquakes can only be observed indirectly from the surface, so this is a very difficult task. Nevertheless, there are many different sorts of observations made by earth scientists which give clues as to what is happening deep in the earth's crust. For instance, the creeping or locking of faults, tilting and bulging of the ground, clustering or unusual absence of small earthquakes, and changes in groundwater levels or in the force of gravity at a particular place can all be measured. Perhaps it will be possible to recognize some regular patterns in these observations which will allow us to detect a big earthquake in advance. Such telltale patterns are called earthquake *precursors.*

The Chinese have shown that precursors *do exist* by their timely evacuations of vulnerable areas before two large damaging earthquakes in 1975 and 1976. They did this by identifying several different precursors after a massive effort in collecting and analyzing observations of the kind mentioned above.

In this chapter we examine the sort of information that can be gleaned from a single kind of precursor that has been proposed. The

value of this precursor for prediction purposes has yet to be demonstrated, and just how such a demonstration might be made is one of our chief interests. Also, we want to show how statistical ideas are useful in making precise the statement of any forecast that might be made using this precursor.

PRECURSORY SWARMS

Our example of a proposed earthquake precursor, which comes from the catalog of historical earthquakes in New Zealand, was recognized by Evison, a geophysicist at Victoria University of Wellington (Evison, 1977). The general idea is to use patterns in the occurrence of small earthquakes to predict big ones. The pattern begins with a cluster of small earthquakes of rather similar size occurring in a fairly small area. This cluster is called a *swarm*. The swarm is followed by a *gap*—a period, lasting much longer than the swarm, in which an unusually low rate of earthquake occurrence

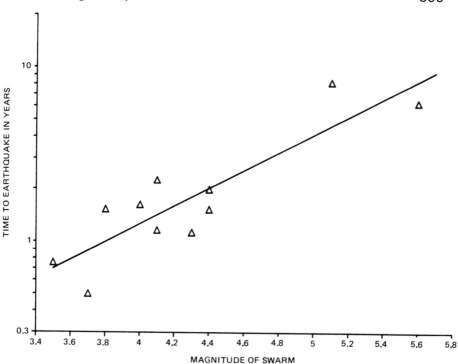

FIGURE 1 Relationship between precursor time and magnitude of precursory swarm for New Zealand historical earthquakes. (After Evison, 1977.) The magnitude of the swarm is defined as the mean of the three largest Richter magnitudes of earthquakes in the swarm.

is observed in the same small area. The gap is then followed by a moderate-to-large earthquake* in the same small area.

F. F. Evison observed that most of the recent moderate-to-large shallow† earthquakes in New Zealand have been preceded by such a

---

*The size of an earthquake is commonly described by its *magnitude* on the *Richter* scale, a quantity that can be determined from the maximum displacement of the trace on a seismograph, and which is closely related to the logarithm of the energy released in the form of elastic ground waves. The term "large" is usually reserved for earthquakes of magnitude 7 or greater. By "moderate to large" we mean earthquakes of magnitude about $5\frac{1}{2}$ or greater. The hazard rate for such events in New Zealand is about 2 per year.

† Shallow earthquakes, for this purpose, are those that occur at a depth no greater than 33 k. Almost all damaging earthquakes are shallow.

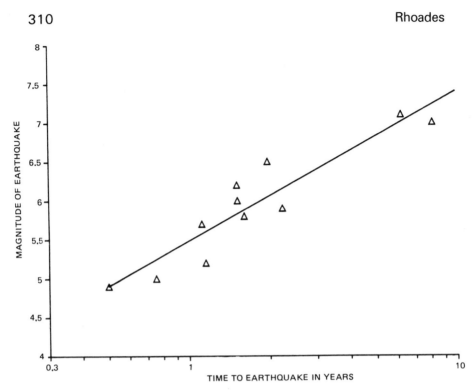

FIGURE 2   Relationship between earthquake magnitude and precursor time for
New Zealand earthquakes and precursory swarms. (After Evison, 1977.)

pattern of activity and that swarms are otherwise quite rare, except near
active volcanos. What is even more interesting, he found that the time
from the swarm to the earthquake is related to the size of the earthquakes
in the swarm (Figure 1) and also to the size of the eventual earthquake
(Figure 2). So the bigger the swarm is, the longer the time to the
earthquake *and* the bigger it will be. These data appear to offer us a
chance of predicting, several years in advance, the location of most
moderate-to-large shallow earthquakes in New Zealand and also of
forecasting approximately their time of occurrence and magnitude.

DISTRIBUTIONS FOR THE TIME OF OCCURRENCE
AND MAGNITUDE

That any forecast of the time of occurrence and magnitude can only be
approximate is evident from the scatter of points about the straight lines

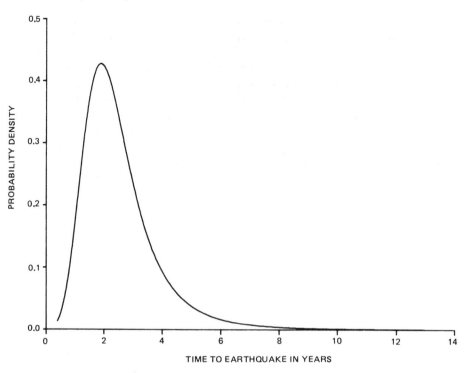

FIGURE 3 Probability density function for the time of occurrence of the earthquake given a precursory swarm of mgnitude 4.5.

of best fit in Figures 1 and 2. For instance, if a swarm of magnitude 4.5 is observed, we should expect to wait about 2 years for the earthquake (Figure 1) and the magnitude of the earthquake should be about 6.0 (Figure 2). But from the scatter of the data points about the line in Figure 1, it is clear that a waiting time as short as 1 year or as long as 4 years is not impossible. However, 2 years would seem to be a more likely time than either 1 year or 4 years. But how much more likely?

To answer the question above in a precise way we need to fit the deviations from the line to a probability distribution. If the relationship between the logarithm of the time to the earthquake and the magnitude of the swarm is truly linear, as it appears to be in Figure 1, except for random variations that follow a normal distribution, statistical theory tells us that the deviations from the fitted line are distributed as a t distribution with $n - 2$ degrees of freedom, where n is the number of data points. If we know the standard deviation of points about the fitted line, we can calculate the *probability density function* for the time to the earthquake given the magnitude of the swarm. This function has been plotted in Figure 3 for a swarm of magnitude 4.5.

Using Figure 3 we can easily answer questions about how much more likely one time of occurrence is than another, just by comparing the values of the probability density function for the two times. For instance, for a time of two years the probability density is 0.42 and for 1 year it is about 0.14. The ratio of 0.42 to 0.14 is 3. Thus the time to the earthquake is three times as likely to be 2 years than it is to be 1 year. Similarly, we can see that it is about $4\frac{1}{2}$ times as likely to be 2 years as it is to be 4 years.

The probability density function is useful for another purpose in that areas under the curve represent probabilities of the time of occurrence falling within a given interval. For instance, we can see that the probability of the earthquake happening between 2 and 4 years is about 1/2 since about 1/2 the total area under the curve lies between these two values. Of course, the total area under the curve must be 1 (the probability that the earthquake occurs within 4 years and has magnitude greater than 7 or the probability that the earthquake occurs during the fifth year and has magnitude between 6.5 and 7.5, and so on.

Just as the data in Figure 1 allow us to compute the probability distribution for the time of occurrence given a particular value of the swarm magnitude, the data in Figure 2 allow us to compute the distribution for the magnitude of the earthquake given a particular time of occurrence. The laws governing conditional probabilities then allow us to compute the whole joint distribution for the time and magnitude of the earthquake. This means that we know all that we need to know about probabilities concerned with the time of occurrence, or magnitude, or both in combination. For example, we could calculate the probability that the earthquake occurrs within 4 years and has magnitude greater than 7 or the probability that the earthquake occurs during the fifth year and has magnitude between 6.5 and 7.5, and so on.

FALSE ALARMS AND FAILURES

So far we have not discussed the possibility that a predicted earthquake might not happen at all. It would be quite foolish for us not to acknowledge this as a possibility. Also, as time goes by with no earthquake following the precursor, common sense demands that we gradually come to the conviction that the prediction was a false alarm and that the earthquake is no longer expected to happen. How soon we reach this conclusion should depend on how confident we were in the first place that an earthquake would occur. Our level of confidence will

TABLE 1   Classification of Alarms and
Earthquakes for Precursory Swarm
Criteria

Total number of alarms	10
Number of valid alarms	8
Number of false alarms	2
Number of shallow earthquakes	11
Number of successes	8
Number of failures	3

depend on how common false alarms have been in the past. Clearly, we need more data than have already been presented.

The recognition of precursors, in this case precursory swarms, was initially carried out by studying what happened before known historical earthquakes. But precursors of future earthquakes have to be recognized by some set of criteria that distinguishes as well as possible the precursory events from other clusters of small earthquakes which may occur from time to time and which are not precursory. For any given set of criteria that we might settle on, it is possible to distinguish four different kinds of event:

1. A *valid alarm:* an event satisfying the criteria that is followed by a (moderate-to-large shallow) earthquake
2. A *false alarm:* an event satisfying the criteria that is not followed by an earthquake
3. A *success:* an earthquake that is preceded by an event satisfying the criteria
4. A *failure:* an earthquake that is not preceded by an event satisfying the criteria

Thus every event satisfying the criteria (an *alarm*) is either a valid alarm or a false alarm according to whether or not an earthquake subsequently occurs, and every moderate-to-large shallow earthquake is either a success or a failure according to whether or not the events preceding it satisfy the criteria. In setting the criteria we would like to obtain as many valid alarms and successes as possible and as few false alarms and failures as possible. It is necessary also to use common sense and to keep the criteria simple and consistent with physical explanations of the phenomenon.

Neither the details of the criteria adopted in the case of precursory swarms nor the physical explanation for them are our concern here. But

Rhoades

TABLE 2   Estimated Proportional Rates
for Precursory Swarm Criteria

Valid alarm rate	0.8
False alarm rate	0.2
Success rate	0.73
Failure rate	0.27

for the criteria that have been adopted, Table 1 gives the number of events in the different classes for New Zealand earthquakes since 1963, excluding those in the volcanic region. (This was the year that the present seismograph network was substantially put in place. Since 1963 the measurement of small earthquakes such as those in the swarms has been much better than it was previously.)

Table 1 covers only swarms with magnitude greater than or equal to 4.0 and shallow earthquakes with magnitude greater than or equal to 5.7, so that not all the data from Figures 1 and 2 are included here. These limits were set high enough so that the ability of the seismograph network to detect the relevant swarms is not in question. These data can be used to make simple estimates of the valid alarm rate (or equivalently, its complement, the false alarm rate) as a proportion of the total number of alarms and of the success rate (or its complement the failure rate) as a proportion of the total number of shallow earthquakes. These estimates are given in Table 2.

These are the rates estimated from the historical data. Remember that we have had the benefit of "data snooping" in setting the criteria, so we cannot be certain that the rates will be so impressive if the criteria are applied to future events. However, if future events are accumulated on the score sheet in Table 1, the estimates of the proportional rates will eventually converge toward the true values for the criteria that have been set.

IS THE ALARM VALID?

The valid alarm rate enters into the forecast in the following way. For an individual alarm it represents an initial estimate of the probability that the alarm is valid. If, as time passes, the earthquake fails to occur, this probability will diminish and eventually tend to zero. The precise way in which this probability changes is governed by the rules for manipulating conditional probabilities. In particular, a well-known theorem named after the eighteenth-century clergyman and statistician Thomas Bayes

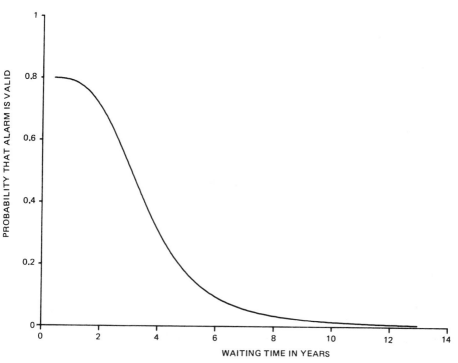

FIGURE 4   Variation of the probability that an alarm associated with a precursory swarm of magnitude 4.5 is valid. The graph shows how this probability would decrease as the waiting time increases without an earthquake occurring.

comes into play. Let V denote the event that the alarm is valid and $\bar{V}$ the complementary event (i.e., a false alarm). Also, we shall denote probabilities by Pr, so that $Pr(V)$ is the probability that the alarm is valid [i.e., the valid alarm rate and $Pr(\bar{V})$ the false alarm rate]. Now let T be the time to the earthquake and suppose that we have already been waiting for some time t during which the earthquake has not occurred. We know that $T \geqslant t$. What we want to calculate is the probability that the alarm is valid given this information [i.e., $Pr(V \mid T \geqslant t)$]. According to Bayes' theorem,

$$Pr(V \mid T \geqslant t) = \frac{Pr(V)\ Pr(T \geqslant t \mid V)}{Pr(V)\ Pr(T \geqslant t \mid V) + Pr(\bar{V})\ Pr(T \geqslant t \mid \bar{V})}$$

We can calculate all the quantities on the right-hand side of this equation. $Pr(T \geqslant t \mid V)$, the probability that no earthquake occurs before time t given that the alarm is valid, is just the area under that part of the probability density curve which lies to the right of t (Figure 3). On the other hand, $Pr(T \geqslant t\ \bar{V})$, the probability that no earthquake occurs before time t given that the alarm is false, is clearly 1 no matter how big t is.

The way that $Pr(VT \geqslant t)$ changes as a function of t when the swarm magnitude is 4.5 has been plotted in Figure 4. It can be seen that the probability, which starts out at 0.8, remains fairly steady for about 2 years and then begins to drop fairly sharply and is only about 0.1 by the time 6 years have elapsed.

THE HAZARD DUE TO FAILURES

The impact of the success and failure rates is in affecting what can be forecast in the event that there is no alarm. The fact that the success rate is not 1 (and in practice it never could be) means that we acknowledge that there are some earthquakes that we shall fail to predict. In fact, the failure rate is our estimate of the proportion of earthquakes that we shall fail to predict using the criteria. This means that at any particular time there will be two kinds of earthquake hazard that we need to be concerned about—a hazard associated with the current alarms—which affects only those areas in which recursors have been observed, and another hazard associated with the possible occurrence of failures— which affects all areas regardless of whether or not a precursor has been observed. Failures will occur randomly but at a constant rate. That is the *hazard rate* for failures, expressed as the number of earthquakes per year exceeding a given magnitude, is always the same, being simply the failure rate multiplied by the long-term historical average rate of occurrence of earthquakes exceeding the given magnitude in the area concerned. The lower the failure rate is, the lower the hazard rate will be in absence of an alarm. To put it another way, a low failure rate means that the absence of alarms implies the absence of earthquakes. A high failure rate means that the absence of alarms implies very little.

IS THE HAZARD RATE UNUSUAL?

The estimation of average historical rates of occurrence of earthquakes is itself a fascinating and rather complex exercise which is of great importance for earthquake-resistant design of engineering structures. For earthquake forecasting it is important because it gives us something to compare our prediction with.

An analogy with weather forecasting is perhaps instructive here. People have often tried to test the skill of a weather forecasting service by

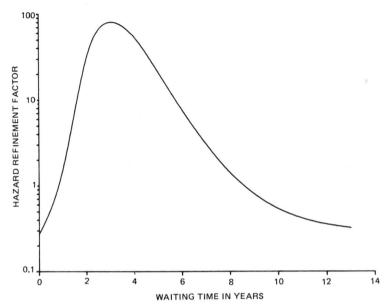

FIGURE 5 Variation of the hazard refinement factor for earthquakes of magnitude 6.5 inside an alarm area in which a precursory swarm of magnitude 4.5 has been observed. The hazard refinement factor expresses the hazard rate relative to the mean historical hazard rate.

comparing the forecasts with those obtained by some simple "low skill" method. For example, one simple method would be to forecast that tomorrow's weather will be the same as today's. Another would be to forecast that the maximum temperature tomorrow will be the same as the average maximum temperature for tomorrow's date over the past 20 years. A good forecaster should be able to improve on these methods. The idea behind such tests is that the forecaster can only demonstrate his skill by predicting some unusual feature of the weather—the sudden change, or the day that is much hotter, or wetter, or windier than usual.

So it is that in earthquake forecasting we can demonstrate skill only by predicting something different from the long-term historical average hazard rate. Since we are interested in testing the value of the forecasting method, it would be useful to express the forecast in a form that highlights any differences from the historical average rate. This is what we have done in Figure 5. The *hazard refinement factor* is just the ratio of the hazard rate under the forecasting method to the average historical hazard rate. Figure 5 refers to an alarm area in which a precursory swarm of magnitude 4.5 has been observed and describes the hazard for an earthquake of magnitude 6.5. The alarm area has been assumed to occupy some 2000 km² of the main seismic region of New Zealand, a region of about 200,000 km². This entire region, which can be considered for this purpose to have uniform seismicity, has an average historical hazard rate of one event in $3\frac{1}{2}$ years for shallow earthquakes of magnitude 6.5 and above. The historical hazard rate in the alarm area is therefore one event in 350 years for earthquakes of magnitude 6.5 and above.

Several factors are interacting to produce the shape of the curve in Figure 5. First, note that the hazard refinement factor is at no time less than the failure rate (0.27). The constant contribution of the possibility of failures to the hazard rate ensures this. The component of the curve above 0.27 is derived from the alarm and is constantly changing. For each waiting time it is adjusted for the fact that the predicted earthquake has not occurred up to that time. The shape of the curve is then determined by the probability density fuction for the time of occurrence (Figure 3), the increasing expected magnitude as the waiting time increases (Figure 2) and the decreasing probability that the alarm is valid (Figure 4). Because the expected magnitude increases with time, the curves for different magnitudes will peak at different times. For instance, the magnitude 7.0 curve would reach its peak later than the magnitude 6.5 curve that we have plotted here. The full statement of the forecast would involve giving the curves for all possible magnitudes in all alarm areas.

ASSESSING THE PERFORMANCE OF THE METHOD

Figure 5 shows that for earthquakes of magnitude 6.5, the hazard rate under the forecast is more than 10 times the historical average rate for a period of about 4 years and reaches a peak of 80 times the historical rate about 3 years after the swarm. If a magnitude 6.5 earthquake were to occur between 2 and 6 years after the swarm, it would represent strong evidence that this method of forecasting is useful. An earthquake occurring 7 years after the swarm would be weaker evidence in favor of the method, since the hazard refinement factor is then only about $2\frac{1}{2}$. An earthquake occurring 9 or more years after the swarm would be evidence *against* the method, because the hazard refinement factor is then less than 1. Any earthquake occurring outside an alarm area would similarly count against the method.

It is not only the occurrence of earthquakes that can be used to demonstrate the worth, or otherwise, of the method. The nonoccurrence of earthquakes can also be used. In an alarm area, the nonoccurrence of a predicted earthquake up to some time t after the precursor will count against the method to the extent that the probabilty of this event is less under the prediction method than under historical estimates of hazard. Similarly, the nonoccurrence of earthquakes outside alarm areas will be evidence in favor of the method, since such events have a higher probability of occurring under the forecasting method than under the historical estimates of hazard. It is quite possible that the nonoccurrence of earthquakes could be the decisive factor in the evaluation of the method. One scenario is that a set of criteria lead to many alarms, all of which turn out to be false. In that case, one would eventually come to the conclusion that the method was worthless, even if no failures occurred.

So whatever earthquakes occur or fail to occur after we begin forecasting, it is possible to evaluate the performance of the method by comparing the probability of all those events under the forecasting method to the corresponding probability under historical estimates of earthquake hazard. The ratio of these two probabilities is a quantitative measure of the value of the method for forecasting. This is how the swarm hypothesis will be evaluated.

At the time of writing it is not known whether the method of precursory swarms is useful in forecasting. Only time and observation of future earthquake occurrences and nonoccurrences will provide the answer.

REFERENCE

Evison, F. F. (1977). The precursory earthquake swarm. *Phys. Earth Planet. Inter.* *15*. 19–23.

24

Genetic Evaluation of Bulls for Milk Production

Robert D. Anderson
Massey University
Palmerston North, New Zealand

THE PROBLEM

Artificial insemination of cattle provides animal geneticists with a convenient and powerful means for increasing the genetic average of a national dairy herd for milk production. Using the technique, an extraordinarily large number of cows can be mated with just a few genetically outstanding bulls. For example, in the 1981–1982 dairying season in New Zealand, 54% of the nation's dairy herd of about 2 million cows was artificially mated, in essentially an 8-to 12-week period, using semen from 38 highly selected bulls, the most extensively used bull having provided 158,000 inseminations. Given that, on average, 51% of calves born are male, but that, through artificial insemination, just a few bulls are required for mating, a dairy cattle geneticist can, of course, be extremely selective in deciding which bulls finally enter the proven bull team for subsequent widespread usage. The need for procedures enabling accurate determination of genetically superior bulls is clearly apparent.

Bulls normally do not produce milk; nevertheless, they possess genes which, following transmission via mating and subsequently, fertilization, partly determine the genetic potential of a daughter for milk production. Given the state of current technology, the evaluation of the genetic merit of a bull for milk production must be based on the lactation records of his female relatives. The statistical issues involved in the

evaluation include the proper collection and analysis of appropriate lactation records in a way that maximizes the accuracy for identifying genetically elite bulls in the population. In general terms, the genetic gain from selection is determined by the selection pressure (number of selected animals as a proportion of the total available) and the accuracy of selection (choosing animals which are, in fact, genetically superior).

DATA COLLECTION

Throughout the world, many schemes have been devised for evaluating dairy bulls, but invariably progeny testing, that is, the analysis of lactation records from daughters that are representative of the bull, has a central role. Conceptually, the idea of utilizing the average performance of progeny in judging the relative merit of bulls is straightforward, but in practice its successful application demands that certain statistical principles be adhered to.

Milk secretion is the culmination of innumerable biochemical and physiological events within an animal, these events being influenced by genetic as well as nongenetic factors (feeding, management, climate, etc.).

The currently accepted genetic model for explaining the mechanism of inheritance of a quantitative trait, such as milk production, is that a very large number of genes, of varying desirability but each with small effect, collectively control the trait. Given the large number of genes involved, together with the modifying or obscuring influence of environmental factors, the effects of individual genes for milk production cannot be distinguished. Moreover, it is not possible to determine which genes the progeny actually have.

The cows to which bulls are mated have varying genetic ability and accordingly, the pattern of mating bulls and cows needs careful thought if the genetic average of resultant daughters is to reflect accurately the genetic merit of the bull concerned. Nonrandom mating of bulls and cows will bias the progeny test results and lead to an erroneous ranking of the bulls of interest. In short, the principle of randomization must be adopted, both in connection with the cows to which bulls are mated and the standard of husbandry accorded to the resultant daughters.

How can a progeny testing scheme be organized in the context of a commercial dairy cattle population such that the genetical and statistical principles already mentioned are given due consideration? One strategy is to establish a tester population specifically for bull-proving purposes. Within such a scheme, approximately 330 inseminations from each young (yearling) unproven bull to be tested are collected and randomly allocated among as many of the participating herds as possible. The bulls are then put aside for 3 years until the daughters of these matings have completed their first lactations. Farmers participating in the scheme are paid by the organizing agency not only for using semen from unproven bulls, but also for recording the production of all resultant daughters that come into milk. Of course, the need for uniform husbandry of the daughters has to be impressed upon those farmers. Taking account of factors such as (1) not every insemination generates a successful conception, (2) abortion, (3) approximately 49% of progeny born are females, and (4) progeny deaths and other losses up to the point at which they themselves calve and complete a first lactation; the 330 inseminations from each young bull lead to first lactation records of approximately 50 daughters 4 years later. An alternative approach would be to dispatch semen from unproven bulls randomly throughout all herds in the country as part of the day-to-day use of artificial insemination. Nevertheless, due to the lesser degree of organizational control that results, experience has shown that roughly three times the foregoing number of inseminations would be needed to generate 50 tested daughters for each young bull. Any increase in the number of inseminations from unproven bulls to generate the required number of tested daughters leads to fewer bulls being progeny tested, but with the size of the proven bull team remaining the same, the resultant selection pressure would be diminished. Moreover, the volume of semen used from unproven bulls, some of whom are genetically inferior, is also increased. Overall, the latter approach results in a lower rate of genetic gain in the population. It should be noted that not all farmers using artificial insemination also herd test their cows.

Why is it necessary to obtain 50 daughter records per young bull? A bull contributes a random sample half of its genes to each daughter, the samplings being independent; the remaining genes, of course, are contributed by the cows to which he is mated. Since the sampling of the bull's genes are independent, two progeny of the same bull, but unrelated cows, will, on average, have $\frac{1}{4}$ of their genes in common. If a bull is mated to three unrelated cows and is carrying one copy of a particularly desirable gene, the probablitity that all three progeny will receive that gene is $\frac{1}{2} \times \frac{1}{2} \times \frac{1}{2} = \frac{1}{8}$, and this argument could be extended if more cows are involved. To minimize the opportunity for sampling effects to result in a group of daughters whose genetic average does not accurately represent the genetic merit of a bull, it is important that as large a number of daughters as possible be achieved. Since the reliability of predicted genetic merit of a bull under progeny test increases only very slowly beyond 50 daughters, that number is adjudged sufficient in the first instance. Further lactation records accumulate should the bull finally enter routine use.

DATA ANALYSIS

Having established the point that the genetic evaluation of a bull for milk production involves the use of lactation records from female relatives of various kinds, it is now of interest to address the mathematical analysis of the recorded data. Many current animal recording schemes are based on the selection index approach. Although out of date in one important respect, it will form the basis of this discussion, as it is relatively simple and straightforward.

Suppose that lactation records from a variety of female relatives of a bull are available; for example:

x = average of the first-lactation records of n daughters of the bull

y = average of k lactation records of the bull's dam

z = average of the first-lactation records of m daughters of the bull's sire (i.e., paternal half-sisters of the bull)

Actually x, and y, and z are deviations from the true mean production for the whole population which may involve the use of adjustments for nongenetic sources of variation.

How is this information utilized to predict the unkown genetic merit, g, of the bull for milk production? The selection index approach leads to the following linear prediction equation:

$$P = ax + by + cz$$

where P is the bull's predicted merit and a, b, c are the appropriate weighting factors to be determined. How are the weighting factors determined? A logical requirement is that the average prediction error, that is, the average of the difference $P - g$, be a minimum, but this leads nowhere mathematically. An alternative principle is that the average of $(P - g)^2$ be minimized, that is, the weighting factors are such that the variance of prediction errors is a minimum. (Note that as P resembles g more closely, on average, the variance of prediction errors tends to zero.) The net result is a set of simultanious equations, akin to, but nontheless different from, multiple regression equations, the details of which are unimportant here.

To simplify the discussion, attention will be confined to the case where a bull is to be evaluated solely on the basis of the average first-lactation records of his daughters in a progeny test. Using g and $\bar{x}$ to denote the unknown genetic merit of the bull and the average of the first lactations of two daughters of the bull, respectively, it is particularly informative to use a path diagram to illustrate the key elements in the biological system through which g and $\bar{x}$ are connected. The diagram could easily be extended to include more daughters.

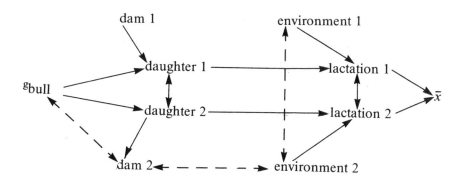

A double-headed arrow signifies the existence of a relationship between the variables involved. The genetic structure is detailed on the left-hand side of the diagram and on the right-hand side, the role of nongenetic effects in influencing milk production is recognized. Arrows of the form $\leftarrow \ldots \rightarrow$ designate a "possible but often avoidable" correlation. For instance, if the bull is related to a dam to which he is mated, there will be a correlation between their genetic values. If the progeny of the bull are managed preferentially in relation to the progeny of some other bull, a correlation between the progeny environments will result. If superior husbandry is accorded the daughters of particular dams, correlations designated by dam $\leftarrow \ldots \rightarrow$ environment will exist. Should the correlations represented by the dashed arrows be nonzero, the accuracy of the progeny test will be suboptimal. The progeny testing scheme described earlier in this article is designed to minimize the opportunity for such correlations to exist.

The selection index in this case is

$$P = b\bar{x}$$

where it can be shown that the weighting factor b is a function of (1) the number of daughters of the bull, (2) the number of other animals whose average performance was compared with the daughters' records, (3) the degree to which production differences between animals are genetically based (i.e., the heritability of the trait), and (4) the existence or otherwise of the correlations designated by $\leftarrow \ldots \rightarrow$. A number of studies have shown that approximately 25% of the variability between cows in milk production can be attributed to varying genetic merit, that is, the heritability of the trait is 0.25. In the most simplistic situation, the weighting factor, b, in a progeny test for milk production can be shown to be $2n/(n + 15)$, where n is the number of daughters in the test. The corresponding reliability (the correlation between P and g) of the progeny test in that situation is

$$\text{Reliability} = \sqrt{\frac{n}{n + 15}}$$

(This formula for quantifying the reliablity of a progeny test for milk production assumes that none of the "avoidable" correlations exists.)

Numerical Example

First-lactation records (i.e., total weight of milk produced) of daughters of three bulls were generated under the conditions of an "ideal" progeny test. The average deviations from the population mean and number of daughters for each bull were:

Bull	Average deviation (kg)	Number of daughters
1	+20	45
2	+20	60
3	+60	5

Assuming the heritability of milk production to be 0.25, the selection index values, with accompanying reliability, for each bull are:

Bull	Index (kg)	Reliability
1	+30[a]	0.87[a]
2	+32	0.89
3	+30	0.5

[a]For bull 1, $b\bar{x} = 2n/(n + 15)\bar{x} = [2(45)/(45 + 15)](+20) = +30$; with reliability

$$\sqrt{\frac{n}{n + 15}} = \sqrt{\frac{45}{45 + 15}} = 0.87$$

How are the index values interpreted? If, for instance, bull 2 was mated to a very large random sample of cows in the same population, the daughters from those matings would be expected to yield, on average, $\frac{1}{2}(32) = 16$ kg more milk in their first lactations than the daughters of a genetically average bull mated to the same sample of cows. On the basis of both the index and reliability, it would be advisable for a farmer to use bull 2. If, for some reason, the farmer did not want to use bull 2, the reliability figures would point to bull 1 as being the next best choice (as the reliability is higher than for bull 3).

DETERMINING HERITABILITY

Having made use of a heritability value of 0.25 in the example, an indication of one method for determing that quantity will now be given. If production differences between animals are not partially heritable, no gain will accrue from applying selection; hence the heritability of a trait needs to be determined at the outset.

Suppose that the conditions for the "ideal" progeny test prevail, that is, sires are randomly mated to dams, progeny are randomly treated with respect to husbandry, and so on. Designating the variability of production of all daughters across the complete set of sire groups as var(x), it is clear that if there are genetic differences between sires, the variability of these differences will contribute to var(x). Symbolically,

$$\text{var}(x) = \text{var}(s) + \text{var}(\varepsilon)$$

where var(s) represents the between-sire variability and var(ε) is the variability in production which is due to all other causes. Using appropriate statistical techniques, it is possible to partition the overall variability into the components mentioned.

For the case being considered, let the average genetic contribution of the ith sire to his group of daughters be designated s_i. The variation between the s_i is, of course, measured by var(s) and this variation contributes to differences in production between daughters in different sire groups. However, within a sire group, s_i is responsible for the likeness between the daughter records of that group. Thus, remembering that the covariance is a measure of the degree to which variables vary together, it is a simple matter to use basic variance-covariance theory to show that var(s) is also the covariance between the records of daughters sired by the same sire. However, under the ideal setting for a progeny test, the sole reason for daughter records tending to resemble one another is the inheritance, and subsequent expression, of a random sample half of the genes of a common sire. Thus var(s) represents $\frac{1}{4}$ of the genetic variability of the trait. Accordingly, determining var(s) leads to an estimate of genetic variability, which, in turn, is used in finding the heritability of a trait. (It stands to reason that if members of the same family do not resemble one another to a greater degree than do individuals chosen at random from the whole population, there will be no genetic basis to the observed variation.) By the earlier definition of heritability, it follows that the quantity is estimated, in this case, by taking 4[var(s)/var(x)].

CONCLUSION

This discussion has been a brief and somewhat superficial overview of the application of statistical methods in one aspect of the dairy cattle industry. Data recorded under field conditions are usually, from a statistical analysis standpoint, less than ideal and the need to analyze such data has meant that animal geneticists have often been at the forefront of the development of statistical techniques capable of generating meaningful information from field data. Furthermore, the sheer volume of data involved in typical analyses require the application of sophisticated computing and data processing strategies. The herd improvement movement employs many people: herd recorders, data entry and computing personnel, farm consultants, artificial insemination technicians, and so on. At the heart of the activity of all these people is the collection and analysis of data according to basic principles of statistics.

VII
Modeling

The next six chapters share the common theme of statistical modeling. We are all familiar with scale models of trains and airplanes, and we are often surprised at how closely the models resemble the real thing, particularly if they are working models. Statistical models, however, aim to strip away unnecessary details to capture the essence of the process and can be though of as having a similar relationship to the real thing that a line drawing has to a photograph. In situations where the real world seems very complex, a statistical model can be developed that involves much fewer variables than real life. At first sight, these may not seem as sophisticated or as interesting as scale models, but they can, in fact, be useful in different ways. Indeed, the very simplicity of the model often allows insights into general patterns exhibited by the data.

Bryan Manly, in his chapter on the survival of the fittest, takes us into the arena of evolution, which has fascinated the public since the views of Charles Darwin were widely aired at the end of the nineteenth century. The model may be set up, as in this chapter, to aid in the understanding of the process of evolution. To various degrees, the other chapters illustrate this same point—that even though the models concentrate on a few of the many variables involved, they give a valid idea of the underlying forces at work. Indeed, this is one of the main reasons for searching for a suitable model.

The chapter by David Dickey on the stock market illustrates quite well how one can go about testing different models. In this case, the general form of the model is kept the same, but possible values of an important parameter are tested to see if they are compatible with the data.

Chapter 27 by Brent Wheeler on the spread of diseases and the growth of cities employs an interesting procedure called simulation, in which the assumptions of the model are used to generate hypothetical outcomes that can throw light on the implications of these assumptions. The speed of computation of modern computers enables such large simulation studies to be carried out to check that the model is a valid one.

Tourism is big business these days, especially for a small country like New Zealand which is rich in spectacular scenery. Peter Thomson is motivated by the tourist potential of the country to analyze the pattern of visitors arriving by air. Graphical displays are perhaps more important in the area of time series than in any other, and Dr. Thomson uses graphs to great effect in this chapter. The power of the computer is also seen in its ability to transform variables and to draw graphs very quickly.

Natural hazards such as floods are very elusive and difficult phenomena to model, for they occur so infrequently that they can be forgotten for long periods of time. Often when they are least expected, they suface with terrifying power and devastation. Richard Heerdegen explains some of the ways that have been employed to try to model such hazards.

Once a statistical model has been devised and tested, it can be used to predict future events. We noted this in the introduction to Part VI, but this aspect does not appear as important for the chapters in this part although the material on predicting the behavior of the stock-market, visitor arrivals, and natural hazards develops models with the ultimate goal of prediction.

Underlying the modeling process is a mathematical basis that is most evident in the Paul Van Moeseke article on the instability of political power. Beginning with a realistic model, he sets out to prove that power is unstable! His article is somewhat different from others in this book, as he uses a theoretical framework of probability to reach a conclusion, whereas other articles consider a set of data.

25

The Survival of the Fittest

Bryan F. J. Manly
University of Otago
Dunedin, New Zealand

NATURAL SELECTION

It is well known that in 1859 when Charles Darwin published his book *On the Origin of Species by Means of Natural Selection, or the Preservation of Favoured Races in the Struggle for Life*, the immediate result was to start a considerable controversy. It was regarded by many as being an attack on the foundations of Western culture.

Scientists generally were convinced by the mass of circumstantial evidence provided by Darwin to justify his thesis that evolution was largely due to natural selection operating on the differences between individuals in populations. However, several of Darwin's supporters did not accept his idea that selection could operate on quite small differences. They felt, rather, that the main targets for selection were the occasional very unusual "sports" that were observed to occur.

A major problem was the lack of data relevant to the question of how natural selection works. Thus Hermon Bumpus writing in 1898 had this to say:

> We are so in the habit of referring carelessly to the process of natural selection, and of invoking its aid whenever some pet theory seems a little feeble, that we forget we are really using a hypothesis that still remains unproved, and that specific examples of the destruction of animals of known physical disability are very infrequent.

331

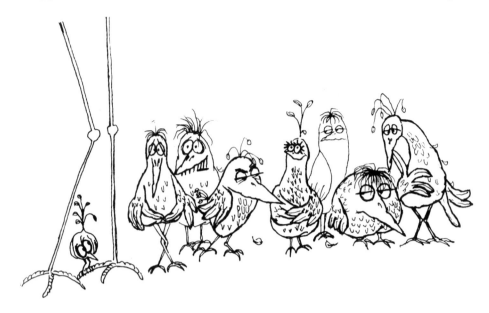

He went on to describe an excellent test of Darwin's theory. As a result of an unusually severe storm in February 1898, 136 exhausted English sparrows were picked up and taken to Bumpus's laboratory. Subsequently, 72 of the birds revived and 64 died. Bumpus weighed the birds and took eight measurements on their skeletons. He then considered the interesting question of whether there were any differences between the survivors and nonsurvivors with respect to these variables. This then amounts to seeing whether birds with certain measurements were able to survive the stress of the storm better than birds in general.

Bumpus reached the conclusion that there were indeed differences between the survivors and nonsurvivors, principally related to sex, total length, weight, the length of the humerus, and the length of the femur. Particularly, he concluded that "average" individuals survived better than those with very small or very large measurements. Bumpus included all of his data as an appendix to his published lecture. These data have been reanalyzed many times since 1898 using more sophisticated statistical methods than were available to Bumpus. However, his conclusions have been verified in general.

Bumpus was not alone in his search for evidence of natural selection taking place. In fact, a new statistical journal, *Biometrika*, was started in 1901 with a primary purpose being to publish the results of investigations in this area. These studies have continued to the present day. In practice it

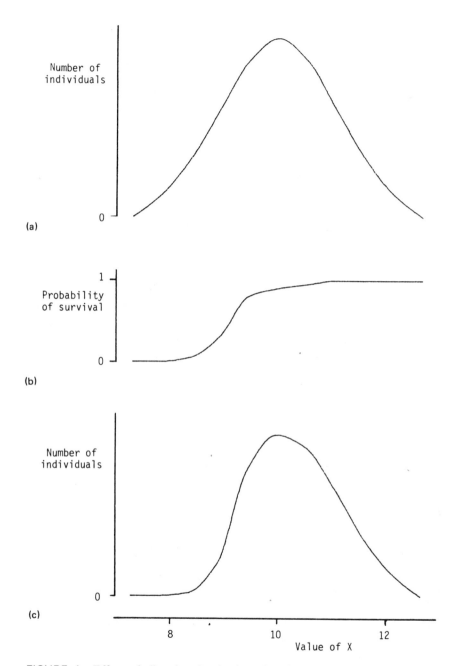

FIGURE 1　Effect of directional selection that favors individuals with large values for X. (a) Distribution of different values of X before selection. (b) A directional fitness function. (c) Distribution of different values of X after selection.

333

has proved to be rather difficult to obtain valid data because of the difficulties involved in observing natural populations. Therefore, the number of good sets of data available to date is still not large.

FITNESS FUNCTIONS

There is a certain amount of difficulty in defining exactly what "fitness" means. A reasonable definition for some purposes is that the fitness of an individual is equal to the number of progeny that it produces. However, it is seldom possible to determine progeny numbers in natural populations, so biologists have to be content with determining just components of fitness. Thus it is common to refer to the probability of an animal surviving a certain period of time as its "fitness," recognizing that this is a somewhat loose use of the term. A fitness function is then a mathematical function which purports to show the relationship between the probability of survival and one or more measured characters.

To be more precise, suppose that the individuals in a population all have values for a certain variable X. This could, for example, be the total body length. The assumption is made that the probability of an individual surviving a certain period of time depends on the value that it has for X, so that a fitness function

$$w(x) = Pr(\text{survival, given } X = x)$$

exists. For the present the question of what exactly is the mathematical form of this function can be left aside. The important point is the assumption of an exact relationship between the survival probability and X.

It is interesting at this point to consider what the likely effects will be for different types of selection operating on a population. Figure 1 illustrates this for *directional selection* where individuals with large values for X are fitter than individuals with small values. Before selection the distribution of X in the population is bell-shaped, with values around 10 being most common. Selection mainly removes individuals with small values of X so that the left-hand "tail" of the distribution is lost.

Figure 2 illustrates the situation with *stabilizing selection*. Here individuals with moderate values for X are fitter than those with either very small or very large values. The effect of selection is then to remove both "tails" of the distribution. After selection the distribution of X is less variable than it was before. Stabilizing selection is thought to be fairly common in nature.

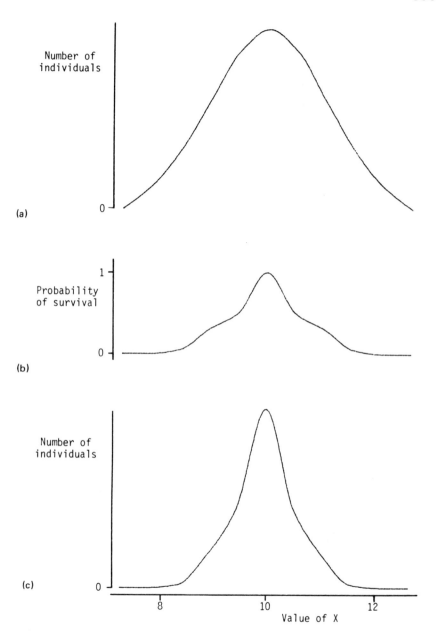

FIGURE 2 Effect of stabilizing selection that favors individuals with values of X around the population average of 10. (a) Distribution of different values of X before selection. (b) A stabilizing fitness function. (c) Distribution of different values of X after selection.

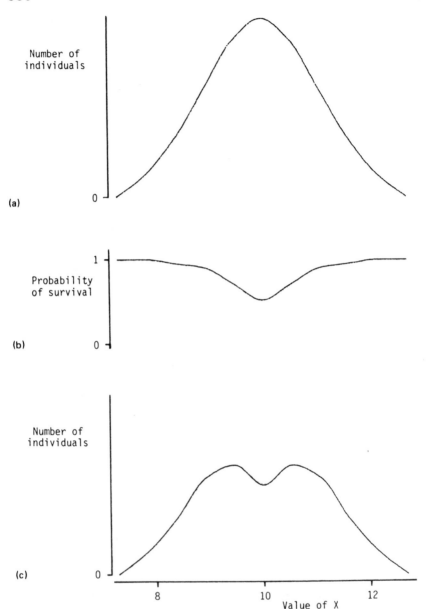

FIGURE 3 Effect of disruptive selection that acts against individuals with average values of X. (a) Distribution of different values of X before selection. (b) A disruptive fitness function. (c) Distribution of different values of X after selection.

Disruptive selection is the opposite of stabilizing selection and is illustrated in Figure 3. Individuals with moderate values for X are less fit than those with small or large values. The effect of selection is then to remove the center of the distribution and increase the variation in the population. In the case shown this results in the distribution becoming bimodal after selection. It is thought that disruptive selection is extremely rare in nature, if it occurs at all.

The idea of directional, stabilizing, or disruptive selection can easily be generalized if there is more than one X variable involved. With two variables it is even still possible to represent a fitness function graphically. For example, Figure 4 shows a fitness surface that has been estimated from Bumpus's data on the survival of female sparrows for the 1898 storm. The height of the surface for any combination of total length and humerus length gives the probability of survival. The surface shows a rising ridge, with directional selection favoring small values of the total length and stabilizing selection favoring middle values of the humerus length. A high intensity of selection is indicated, wtih the probability of survival ranging over almost the full possible range from 0 to 1.

SOME DATA FOR DETERMINING FITNESS FUNCTIONS

It is interesting to look at some specific examples of data relevant to the problem of estimating fitness functions. The first concerns the size of snails and their ability to survive without food. To investigate this, a sample of the species *cepaea nemoralis* was collected and then left in an unheated room in Manchester, England, from the end of September 1968 until the middle of June 1969. Over this period, 80% of the snails died. Table 1 shows the initial distribution of the maximum shell diameter and also the distribution for survivors(Cook and O'Donald, 1971). The highest survival (28%) was for snails with a shell diameter of 26mm, which is somewhat higher than the population mean. There appears in this case to be something of a mixture between directional selection (Figure 1) and stabilizing selection (Figure 2).

Of course, it can be argued that this is not really an example of natural selection because the snails were placed in an unnatural environment. However, the hope of the experimenters was that their results are indicative of what happens naturally.

Bumpus's sparrow data have already been referred to. It may be recalled that in this case there are eight morphological measurements and weight recorded for each bird. If weights are ignored on the grounds that these were likely to be very dependent on the temporary physical condition of the birds, this still leaves eight variables to which survival

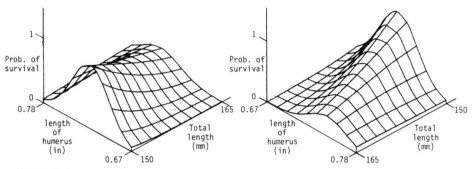

FIGURE 4   Example of a two-variable fitness surface. Both graphs are of the same surface, but viewed from different angles.

may be related. There are ways of handling (or attempting to handle) eight variables at a time. However, for a second example, attention will be restricted to the two variables shown in Figure 4, the total length and the length of the humerus (a wing bone), for female birds only. Figure 5 shows a plot of the data, with survivors and nonsurvivors indicated. For some reason Bumpus measured total length in millimeters and the length of the humerus in inches. This creates no problem since the units of measurement are irrelevant to the question of whether or not selection has taken place.

At first sight Figure 5 does not indicate much difference in the distributions of survivors and nonsurvivors. Notice, however, that there is low survival for individuals on the outside edge of the swarm of points except when the total length is low and the length of the humerus is about average. It is this effect that has caused the fitness surface of Figure 4 to take the particular form that it has.

For a final example, some long-term population changes can be considered. At about the start of this century archaeologists put a good deal of effort into digging up skulls in ancient cemeteries and measuring various dimensions on them. Much of the collecting was done in the area of Thebes in Egypt, so that data are available from this location on large numbers of skulls, dating from around 4500 B.C. to around A.D. 300 (Thomson and Randall-Maciver, 1905). For the present purpose, atten-tion will be restricted to a sample of 89 male skulls from the late pre-dynastic period (circa 3600–3100 B.C.) and another sample of 123 male skulls from the Ptolemaic and Roman periods (circa 330 B.C.– A.D. 320). A number of measurements are available for these sample skulls. Only the maximum breadth and the basibregmatic height will be considered

TABLE 1   Over-Winter Survival of *Cepaea nemoralis*

Maximum shell diameter (mm)	Initial number on 9/29/68	Survivors on 6/14/69	Percent survival
21	21	1	5
22	93	6	6
23	255	44	17
24	343	70	20
25	289	62	21
26	128	36	28
27	29	7	24
	1158	226	20

here, the latter being (more or less) the height of the skull when it is placed on a table.

This example differs from the other two in one rather important respect. In the other examples there has been an initial population of individuals from which some have been removed, possibly in a selective manner. Now there is a first sample from the population of Egyptian skulls (whatever that might mean), and another sample taken from the same population 3000 years later. In the intervening period the population will have been affected by innumerable factors influencing reproduction, survival, and migration. It is still possible to examine changes in the population. Unfortunately, however, it is not possible to be sure what caused the changes. It can, in fact, be argued that the most likely explanation for any consistent changes in skull dimensions is that there was a continuing flow of migrants into Egypt, particularly from the south. If this is true, the population changes have nothing at all to do with different types of skull having different fitnesses.

Subject to this reservation, it is possible to estimate a "fitness" function for two variables, maximum breadth and basibregmatic height. This can be thought of as measuring the amount of selection, assuming that selection was responsible for the population changes.

There is yet another problem with this example that does not occur with the other ones. Evolutionary theory says that over many generations all populations will gradually change because of random genetic drift. Thus a change in the mean of a skull dimension over a period of 3000 years, which is about 120 generations, could easily be due to chance effects rather than selection or migration. The magnitude of genetic drift cannot be calculated in any easy way. There is therefore the possibility

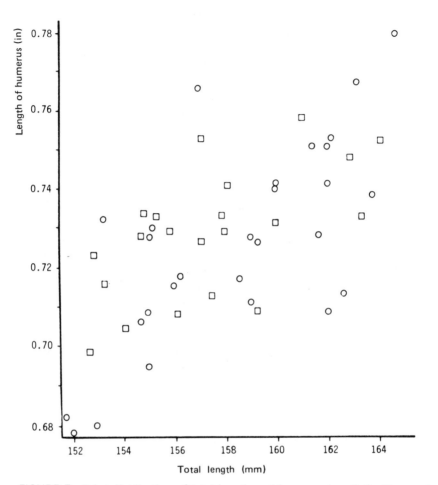

FIGURE 5   Joint distribution of total length and humerus length for Bumpus's female sparrows. □, survivors. O, nonsurvivors.

TABLE 2 Distribution of Maximum Breadth and Basibregmatic Height for Two Samples of Male Egyptian Skulls[a]

Maximum breadth (mm)	Basibregmatic height (mm)						Totals
	120–123	125–129	130–134	135–139	140–144	145–149	
120–124	0 / 0	0 / 0	3 / 0	1 / 0	0 / 0	0 / 0	4 / 0
125–129	1 / 0	4 / 7	5 / 4	1 / 3	1 / 0	0 / 0	12 / 14
130–134	1 / 3	6 / 14	12 / 12	11 / 7	1 / 2	1 / 0	32 / 38
135–139	0 / 1	4 / 7	11 / 20	14 / 12	0 / 1	0 / 0	29 / 41
140–144	0 / 3	1 / 6	2 / 6	5 / 9	2 / 0	0 / 0	10 / 24
145–149	1 / 0	1 / 4	0 / 2	0 / 0	0 / 0	0 / 0	2 / 6
Totals	3 / 7	16 / 38	33 / 44	32 / 31	4 / 3	1 / 0	89 / 123

[a]In each cell of the table the top value is for the first sample (pre-dynastic) and the bottom value is for the second sample (Ptolemaic and Roman).

that any apparent differences between samples taken from a population many generations apart have no cause at all. However, in the present case several samples of skulls of different ages between the pre-dynastic and Roman periods all indicate a tendency for the basibregmatic height to get smaller and the maximum breadth to get bigger with time. The effect of genetic drift is random changes in each generation, so that this can be ruled out as the explanation for any consistent long-term trends. Thus it does seem that the changes in basibregmatic height and maximum breadth must have been due to something else.

Table 2 shows the bivariate distribution of basibregmatic height and maximum breadth for the two samples being considered. It is not particularly easy to pick out differences between the two samples from a table like this. However, some idea can be gained by recognizing that if both samples came from the same distribution, the ratio of the first sample frequency to the second sample frequency should be about 89/123 for all the cells in the table. This should also be true for row and column totals. Viewed from this point of view the first sample seems to have too few skulls with small basibregmatic heights and large maximum breadths. It seems that as time progressed the mean basibregmatic height got smaller and the mean maximum breadth got bigger. As mentioned above, this general trend is confirmed by intermediate samples.

THE ESTIMATION OF FITNESS FUNCTIONS

In the examples that have been considered it seems that the fitness of individuals is related to the values that they possess for the variables measured. It is therefore worth considering how the relationship can be determined more precisely. In other words, how can fitness functions be estimated?

Strictly speaking, it would be appropriate to ask first whether the apparent relationships between fitness and the variables are genuine or whether they can be explained easily enough by chance happenings. For example, suppose that the probability of surviving 9 months without food is the same for all *Cepaea nemoralis* snails and, in particular, has nothing to do with size. It is interesting to calculate how likely it is that a set of data such as that shown in Table 1 will occur under these conditions. If it is quite likely, then clearly there is no need to introduce the idea of selection and no point in estimating fitness function.

The question of testing the statistical significance of selective effects is not particularly straightforward. Here it suffices to say that in all the examples mentioned, the apparent selection is strong enough so that it is unlikely to be merely due to chance. Hence the estimation of fitness functions becomes a sensible problem.

There are a number of different approaches that have been suggested for determining fitness functions. The statistically "best" methods need the use of an electronic computer to carry out rather lengthy calculations. However, a simple approximate method is available in cases where there is only one X variable and the distribution of this is roughly bell-shaped before and after selection (O'Donald, 1971). That is, before and after selection the distribution of X should have the shape shown in part (a) of Figures 1 to 3. More precisely, X should approximately at least follow what is called a normal distribution.

Apart from making assumptions about the shape of the distribution of X, it is also necessary to make an assumption about the form of the fitness function. Here the somewhat arbitrary form

$$w(x) = Pr(\text{survival, given } X = x)$$
$$= W \exp[-K(\theta - x)^2] \tag{1}$$

will be assumed, where the constant W has to be positive. If K is positive, then individuals with $X = \theta$ have the maximum possible survival probability of W. Hence if θ is large compared to the values of X in a population, equation (1) will provide an approximation for a directional fitness function such as is shown in Figure 1b. The individuals with the largest X values in the population will have highest survival. If K is positive and θ is about equal to the average value of X in a population, then equation (1) will approximate a stabilizing fitness function such as that shown in Figure 2b. Individuals with average values of X will have highest survival. On the other hand, if K is negative, individuals with $X = \theta$ will have the lowest possible survival probability of W. Hence if θ is about equal to the average value of X in a population, equation (1) will approximate a disruptive fitness function, such as that shown in Figure 3b. In general, the equation gives a flexible way of representing most realistic types of fitness function. The estimation of the parameters W, K, and θ is quite straightforward and is discussed in the appendix to this chapter.

When there are two X variables equation (1) can be generalized by making the argument of the exponential function depend on both of them. This makes estimation more complicated (Manly, 1981) but apart from that does not create any new difficulties.

ESTIMATED FUNCTIONS FOR THE EXAMPLES

The calculations for determining a fitness function from the data of Table 1 are given in detail in the appendix. This function has the equation

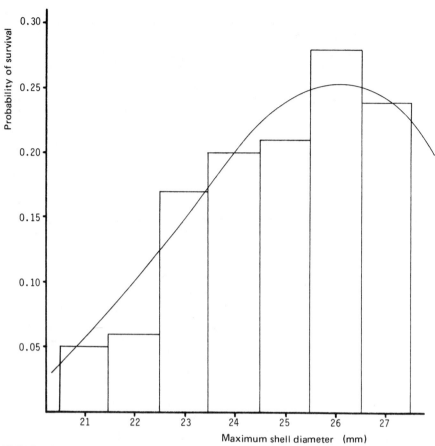

FIGURE 6 Observed proportions surviving for snails with different maximum
shell diameters (vertical bars) and the estimated fitness function (the curve).

$$w = 0.256 \exp\,[-0.0598(26.03 - x)^2]$$

which says that snails with a shell diameter of 26.03 had the maximum
survival probability of 0.256. A plot of the function and the original data
are shown in Figure 6. The function seems to give fair approximations for
true fitness values. Clearly, there was a good deal of selection in this
experiment, with the survival of snails ranging from the maximum of
about 0.26 down to about 0.05.

Using an extension of the approach outlined in the appendix, the
fitness function estimated from Bumpus's female data (Figure 5) is found
to be

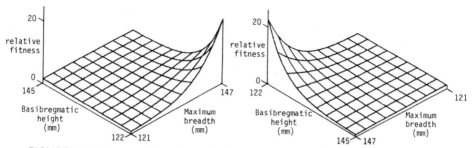

FIGURE 7   Fitness surface for male Egyptian skulls between late pre-dynastic and Roman/Ptolemaic times.

$$w = 0.53 \exp(-0.380z_1 + 0.326z_2 - 0.034z_1^2 + 0.122z_1z_2 - 0.458z_2^2)$$

where

$$z_1 = \frac{\text{total length} - 157.8}{3.62}$$

and

$$z_2 = \frac{\text{humerus length} - 0.727}{0.0219}$$

Here z_1 is the total length coded, so that the mean is zero and the variance is 1 for the sparrows before selection. Similarly, z_2 is the humerus length coded to have a mean of zero and a variance of 1 before selection. (The mean and variance are defined in the appendix.) This fitness function is the one that gives the surface shown in Figure 4. Apparently, sparrows with a moderate humerus length and a small total length had the highest probability of survival. Low probabilities are particularly associated with either very large or very small humerus lengths.

When it comes to the example on Egyptian skulls, a modification of the definition of fitness is required. What a fitness function needs to do in a case like this is to express the fact that between the time of the first sample and the time of the second sample, certain skull types increased in frequency while other types decreased in frequency. Therefore, the fitness function needs to show multiplication factors for frequencies. In other words, the fitness of a skull with maximum breadth x_1 and basibregmatic height x_2 can be defined as

$$w = C \frac{\text{frequency of skulls with these measurements in second period}}{\text{frequency of skulls with these measurements in first period}}$$

where C is any suitable scaling constant. The value of C does not matter since it is only the relative fitness values of different skulls that is of interest. A convenient choice for C is the value that makes w equal to 1 for a skull that has mean values for x_1 and x_2 in the first sample.

On this basis the same method of estimation that was used with Bumpus's data produces the function

$$w = \exp(0.4117z_1 - 0.3910z_2 - 0.0556z_1^2 - 0.1149z_1z_2 + 0.0033z_2^2)$$

where

$$z_1 = \frac{\text{maximum breadth} - 133.7}{5.22}$$

and

$$z_2 = \frac{\text{basibregmatic height} - 133.0}{4.59}$$

are coded values of the maximum breadth and the basibregmatic height. This function gives the fitness surface shown in Figure 7. The changes between the first and second sample times suggest that skulls with large maximum breadths and small basibregmatic heights were fitter than average skulls by a factor of up to 20 or more. There therefore seems to have been strong directional selection.

DISCUSSION

These three examples show that individuals can apparently differ substantially in their Darwinian fitnesses according to the values that they possess for easily measured characters. These and similar cases indicate that natural selection does not operate only on grossly unusual "sports."

It must be mentioned, however, that nonsense relationships can be found for fitness just as they can for any other measurements. A well-known case occurs with sternopleural bristle numbers of the fruit fly *Drosophila melanogaster*. These bristles are small hairs on the sides of the thorax of the adult flies. It has been found that when the flies are reared in crowded conditions (i.e., under strong selection) the distribution of the number of bristles is quite different from what it is for flies reared in uncrowded conditions (i.e., under weak selection). Indeed, it looks remarkably like an example of stabilizing selection, as shown in Figure 2. Flies with moderate bristle numbers seem to survive best when selection is intense. However, the interesting thing is that the bristles do not develop until after the selection has taken place. Therefore, the number of bristles cannot in itself be a *cause* of differential survival. The real cause of different fitnesses must be something else that is partly expressed in bristle numbers.

This raises a very important point. A fitness function is merely a convenient representation of a relationship between fitness and one or more characters. It says nothing about what causes natural selection or why there is a relationship. Interestingly, this does not necessarily affect the usefulness of fitness functions because the biologist is usually just concerned to discover whether any selection at all is going on. In that case the biologist is quite happy to regard any measurement as merely an index of some more fundamental physiological variable that he or she is not able to measure directly.

REFERENCES

Bumpus, H. C. (1898). The elimination of the unfit as illustrated by the introduced sparrow *Passer domesticus*. Biological Lectures from the Marine Biology Laboratory, Woods Hole. Eleventh lecture, pp. 209–226.

Cook, L. M., and P. O'Donald (1971). Shell size and natural selection in *Cepaea nemoralis*. In *Ecological Genetics and Evolution* (R. Creed, ed.). Blackwell Scientific, Oxford.

Manly, B. F. J. (1981). The estimation of a multivariate fitness function from several samples from a population. *Biom. J. 23*, 267–281.

O'Donald, P. (1971). Natural selection for quantitative characters. *Heredity 27*, 137–153.

Thomas, A., and R. Randall-Maciver (1905). *The Ancient Races of the Thebaid*. Oxford University Press, Oxford.

APPENDIX: ESTIMATING A FITNESS FUNCTION

Suppose that a sample consists of values $X_1, X_2, \ldots, X_n$. Then the mean is defined as

$$\bar{X} = \frac{X_1 + X_2 + \cdots + X_n}{n}$$

and is a measure of the "center" of the distribution. The variance can be defined as

$$V = \frac{(X_1 - \bar{X})^2 + (X_2 - \bar{X})^2 + \cdots + (X_n - \bar{X})^2}{n}$$

and this is a measure of the "spread" of the distribution.

Assuming a bell-shaped distribution for X before selection, as in part (a) of Figures 1 to 3, and a fitness function of the form

$$w = W \exp[-K(\theta - x)^2]$$

the parameters K and θ can be estimated by seeing how the mean and variance change by selection. Thus if $\bar{X}$ and V are the mean and variance of the distribution before selection, while $\bar{X}'$ and V' are the values after selection, estimates of K and θ are given by

$$\hat{K} = \frac{V - V'}{2VV'}$$

and

$$\hat{\theta} = \frac{V\mu' - V'\mu}{V - V'}$$

Having estimated these two parameters, W can be estimated by making the expected survival rate equal to the observed one, as shown below.

Example

Consider the snail data of Table 1. Here the mean and variance for the initial population are $\bar{X} = 24.11$ mm and $V = 1.61$, while for the survivors $\bar{X}' = 24.42$ mm and $V' = 1.35$. Substituting into the equations above gives

$$\hat{K} = \frac{1.61 - 1.35}{2 \times 1.61 \times 1.35} = 0.0598$$

and

$$\hat{\theta} = \frac{1.61 \times 24.42 - 1.35 \times 24.11}{1.61 - 1.35} = 26.03$$

so that

$$w = W \exp[-0.0598(26.03 - x)^2]$$

Now W can be chosen so that the expected overall survival rate is equal to the observed rate of 226/1158. This is done by taking $W = 1$

TABLE A1   Calculation of a Fitness Function

Maximum shell diameter (*mm*)	Initial number of snails	Taking *W* = 1 Fitness function *w(x)*	Taking *W* = 1 Expected number of survivors	Taking *W* = 226/881.8 = 0.256 Fitness function *w(x)*	Taking *W* = 226/881.8 = 0.256 Expected number of survivors	Observed number of survivors
21	21	0.220	4.6	0.056	1.2	1
22	93	0.379	35.2	0.097	9.0	6
23	255	0.578	147.3	0.148	37.8	44
24	343	0.782	268.1	0.200	68.7	70
25	289	0.939	271.2	0.241	69.5	62
26	128	1.000	128.0	0.256	32.8	36
27	29	0.945	27.4	0.242	7.0	7
			881.8		226.0	226

initially and working out the corresponding overall survival rate. Then W can be adjusted to give the correct overall survival rate, as shown in Table A1.

Clearly, there is good agreement between the expected and observed numbers of survivors, as shown in the last two columns of this table. Therefore, the fitness function

$$w = 0.256 \exp[-0.0598(26.03 - x)^2]$$

gives a good representation of the relationship between survival and the shell diameter.

This approach can be extended to cover situations with more than one variable. In that case the correlation between variables needs to be taken into account and the equations become a good deal more complicated. With more than two variables an electronic computer is needed to carry out the calculations.

26

Forecasting Visitor Arrivals to New Zealand: An Elementary Analysis

Peter J. Thomson
Victoria University
Wellington, New Zealand

INTRODUCTION

The need to predict the number of visitors arriving in a country such as New Zealand is self-evident. The tourist industry in New Zealand is large, growing, and is a major earner of precious overseas currency. Any reasonable prediction of future trends in visitor arrivals will assist airlines with flight scheduling and the purchase of aircraft. It will also help financiers to make decisions about the building of new tourist hotel complexes, local tourist attraction operators in forward planning, and so on; the list is extensive.

In this study the objective is to forecast the number of short-term visitor arrivals to New Zealand between April 1982 and March 1986 given only the number of visitor arrivals that have arrived each month during the period April 1956 to March 1982. Approximately 50% of such visitors come from Australia, 20% from North America, 15% from the United Kingdom and Europe, with the remainder coming mainly from the Pacific Islands and Asia. In practice the use one makes of such forecasts will, in many cases, dictate the analysis undertaken. For example, if one were attempting to determine whether there were any factors that might limit the flow of tourists, then concomitant measurements such as the maximum number of tourist hotel beds and the maximum number of airline passengers that can arrive in the country over time would be

needed. Here the emphasis will be on simple data analysis combined
with plenty of graphs.

THE DATA

The data are given in Table 1, which compactly records the number of
visitors arriving in New Zealand each month during the period April
1956 to March 1982. Note the convenient year-by-month form. Here
short-term visitor arrivals are visitors who intend staying up to 1 year in
New Zealand, excluding through passengers (i.e., those who are in transit
and stay at the airport or on board their ship for only a short time prior to
moving on to their intended destination). The data are collected from the
ubiquitous arrival cards that every visitor must fill in on arrival to New
Zealand.

   In any statistical investigation of this sort one should always
determine the source of the data and also any qualifications that might
apply to it. In this case it turns out that before March 1975 all cards were
analyzed, between April 1975 and June 1976 only half of the cards were
analyzed, and after July 1976 one-fourth of the cards were analyzed. In

TABLE 1 Total Visitor Arrivals to New Zealand by Months (Excluding Through Passengers), Years Ended March 31, 1957–1982

Year ended March 31	Apr.	May	June	July	Aug.	Sept.	Oct.	Nov.	Dec.	Jan.	Feb.	Mar.	Total
1956–57	2,171	1,398	1,061	1,150	1,232	1,761	2,441	3,903	3,277	3,368	3,162	2,955	27,879
1957–58	2,062	1,330	1,271	1,373	1,540	2,056	2,714	2,824	4,407	3,318	3,619	3,258	29,772
1958–59	2,687	1,413	1,202	1,262	1,597	2,377	2,371	2,753	4,923	3,770	3,817	3,001	31,173
1959–60	2,238	1,551	1,351	1,515	1,787	2,931	3,367	3,482	5,395	4,391	4,598	3,951	36,557
1960–61	2,453	1,783	1,568	1,599	1,944	2,874	3,862	3,952	6,159	5,213	5,171	5,346	40,924
1961–62	2,921	2,504	1,966	2,128	3,166	3,682	5,567	5,184	7,181	6,185	6,601	4,603	51,688
1962–63	3,351	2,936	2,336	2,485	3,026	4,107	6,172	6,798	8,325	7,421	6,584	5,524	58,885
1963–64	3,754	3,268	2,690	2,720	4,071	5,063	6,959	7,142	10,504	8,139	8,552	6,842	69,704
1964–65	4,195	4,163	3,205	2,993	4,308	5,469	7,750	8,672	11,641	9,716	11,456	8,467	82,035
1965–66	5,430	4,992	3,478	3,955	5,725	7,109	9,266	9,789	14,859	12,187	11,207	10,019	98,016
1966–67	6,678	6,692	4,377	5,105	7,678	7,285	10,992	10,146	15,729	14,587	12,457	11,145	112,871
1967–68	7,998	6,875	6,285	7,487	7,701	7,574	9,441	10,765	17,169	14,541	14,265	13,087	123,188
1968–69	7,891	7,368	6,059	7,052	8,550	7,328	10,409	12,176	21,328	13,809	15,323	14,609	131,902
1969–70	9,314	8,645	7,192	8,432	10,029	8,701	11,678	15,962	24,415	16,423	16,902	17,298	154,991
1970–71	12,086	9,974	8,837	10,463	12,917	10,338	15,552	18,597	28,193	19,761	22,278	21,873	190,869
1971–72	13,623	17,055	9,995	12,554	14,615	12,508	18,974	22,218	34,035	24,842	22,372	24,789	227,580
1972–73	16,299	13,944	11,207	14,674	16,204	13,593	21,782	23,380	39,376	27,925	26,995	29,265	254,644
1973–74	21,897	17,684	14,085	18,839	21,250	17,313	23,174	28,945	50,209	39,530	30,791	34,527	318,244
1974–75	25,443	21,898	14,890	18,221	24,627	20,779	28,630	32,624	56,982	44,209	35,180	37,711	361,194
1975–76	26,958	23,752	15,424	17,732	27,650	20,354	30,404	36,390	61,770	47,030	39,070	38,052	384,586
1976–77	27,212	24,134	13,312	17,444	27,632	21,404	31,572	36,472	60,696	48,176	35,792	36,376	380,222
1977–78	29,784	21,296	17,032	22,804	27,476	21,168	29,928	37,516	62,156	44,672	40,500	36,608	390,940
1978–79	28,524	23,060	15,760	20,892	28,992	23,048	35,052	40,564	69,304	49,968	42,068	41,512	418,744
1979–80	29,272	25,868	18,216	23,166	29,808	25,232	33,780	43,916	69,576	48,224	51,353	46,784	445,195
1980–81	31,284	26,681	22,817	26,944	32,902	25,567	37,113	44,788	70,706	49,699	48,783	46,172	463,456
1981–82	33,932	29,105	24,159	29,089	33,134	29,011	39,303	44,924	70,726	50,300	46,463	42,435	472,581

Source: New Zealand Visitor Statistics, published by the New Zealand Tourist and Publicity Department.

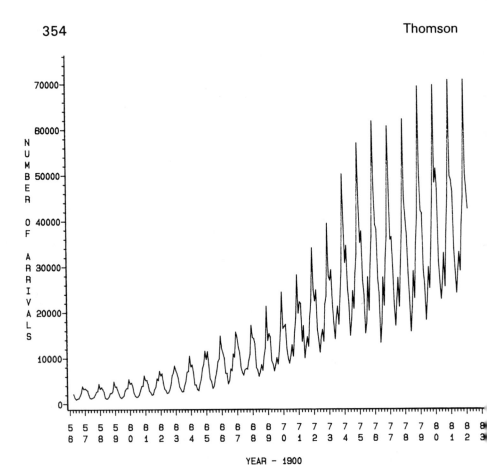

FIGURE 1   Visitor arrivals to New Zealand by year.

the latter cases random selection schemes were used and the number of visitor arrivals obtained were multiplied by 2 or 4 as the case may be. We shall ignore the effects of this sampling except to note that it should not affect the number of arrivals, on average, but will affect their variability. Those of you with a good grounding in the theory of probability may care to determine the precise effect on the standard deviation of any monthly figure for the number of arrivals. It should be noted that the data conclude in March 1982. In fact, one can readily obtain more recent data. However, for the purposes of this study, it will be instructive to use these additional data as a means of assessing the accuracy of any predictions we might make. In practice, of course, we would use all the information that we can possibly lay our hands on.

Although Table 1 is neatly and compactly laid out, it is not easy to

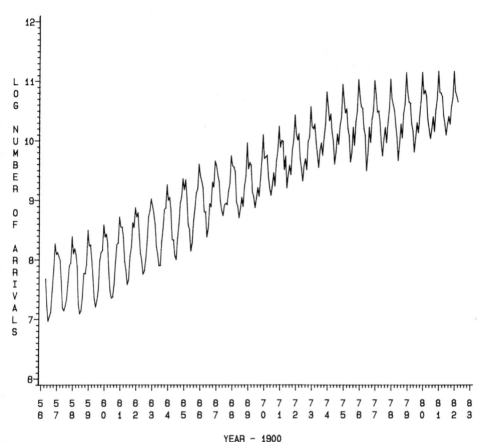

FIGURE 2   Log visitor arrivals to New Zealand by year.

assimilate all the information in the figures, nor is it easy to see any underlying regularity or trends that might be helpful in predicting future arrival patterns. Another display is the simple graph given in Figure 1. The series appears to comprise a generally increasing trend with a superimposed seasonal fluctuation which is increasing in amplitude as the level of the series increases. This display of the data is much more useful and informative than that given in Table 1.

TRANSFORMING THE DATA

For good prediction one needs to see fairly regular patterns in the data which can be projected forward into the future. Figure 1 has structure but it does not have the regular repeatable pattern we seek. At this stage we seek to transform the data in some way so that we retain the structure but make the pattern more predictable. Now, at any instant of time, the rate of increase in visitor arrivals looks to be proportional to the actual number of visitor arrivals at that time. This suggests that a log transformation might be appropriate. The graph of log visitor arrivals is given in Figure 2 [All logarithms in this chapter are natural logarithms (i.e., to base e).]

The log data exhibits the regularity we are seeking. On the whole there appears to be a steadily increasing trend which is linear for the most part. Superimposed on this is a seasonal fluctuation which now has a fairly constant amplitude over time. However, there does appear to be a slight decrease in the amplitude of the seasonal fluctuation between 1967 and 1974. Moreover, the general linear increase appears to flatten toward the end of the graph.

Now consider the evolution of the log arrivals over time for each of the months in turn. What would we expect to see? Presumably, the general trend should be present plus the effect due to the month concerned. If this monthly effect is roughly constant over time, the overall picture would consist primarily of the trend displaced either upward or downward, depending on the monthly effect. A typical selection of the graphs of these monthly series is given in Figure 3.

On the whole the graphs live up to expectations. Note the big jump in May 1971. This could be due to an error in coding the data, or it may be a genuine effect. A rogue observation such as this, called an "outlier" in statistical terminology, warrants further investigation. Running an eye down the May column of Figure 1, it is clear that there was a dramatic increase in the number of arrivals in 1971. Why? By delving further into the source document it was ascertained that this hiccup was due to a Rotary conference held in Sydney at that time. It was obviously a huge conference to cause such a large effect in New Zealand. Indeed, the magnitude of the increase is sufficiently great as to cast further doubt on this value. Further enquiry is clearly needed, but we shall not follow this up here. It will be sufficient to record that the observation is atypical and should be discounted or ignored in any subsequent analysis.

FORMULATING A MODEL

The log series appear to consist of an annual cyclical shape super-imposed on an increasing trend that is smooth and locally linear. So far we have used the word "trend" rather loosely to mean the general direction of the graph. Let us now define it to be a smoothly varying function of time which depicts the general movement of the series.

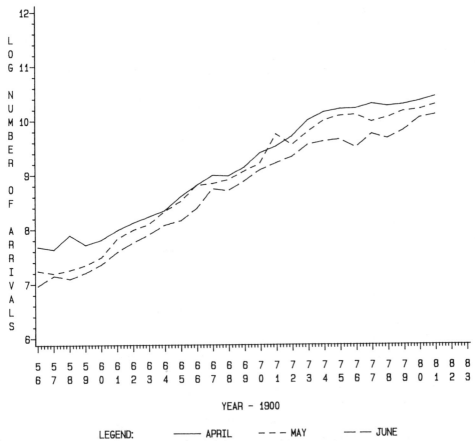

FIGURE 3   Log visitor arrivals to New Zealand by year.

Around the trend there will, in many cases, be a seasonal or cyclical fluctuation. The combination of these two components will explain the systematic part of many time series met in practice, with the remainder left over after taking out these components being nonsystematic errors. It would appear that the log visitor arrivals series can be modeled in this way.

First some notation. We shall denote the number of visitor arrivals in month t as $N(t)$ and set $Y(t)$ equal to $\log N(t)$. Here t runs from 1 to 312 with $t = 1$ denoting April 1956, $t = 2$ denoting May 1956, and so on. Thus our model is now

$$Y(t) = \log N(t) = T(t) + S(t) + E(t) \qquad (t = 1, 2, ...) \qquad (1)$$

where $T(t)$ and $S(t)$ denote the trend and seasonal components, respectively. $E(t)$ encapsulates all that cannot be explained by $T(t)$ and $S(t)$; therefore, if the model is reasonable, it should comprise nonsystematic errors with no structure. Note that $S(t)$ is cyclical, repeating itself every 12 months [i.e., $S(t) = S(t + 12)$ for any month t]. Thus, given 12 consecutive values of $S(t)$, we can readily generate the remainder. We shall think of $S(t)$ in month t as the deviation from the trend due to the particular calendar month concerned. This means that the sum of the deviations over 12 consecutive months should add up to 0. (If they did not, we would simply adjust the trend until they did.) Hence, in addition to equation (1), we add the constraints

$$S(t) = S(t + 12)$$

$$S(t + 1) + \cdots + S(t + 12) = 0 \qquad (t = 1,2,...) \qquad (2)$$

Let us now attempt to extract the trend and seasonal components.

ESTIMATING THE TREND

To estimate the trend we shall form a moving average of the log series. The concept of a moving average will become clear as we proceed. Consider the simple arithmetic average of 12 observations. Form a new series of averages from the log arrival series by first averaging the values of the series at time points 1 to 12, then 2 to 13, then 3 to 14, and so on. Because the average of one set of 12 points will be much the same as the average of the next set of 12 points, the resulting series will appear smoother than the original series. In time-series jargon we say that the original input series has been filtered by a moving-average filter to produce a filtered output series. In this case the filter is a smoothing filter since is produces a smoothed output series. Clearly, any finite (possibly weighted) average would qualify as a moving-average filter. We have chosen the ordinary arithmetic average of length 12 for a particular reason. Let us denote this particular filtering operation by the operator MA12. Since the seasonal component repeats itself every 12 months irrespective of starting point, a 12-month moving average of the log series (1) will yield

$$MA12(Y(t)) = MA12(T(t)) + MA12(E(t)) \qquad (3)$$

since $MA12(S(t))$ is 0 by equation (2). Moreover, if the trend is roughly a constant or a straight line over any 12-month period (this is consistent with our definition of the trend), $MA12(T(t))$ will be the values of the

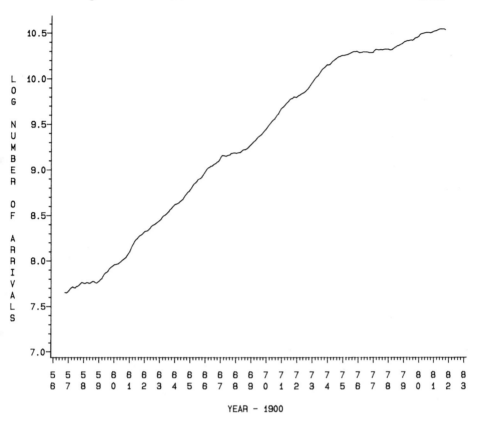

FIGURE 4   Log visitor arrivals to New Zealand by year, the trend adjusted for monthly effects.

trend at the successive time points $(1 + 2 + \cdots + 12)/12 = 6.5$, $(2 + 3 + \cdots + 13)/12 = 7.5$, $(3 + 4 + \cdots + 14)/12 = 8.5$, and so on. We need to bring this series into line with our prevous time points 1, 2, 3, ..., 312. One simple way of doing this is to apply another moving-average filter to the filtered series (3), where this time the average used is the simple average of just two observations. Then the averages of the averaged trend values $MA12(T(t))$ will be values of the trend at the time points 7, 8, 9, ..., 306. Verify that this is the case and that the last time point if 306. This is the approach we shall take. Note that the combined operation of these two moving averages is just another moving average, in this case a 13-point weighted moving average where the first and last observations have weights 1/24 and the remaining observations have weights 1/12. Thus each average is computed as

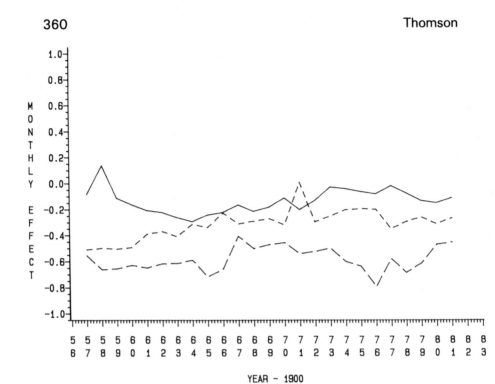

FIGURE 5   Log visitor arrivals to New Zealand by year, adjusted for trend.

$$\text{Average} = \frac{\text{1st obsn} + 2(\text{2nd obsn} + \cdots + \text{12th obsn}) + \text{13th obsn}}{24}$$

Check out the development in this paragraph by applying these moving averages to the first few points of the log arrival series.

In summary, the 13-point moving average filter (which we shall denote by MA) applied to the original log series will yield (if the model is appropriate)

$$MA(Y(t)) = MA(T(t)) + MA(E(t)) = T(t) + MA(E(t))$$

The averaging operations should reduce the absolute size of the errors; i.e., $MA(E(t))$ should be smaller in magnitude than $E(t)$. (Those of you familiar with the statistical properties of sums of variates should be able to quantify this reduction in size, especially for the case where the errors are independent or unrelated to one another.) Thus $MA(Y(t))$ should provide a good estimate of the trend at the time points 7, 8, ..., 306.

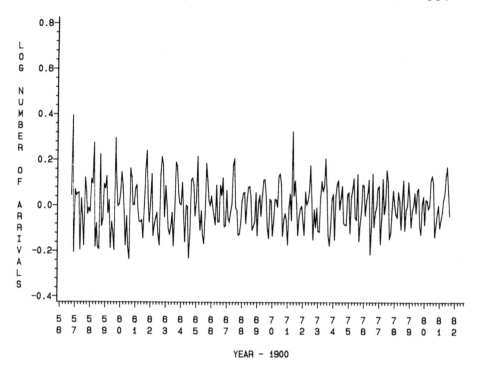

FIGURE 6   Log visitor arrivals to New Zealand by year: residuals after trend and
month effects removed.

A graph of the estimated trend is given in Figure 4. Note the decrease
in the number of arrivals between 1967 and 1968 and also from 1975 on.
Is this an indication of some other limiting factor, such as hotel
accommodation or airline seats available?

ESTIMATING THE CYCLIC EFFECT

Let us now subtract the estimated trend from the original log arrival
series. The resulting series should now consist mainly of the seasonal
effect plus error. It is instructive to graph this series or, more appro-
priately, the individual monthly series. A typical selection of these graphs
is given in Figure 5.

If our model is appropriate, the individual graphs should look flat,
any deviations being due to the nonsystematic error. In this case,
however, it would seem that the graphs are not quite flat, but exhibit a
slowly changing level. Consider the month of May, for example. In

addition to the obvious effect of the Rotary conference, the May level appears to be slowly increasing over time. Ignoring, for the moment, these slow changes, we now average each of these monthly series to obtain an estimate of the seasonal effect for each particular month.

RESIDUALS

Finally, we subtract the cyclic effect and the trend from the series leaving (we hope!) essentially nonsystematic error. This should exhibit irregular and unpredictable behavior with no discernible patterns (see Figure 6).

The graph certainly looks irregular. Note that if we were to draw a histogram of the values of the residuals, the resulting distribution would be positively skewed. [One quick and easy way of doing this is to rotate the graph of the residuals through 90° and then imagine the shape of the histogram that would result when all the points were allowed to fall onto the (new) horizontal axis.]

Checking that the series really is unpredictable is not as straightforward as it might seem. One tool to use is the sample autocorrelation function (acf). This is formed by correlating the series with a lagged version of the series via the formula

$$r(t) = \frac{c(t)}{c(0)}$$

where

$c(t)$ = sample covariance between the series of observations $Y(s)$ $(s = 1,2, \cdots)$ and the series of observations $Y(s + t)$ $(s = 1,2, \cdots)$

$$= \sum_s \frac{[Y(s) - \bar{Y}][Y(s + t) - \bar{Y}]}{312} \qquad (t = 0,1, \cdots)$$

and $\bar{Y}$ is just the sample mean of the $Y(t)$. Verify that $r(t)$ is indeed a sample correlation; that is, it is of the form

$$\frac{\text{sample covariance}}{(\text{sample variance} \times \text{sample variance})^{1/2}}$$

To get a feeling for this operation, compute $r(12)$ and check that the value obtained looks reasonable by inspecting the scatter diagram of points $(Y(s), Y(s + 12))$, $s = 1, 2, ...$ A graph of the first few autocorrelations is given in Figure 7. Clearly, points a multiple of 12 months apart are highly

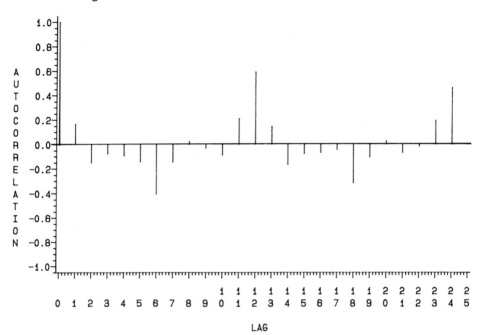

FIGURE 7   Log visitor arrivals to New Zealand by year: autocorrelation function
of residuals.

correlated. This indicates that our model is inadequate. A structureless
error process should have no significant autocorrelations, apart from $r(0)$,
which is exactly 1. How has this eventuated? It is clear that we have not
taken account of all the structure in our series. Since the residuals show
no obvious trend, but do have significant 12-month autocorrelations, the
deficiency would appear to lie in the estimation of the seasonal effects.
Looking back at the graphs of the monthly series adjusted for trend given
in Figure 7, we see that what is present is a slowly changing seasonal
profile. This failure to adhere to the assumption that seasonal effects are
constant over time lies at the root of the problem. However, all is not lost,
as we shall see later when we formulate our predictions.

By way of a digression, consider the graph in Figure 8. This graphs
the trend, seasonal, and residual components on the same scale. It is clear
which phases of the analysis are important. The trend is the most
important component in terms of explaining the movement of the series,
followed in importance by the seasonal component and then the residual.
Let us now model the data from a slightly different point of view.

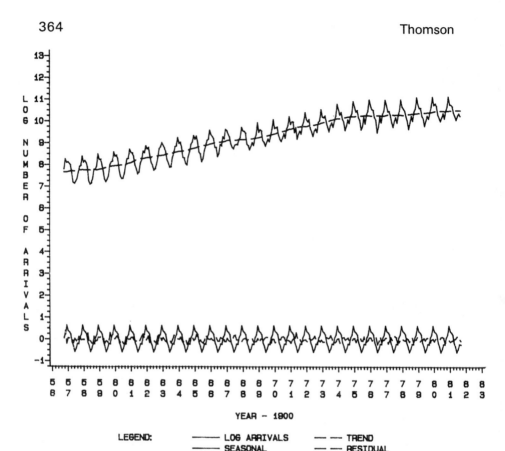

FIGURE 8   Log visitor arrivals to New Zealand by year: log arrivals = trend + seasonal + residual.

AN ALTERNATIVE APPROACH

A natural variable of interest (at least to the tourist industry) is the increase of arrivals expressed as a proportion of arrivals in the same period of the previous year, that is, the variable

$$P(t) = \frac{N(t) - N(t - 12)}{N(t - 12)}$$

Indeed, all too common expressions of the form "our numbers are so many percentage points up or down on last year" are evidence of the popularity of this measure.

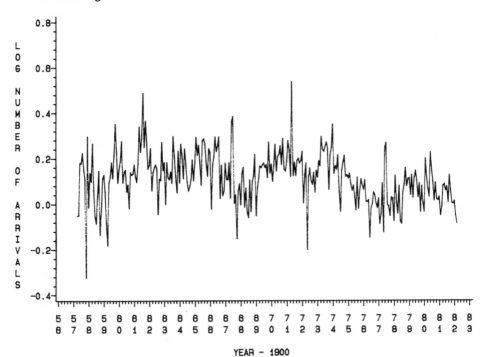

FIGURE 9   Log visitor arrivals to New Zealand by year: twelfth differences.

Now, provided that $P(t)$ is small, it is very close to the twelfth differences of the log series $Y(t)$, that is,

$$P(t) \approx Y(t) - Y(t-12) = \log N(t) - \log N(t-12)$$

for $P(t)$ small. This follows from the fact that

$$Y(t) - Y(t-12) = \log[1 + P(t)]$$

and $\log(1 + x)$ is approximately the same as x when x is small. In our case $P(t)$ and the twelfth differenced $Y(t)$ series are almost identical. As a consequence we shall, for convenience, work with $Y(t) - Y(t-12)$, but interpret this (and refer to it) as $P(t)$, a series of proportionate differences. A graph of the series is given in Figure 9.

What observations can we make about $P(t)$? There seems to be little structure in the series. However, it does look as if there is evidence of a slowly changing level. Let us look at the first few autocorrelations of $P(t)$ (Figure 10). These show a gradual decline, the higher values at the lower

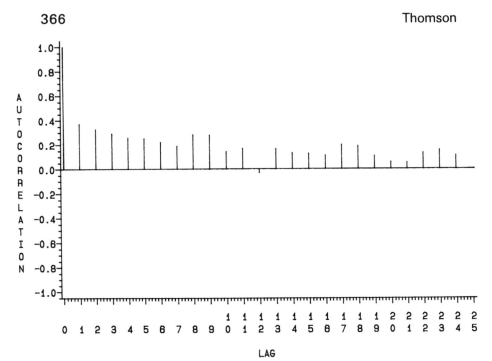

FIGURE 10 Log visitor arrivals to New Zealand by year: autocorrelation function of twelfth differences.

lags being due to the slowly changing level of $P(t)$. It would be useful to remove the effect of the changing level. One obvious way would be to adopt the trend estimation techniques discussed earlier. Another approach would be to difference the series, that is, form the series $P(t) - P(t - 1)$ by subtracting from each observation the value of the preceding observation. This operation will remove any constant level and reduce any linear trend to a constant. (Check these statements by differencing the constant series a and the straight-line series $a + bt$, where a and b are constants and t indexes time.) Since the level of $P(t)$ looks reasonably smooth, this operation (or filter) should effectively do the trick.

However, differencing or, indeed, any form of filtering, will induce structure into (previously) structureless errors. To check that this induced structure is insignificant, we should look at the autocorrelations of $P(t) - P(t - 1)$ which are given in Figure 11. These autocorrelations clearly exhibit structure.

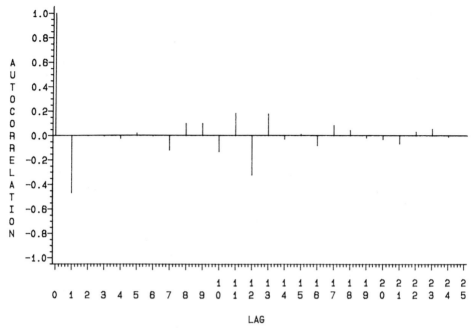

FIGURE 11 Log visitor arrivals to New Zealand by year: autocorrelation function of first differences of twelfth differences.

The alternative method also seems beset by difficulties. Let us now try to reconcile the two approaches and, in so doing, determine the one we should adopt when predicting future values of the arrival series.

SELECTING THE MODEL

As has been mentioned before, any form of filtering will induce structure into previously structureless errors. What sort of structure does it induce? Suppose that $e(t)$ is a series of independent random variables each with zero mean and common variance. This is a fairly natural definition for a structureless error series and is frequently refered to as a "white noise process" by time-series analysts. For such a white noise process it can be shown that the autocorrelations of the series

$$Z(t) = \Sigma_j a(j)e(t - j)$$

Thomson

are given by

$$\rho(t) = \frac{\sum_j a(j)a(j+t)}{\sum_j a(j)^2} \qquad (t = 0, 1, \cdots) \qquad (4)$$

Thus to assess the effect of the difference filter on white noise (i.e., structureless errors) we set $a(0)$ equal to 1, $a(1)$ equal to -1, and the remainder of the $a(j)$ equal to 0. This makes $\rho(0)$ equal to 1, $\rho(1)$ equal to -0.5, and the remainder of the $\rho(t)$ equal to 0. These values are not too dissimilar from those given by Figure 11. However, there is still a little too much structure around lag 12 for us to be totally convinced that the structure in Figure 11 is due solely to the difference operator. What does the autocorrelation structure around the lag of 12 months tell us?

Consider, for a moment, the original model given by equation (1), but where now we allow the seasonal effects to change slowly over years. (This seems to be a little more realistic in the light of our discoveries earlier.) Verify that twelfth differencing the $Y(t)$ series will remove any slowly changing seasonal effect and also the more slowly changing parts of the trend. Since we believe that the seasonal pattern is changing slowly over time, the twelfth difference should effectively remove $S(t)$. Although $T(t)$ is changing slowly, it may not be changing slowly enough for the twelfth differencing to remove it completely. This is borne out by the plot of twelfth differences given in Figure 9, which exhibits a slowly changing level. As mentioned earlier, first differencing should remove any remaining trend effects. However, if the error series $E(t)$ in equation (1) is white noise, the effect of the combined differencing operations (first and twelfth differencing) will yield the autocorrelation structure shown in Figure 12. This can be checked by writing the two differencing operations as one filtering operation and then applying equation (4). Compare the above with the observed autocorrelation pattern given in Figure 11. They are suspiciously close.

This evidence together with our earlier results would seem to indicate that the relevant model for the log arrival series comprises slowly evolving trend and cyclical components together with a superimposed white noise process. This is the model we shall adopt when generating predictions.

PREDICTIONS

Since we have concluded that the error process is (approximately) white noise, this component is unpredictable and hence best predicted by its average value, zero. As a consequence our predictions must be based on the most recent estimates of the evolving trend and seasonal pattern.

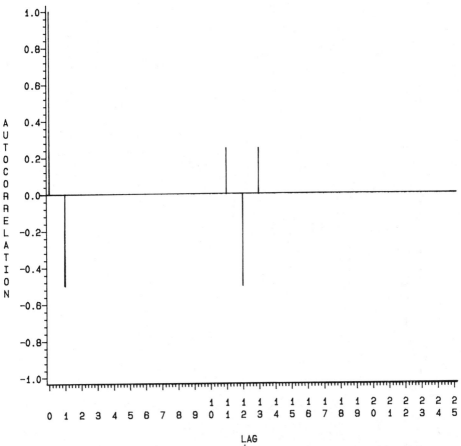

FIGURE 12 Combined first and twelfth difference filters on white noise:
theoretical autocorrelation function.

These are most expeditiously obtained by fitting separate straight-line
regressions to the last data points in each of the monthly series graphed
in Figure 3. (One could also use a robust line-fitting procedure if it was
felt that outliers might be a problem.) Clearly, one should choose the
maximum number of time points that appear consistent with the straight-
line hypothesis. In this case the data from April 1974 to March 1982 were
used. Once the various straight lines have been fitted, they are projected
into the (near) future. The predicted log arrival series is now reconstituted
from the individual monthly projections. Finally, the forecasted arrivals
are obtained by taking antilogs of the projected log arrival series. Graphs

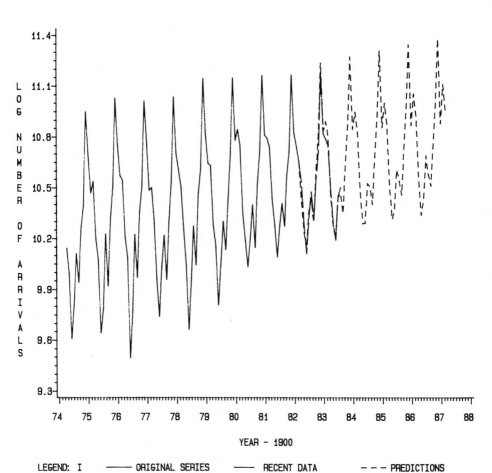

LEGEND: I ——— ORIGINAL SERIES ——— RECENT DATA - - - PREDICTIONS

FIGURE 13 Log visitor arrivals to New Zealand by year: predictions to March 1986.

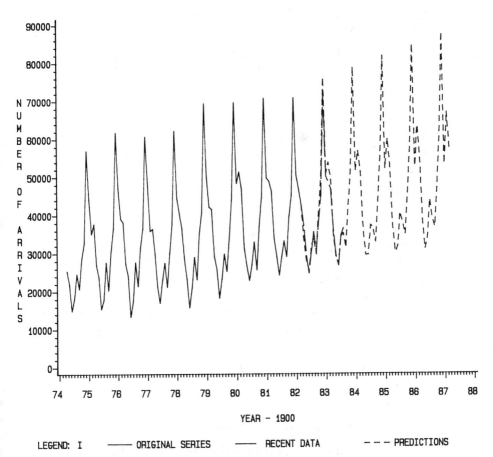

FIGURE 14   Visitor arrivals to New Zealand by year: predictions to March 1986.

TABLE 2 Total Visitor Arrivals to New Zealand by Months (Excluding Through Passengers), April 1982–August 1983

Apr.–Dec. 1982	37,339	29,408	24,699	30,412	35,463	30,226	38,294	44,954	71,733
Jan.–Aug. 1983	50,086	48,477	46,567	35,892	29,343	26,708	34,600	36,401	

of the resulting forecasts for both the log and original series are given in Figures 13 and 14. Included in these graphs are the most recent data points, (i.e., the values for the period April 1982 to August 1983). These values are given in Table 2.

Recall that this stretch of the data was omitted from the study. The prime reason for this was to see how good our forecasts were in actuality. The results do not look too bad. The only major descrepancy between the recent data and the corresponding forecasts occurs in February and March 1983, where we have overestimated the number of visitor arrivals. This is due to the fact that we have not quite captured the recent evolution of these particular months. This can be cured by projecting forward for these months using more recent data than the post-1974 data.

One measure of accuracy is the mean-squared error (MSE) criterion. This is just the average sum of squares of the actual forecast errors; that is,

$$\text{MSE} = \frac{\text{sum from April 1982 to August 1983 of } [F(t) - N(t)]^2}{17}$$

where the $F(t)$ denote forecasted values. In this case a better measure of accuracy is the mean-squared proportionate error (MSPE) given by

$$\text{MSPE} = \frac{\text{sum from April 1982 to August 1983 of } [(F(t) - N(t))/N(t)]^2}{17}$$

Note that

$$\text{MSPE} \approx \frac{\text{sum from April 1982 to August 1983 of } [\log F(t) - \log N(t)]^2}{17}$$

since

$$\log F(t) - \log N(t) = \log \frac{1 + [F(t) - N(t)]}{N(t)}$$

and $\log(1 + x) \approx x$ when x is small. Here

$$\begin{aligned}
\text{MSE} &= 6{,}136{,}633 & (\text{MSE})^{1/2} &= 2477 \\
\text{MSPE} &= 0.0030443 & (\text{MSPE})^{1/2} &= 0.05518
\end{aligned}$$

The square root of these quantities gives some idea of the magnitude of the errors likely to be incurred when forecasting visitor arrivals using these techniques. In particular, the average forecast error is approximately 5.5 percent. These figures could be reduced further by improving, in particular, the February and March projections.

Many questions remain. How would "high" and "low" forecasts of future visitor arrivals be constructed? Given that, in general, future values of the series are not known, how would estimates of the MSE and the MSPE be determined? Under what conditions can the size of these errors be reduced? Would the forecasts be improved by disaggregating the series into Australian and non-Australian arrivals and then applying the projection techniques above to each of these series? If so, would it pay to further disaggregate into yet more categories? Unfortunately, we do not have the space to dwell on these and other such interesting questions, which are left for the reader to ponder and resolve.

CONCLUSION

It is apparent that a reasonably comprehensive analysis of the data and the subsequent generation of forecasts can be undertaken with a minimum of mathematical and statistical tools. Indeed, these simple techniques combined with a little bit of common sense will frequently perform as well as more sophisticated forecasting techniques. Happy forecasting!

27

Spreading Disease by Statistics and Growing Cities with Computers

P. Brent Wheeler
Palmerston North City Corporation
Palmerston North, New Zealand

INTRODUCTION

On the corner of Broadwick and Lexington Streets in Soho, London, stands a public house with a small plaque on the wall recording the site of a water pump. The inn sign is a portrait of one John Snow. The plaque and the portrait commemorate John Snow's discovery of the link between cholera and water. [Snow's work is described in Stamp (1964).] During a cholera epidemic in 1848, Snow mapped the incidence of death by cholera in his Soho practice and noticed a suspicious clustering around the (then named) Broad Street water pump. He also noticed that deaths from the disease appeared to have spread outward from the pump. Snow had the pump handle removed and deaths from cholera ceased almost immediately. While this contribution to medical science was of considerable importance, it also represents one of the earliest spatial diffusion studies ever undertaken.

Diffusion is the process by which phenomena move across space over a period of time. The subjects of diffusion processes are of vastly differing types and involve a wide variety of phenomena. A bottle of ammonia is opened in a chemistry laboratory and the pungent odor gradually diffuses throughout the room. A fashion in haircuts appears in London, then New York and Paris, and the popularity of the fashion diffuses throughout Western society. New inventions, such as the

375

transistor radio, the electric frying pan, or the microcomputer originate at the place of invention and then spread throughout the world.

The space over which phenomena diffuse varies greatly in scale. It may be relatively small, as, for example, in the diffusion of ammonia fumes through a room. At the opposite extreme, the scale involved may be global, as in the diffusion of transistor radios around the world.

The length of time over which diffusion takes place also varies and may be relatively short—a matter of minutes in the case of a smell permeating a room, or extremely long. The migration of Europeans and their subsequent diffusion through North America, for example, has taken place over several centuries.

In spite of this complexity, the wide variety of time and distance scales involved, and our meager knowledge of diffusion processes, certain generalizations may be made. These generalizations allow diffusion processes to be simulated by simple statistical models. In this chapter some very simple probabilistic and deterministic concepts are used to develop two models for simulating diffusion processes.

The first step in building these models is to consider the main factors that influence diffusion. A useful way of considering the factors that

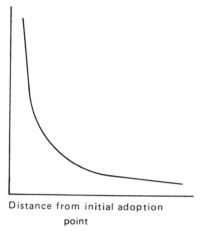
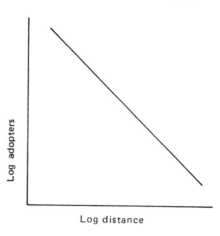

FIGURE 1   Neighborhood effect.

affect diffusion is to examine the process in terms of the diffusion of innovations, since the processes that determine the successful introduction of a new idea or artifact into a given social system are analogous to many other diffusion processes.

THE INFLUENCE OF DISTANCE AND SIZE

The most crucial variable involved in creating the spatial patterns associated with diffusion patterns is distance. Most innovation diffusion studies have shown that the likelihood of people adopting a given innovation decreases as distance between the origin of the innovation and potential adopters increases. Accordingly, the Swedish geographer Hagerstrand's studies of pasture improvement diffusion showed that as distance from the points of initial adoption increased, the number of adopters decreased (see Hagerstrand, 1966). The process whereby adoption is more likely to occur around existing adopters is termed the *neighborhood effect* and is depicted in Figure 1.

Many studies of diffusion suggest that the form of the neighborhood effect is negative exponential and may therefore be expressed as a simple linear equation with distance as the independent variable. Although it is apparent that the probability of contact between adopters and potential adopters depends partly on distance, it also depends on the number of potential adopters at a given distance from existing adopters. It is commonly found that the diffusion of new fashions, for example, begins in the world's larger cities, often London, Paris, or New York, and then

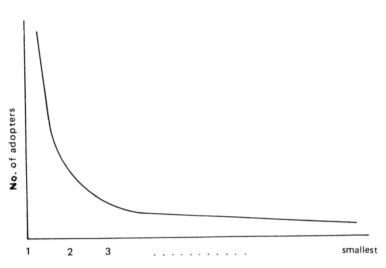

Rank of city in terms of size

FIGURE 2   Hierarchical effect.

spreads down through the hierarchy of different city sizes around the world. This "trickling down" process is known as the *hierarchical effect* and is depicted in Figure 2.

For many innovations it has been observed that both neighborhood and hierarchical effects are important, and these two influences may be combined in a general form as a linear gravity model. This model, which is adapted from Newton's famous law of gravitation, states that the number of adopters is proportional to the product of the size of the two places involved and inversely proportional to the distance between them.

THE INFLUENCE OF TIME

Diffusion is a dynamic process that takes place through time. Studies of innovation diffusion suggest that adoption occurs initially at a slow rate (innovaters) and gradually gathers momentum (early majority) until a peak is reached and the rate of adoption begins to diminish (late majority) at an ever-increasing rate because the bulk of the potential adopters have taken up the innovation, leaving only a small number of late adopters (laggards). This pattern is readily described by the normal

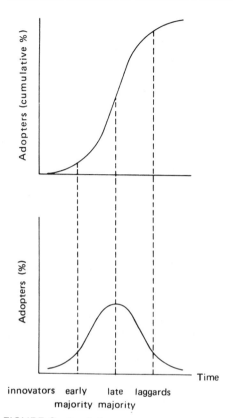

FIGURE 3   Logistic curve.

curve and by accumulating the number of adopters, with an S-shaped or *logistic* curve (see Figure 3).

COMBINING DISTANCE, SIZE, AND TIME

To complete this brief introduction to the diffusion process it is helpful to consider the factors discussed above as an integrated set of influences that take place simultaneously through time. The graph shown in Figure 4 may be thought of as representing either distance or time or size effects. In addition, it may be used in a general sense to represent all three factors concurrently.

Figures 5 and 6 show how such a diffusion process characterized the adoption of radio broadcasting in the United States between 1921 and

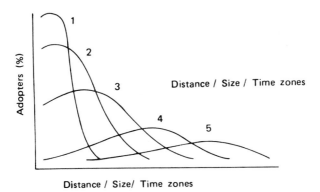

Distance / Size/ Time zones

FIGURE 4   Combining distance, size, and time. (Note: size expressed as ranks with 1 being large.)

1924 (see Bell, 1965). Figure 5 shows that large cities which were spaced relatively close to one another were the earliest adopters—evidence of a neighborhood and hierarchical effects. This pattern continued so that by 1924 (Figure 6) a trickle-down effect led to small cities in the hierarchy adopting. While several cities far to the west of the main area of initial innovation had adopted, the majority of new acceptors were located near existing adopters. The time element in the process is more difficult to discern, but it appears that the period prior to 1921 was an early innovation stage, with the period 1921–1924 making up the early majority.

SIMULATING THE DIFFUSION OF CHOLERA DEATHS

Having identified the major factors involved in diffusion processes it is possible to develop a model to simulate the spread of deaths by cholera which was observed by Snow in 1848. This form of diffusion is termed "expansion" diffusion and is the most simple form of diffusion since it involves the simple spread of some phenomena from a central point. Thus the object of the simulation is to reproduce the pattern of deaths by cholera in Soho. If this can be achieved, the model may be able to be applied in other areas.

As an initial step it is helpful to make several simplifying assumptions. First, assume that the potential victims were evenly spread around the water pump, which was the origin of the infection. Second, assume that the disease may spread with equal ease in any direction around the water pump over the area in which deaths occurred. These

FIGURE 5   Cities adopting radio broadcasting in 1921. (Adapted from Bell, 1965.)

two assumptions, an even distribution of potential victims and equal ease of movement in all directions, are commonly termed "isotropic conditions."

Two further assumptions are useful. One is that the disease was caught only by using water that originated from the Broad Street pump. The possibility of catching the disease from another person is therefore ignored. The other is that the probability of catching cholera was entirely dependent on the distance from the pump.

The next step is to use these assumptions to estimate the probability of contact between potential victims and the pump. One way of doing this is by placing a grid over Snow's map and calculating the average number of deaths occurring in grid squares at a given distance from the water pump (see Figure 7). The data in Table 1 were generated in this manner and the procedure may be thought of as smoothing the pattern observed by Snow to fit the assumption of isotropic conditions.

The probability of contact with the pump occurring for a person in any given cell may now be simply calculated by dividing the total number

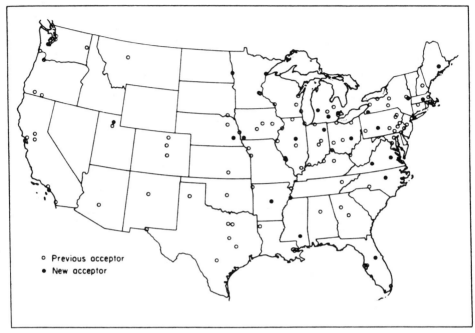

FIGURE 6   Cities adopting radio broadcasting in 1924. (Adapted from Bell, 1965.)

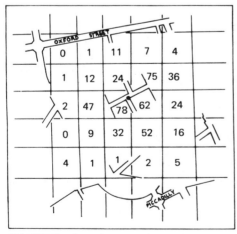

FIGURE 7   Snow's map of cholera deaths in 1848. Total deaths = 500.

TABLE 1   Deaths in Soho, 1848

Observed deaths					Average deaths in grid squares of a given location[a]				
0	1	11	7	4	3	7	10	7	3
1	12	24	75	30	7	37	41	37	7
2	47	78	62	24	10	41	80	41	10
0	9	32	52	16	7	37	41	37	7
4	1	1	2	5	3	7	10	7	3
	Total 500						Total 500		

[a]Calculated by summing values of cells in identical positions on the grid and dividing by the number of such cells; for example, in the case of the outside corner cells, $0 + 4 + 5 + 4 = 13 \div 4 = 3$.

of deaths into each of the cell frequencies. The resulting probability matrix is shown in Table 2. Notice that although the probabilities could have been calculated directly from Snow's data, these would have depended on direction as well as distance, and would therefore have required a more complicated model which would have fitted Snow's data better, but would have provided a less general expression of the distance decay effect, thereby limiting the potential applications of the model.

The probability matrix expresses the distance decay effect involved in the diffusion of cholera deaths from a central point, and in Figure 8 the plot of log distance versus log number of deaths is a straight line that may be summarized with the equation

$$\log(\text{deaths}) = 2.578 - 2.583 \log(\text{distance})$$

with −2.583 being the rate of distance decay. It should be noted that the distance metric, in this case the distance between the cell with the Broad Street pump and all other cells, is rather crude and does not take account of distance within cells.

The probability matrix is used to generate the pattern of deaths by accumulating the probabilities assigned to each cell to express the chances of infection and death as a set of integers between 1 and 500. The interval 0–2, for example, is assigned to the upper left-hand cell and corresponds to the number of chances in 500 of infection occurring in that cell. Table 3 contains the complete set of intervals.

To generate the pattern of deaths, the accumulated interval matrix is placed on this cell on the map and a random number between 0 and 499 is selected from a random number table. The cell on the map over which the interval surrounding the random number lies is then said to have a

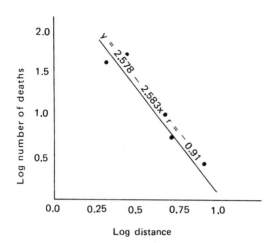

FIGURE 8   Distance decay in the diffusion of cholera deaths.

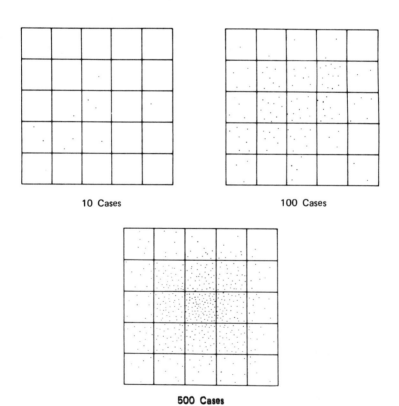

FIGURE 9   Buildup of cholera through time.

TABLE 2   Probabilities of Contact with Pump

0.006	0.014	0.020	0.014	0.006
0.014	0.074	0.082	0.074	0.014
0.020	0.082	0.160	0.082	0.020
0.014	0.074	0.082	0.074	0.014
0.006	0.014	0.020	0.014	0.006

Total = 1.0

TABLE 3   Probabilities Expressed as Accumulated Intervals

000–002	0003–009	010–019	020–026	027–029
030–036	037–073	074–114	115–151	152–158
159–168	169–209	210–289	290–330	331–340
341–347	348–384	385–425	426–462	463–469
470–472	473–479	480–489	490–496	497–499

new victim. If the number 170 were drawn from the random number table, a new victim would be placed in the map cell immediately to the left of the center cell on the accumulated interval matrix.

Figure 9 shows the result of applying this procedure, and a clear distance decay effect is evident. The first 10 and 100 cases are found predominantly around the initial point of infection and when 500 cases have been simulated a clear concentration is found in the center of the area with fewer deaths in peripheral areas. The final result when compared with Snow's data shows that the isotropic assumptions have produced data that appear more uniform. Even this simple simulation, therefore, shows the order of variation that chance could cause and suggests that something more than chance must be operating in Snow's data, and that the cholera spread more in one direction than another.

A GENERAL MONTE CARLO DIFFUSION MODEL

Although the model described above, which is termed a Monte Carlo simulation under isotropic conditions, produces some useful information it has several drawbacks. In the first place, the rate of infection is not well simulated since the model produces geometric growth rather than an S-shaped curve which is more typical of diffusion processes. Second, the model takes no account of either varying population sizes or suscepti-bility to infection. Third, the model deals only with expansion diffusion, and strictly speaking is little more than an exercise in generating patterns.

To overcome some of these limitations it is possible to develop a more sophisticated model which can be run by computer. The most advanced work in this area has been carried out by the British geographer Robson in the context of city growth, and the model developed below is an adaptation of his work (see Robson, 1973).

The simulation problem concerns the diffusion of those technological innovations that lead to rapid growth in urban areas. The major influences modeled include distance (neighborhood effect), size (hierarchical effect), and time (logistic curve) and, as with the cholera model, a number of simplifying assumptions form the basis for the model.

It is assumed at the outset of the simulation that only the largest city in the hypothetical urban system has adopted an innovation, or more realistically a set of innovations, and that their adoption leads to growth in that town. Because innovations are not always adopted immediately, a "level of resistance" to innovation is incorporated in the model and towns are said to adopt only after they have received a certain number of "stimuli."

Adoption of the set of innovations by a town has two results. In the first place, the town adopting experiences an immediate growth increment at the time it adopts, a growth of half that increment in the next time period, and a third, smaller increment in all subsequent periods.

The rationale for this pattern of growth is that an initial adoption causes a growth in employment through the establishment of new factories or plants. A second wave of employment arises in the next generation as the new factories are serviced. The net effect of establishing new factories and their servicing populations is a small but continuing growth increment through reproduction and migration in all subsequent periods.

The second result of adoption is that the towns which have adopted begin to emit innovative stimuli themselves and the wave of innovative activity gradually spreads throughout the urban system. The probability of a town being "hit" is said to be a function of both the size of the towns in the system and the distance between them. As in the Monte Carlo model, a random number is selected for each town and if it falls in the range of probabilities for that town and if the town's level of resistance is passed, the town adopts and grows. One "pass" over all towns constitutes a generation.

To test the capability of the model to simulate spatial diffusion

TABLE 4 Combined Simulation

Town	Initial population	After 10 generations	After 15 generations
A	10,000	59,200	151,450
B	10,000	48,900	108,600
C	10,000	49,750	118,450
D	10,000	46,550	113,700
E	10,000	36,050	73,950
F	10,000	43,900	106,800
G	20,000	130,500	417,250

processes through time, a simple simulation was run. The hypothetical urban system consisted of seven towns, six of which were arranged as a regular hexagon with the seventh located in the middle of the system. Thus the distance between any one of the six outer towns and all other towns was identical, while the distance from the center town to all other towns was shorter.

In the simulation the model was run for 15 generations. The level of resistance was set at 10; that is, each town had to receive 10 stimuli before adopting and growing. The distance decay parameter was set to 2 so that interaction between any two towns was deemed to be proportional to the product of their populations and inversely proportional to the square of the distance between them. Although these values are clearly arbitrary, they are similar to those observed in many diffusion studies. In the simulation the neighborhood and hierarchical effects were combined by assigning the centrally located town (town G) an initial population of 20,000 with all other towns being assigned a population of 10,000.

Having adopted, towns were said to grow by 100 units in the period of adoption, by 50 units in the next period, and by 10 units in all periods subsequent to adoption regardless of hits. Thus as the population of a given town grows, the proportional increase in population is rapid at first, then gradually diminishes in the manner described by the logistic curve. A computer program was written to perform the simulation and to measure the growth rate of towns.

The results of the simulation consist of a table showing the populations of the seven towns after 10 and 15 generations (Table 4), a map with the populations expressed as proportional circles (Figure 10),and a graph showing the rate growth exhibited by the largest and the smallest town (Figure 11).

After 10 generations only town G had grown significantly faster than the other towns, which had roughly equal populations. While a similar

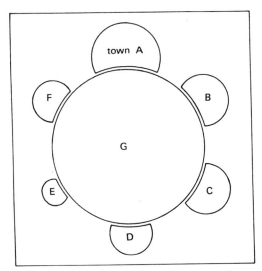

FIGURE 10   Simulation after 20 generations.

situation prevailed at generation 15, town A had grown more than the other outer towns and town G had increased its lead on all other towns. The strong influence of both size and distance is clear in Figure 11, which shows the overwhelming dominance of town G. The dominance of town G from early in the simulation is also shown, as is the considerable difference in growth rates for the fastest and slowest (town E) growing centers. As in the cholera example, this simulation shows the size of difference that can be caused by chance. Towns A and E started equal and had the same chance of growing, but have grown at different rates.

CONCLUSIONS

The models described above, particularly the second, more general model, appear to perform reasonably well and reproduce the patterns noted in most diffusion studies. Even though the applications described are rather artificial, it is easy to imagine more realistic situations in which the simulations would provide useful information.

Obviously, there are numerous contagious diseases whose spread could be modeled under varying conditions. One interesting area for simulating city growth is in estimating the influence of growth promotion and regional development policies.

In principle, the models are applicable to any situation in which the

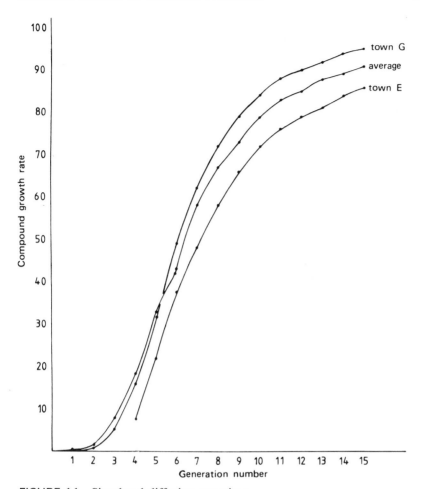

FIGURE 11   Simulated diffusion over time.

diffusion process is controlled by the many deterministic elements of size and distance as well as elements which may reasonably be viewed as random. Examples of size-oriented elements include populations of all kinds, both human and nonhuman, labor pools, cultural artifacts, and various forms of information. Examples of deterministic, distance-oriented elements include physical distance, time, and cost. The probabilistic framework of the models provides a useful way of examining the numerous decision-making and behavior patterns which are nondeterministic in nature.

Finally, a major advantage of the models is that they may be run as computer simulations. This provides the opportunity for considering large-scale processes which take place over long periods of time.

REFERENCES

Bell, W. (1965). The diffusion of radio and television broadcasting stations in the United States. Unpublished M.A. thesis, Department of Geography, Pennsylvania State University.

Hagerstrand, T. (1966). Aspects of the spatial structure of social communication and the diffusion of information. Papers of the Regional Science Association 14, 27–42.

Robson, B. (1973). *Urban Growth: An Approach*. Methuen, London.

Stamp, L. D. (1964). *Some Aspects of Medical Geography*. Oxford University Press, Oxford.

28

Time Series and the Stock Market

David A. Dickey
North Carolina State University
Raleigh, North Carolina

INTRODUCTION

There are many instances in which data are collected at equally spaced time intervals, and the analysis of such data uses the methods of a special branch of statistics known as time-series analysis. Examples include the many monthly series reported by the federal government, such as the unemployment rate, daily environmental measurements such as temperature and rainfall, and industrial measurements such as hourly temperature in a furnace and exhaust gas emissions.

This chapter is an application of time-series analysis to a series of IBM stock closing prices. Interest centers on using the scheme of selling stock when the price is high and buying when the price is low to make a profit. If stock prices are fluctuating about some known, or at least estimable, general mean level, this would appear to be a reasonable strategy. We would observe the series rising and falling as we pass through time, but the existence of an overall constant mean level about which these fluctuations occur would keep our series from rising or falling too far. Thus when the price series gets unusually high, we sell before the anticipated drop toward the mean.

We have theorized one possible type of behavior for stock prices. Here is an alternative. It might be that the value of the series each day is simply equal to the previous value plus a random step either up or down.

In this case, we do not make use of a series mean. The series may behave somewhat like a drunken man staggering, whose position at time t is equal to his previous position plus a random step in any direction. This man has no tendency to return to a straight course but may wander arbitrarily off since, at each time t, he takes a random step. If our stock price series were of this nature, we would not observe a tendency to return to an overall mean level and would be reluctant to try the sell high-buy low strategy. Such a series is called a random walk.

In some areas of time-series analysis, periodic functions such as the sine and cosine functions of trigonometry are used to describe fluctuation in observations. We could pretty well rule out the use of such regular functions for our stock prices since, for example, if the price series had a perfect 30-day cycle, before long, people would notice this and soon would be able to predict well the behavior of the stock prices. Obviously, this has not been done. Thus any realistic conception or model of the price series must contain some element of randomness—but how much?

Let us go to the extreme and ask if the stock prices might consist of a mean plus totally random, independent deviations. In this case, knowing that a price is above the mean level one day tells us nothing about whether it will be above or below on the next day.

One might reason that this independent deviations scenario is not realistic. If stock prices are high today, we would probably expect them to be at least somewhat high again tomorrow. It is this type of behavior, in fact, that makes time-series data different from other types of statistical data. The tendency for positive deviations from the mean to be followed by positive ones and for negative to be followed by negative is referred to as *autocorrelation*. The existence of autocorrelation in data taken over time is observed in a wide variety of research areas.

We now review the facts:

1. The series probably contains some randomness and is thus not perfectly predictable.
2. If the series is fluctuating about an overall mean level, it is probably exhibiting autocorrelation.
3. The series could be a random walk.

A TIME-SERIES MODEL

We now try to formulate a mathematical model that will allow for all of the possibilities mentioned above. It will then be possible to use some real data in the model to decide on the nature of the series. It is a coefficient in the model that will indicate the true nature of the series. The coefficient will be estimated from the data. For one value of the coefficient, the model corresponds to a random walk; for another, it is a mean with autocorrelation; and for another, it is a mean plus completely random independent deviations.

Here we have used a classical approach of statistics. First, we formulate a model that encompasses all of our postulated scenarios. Using data, we estimate the parameters of the model and, based on these estimates, we decide which scenario to choose as our representation of reality.

The price at time t will be denoted Y_t in our model, and the mean of the series will be denoted μ. A model that we will consider at first is

$$Y_t - \mu = \rho(Y_{t-1} - \mu) + e_t$$

where the e_t is part of a completely independent random sequence. The presence of e_t in the model gives a part that is not predictable. The dependence of Y_t on Y_{t-1} shows that a part of Y_t is predictable from the past. For example, if $\mu = 400$, $Y_{100} = 405$, and $\rho = 0.8$, we predict Y_{101} to be 404.

Before going into our data analysis, let's see if the proposed model involves all the possibilities we envision for our series. Notice that if

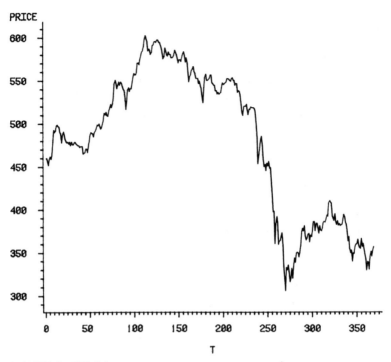

FIGURE 1   IBM data.

$\rho = 0.8$, a value of Y_t above the mean tends to be followed by a value
somewhere around $\mu + 0.8(Y_{t-1} - \mu)$, which is, again, larger than μ.
Similarly, negative deviations tend to be followed by negative ones. This
phenomenon follows for any value of ρ strictly between zero and 1. Thus
our model can exhibit autocorrelation. If $\rho = 1$, our model is $Y_t - \mu =$
$Y_{t-1} - \mu + e_t$ or $Y_t = Y_{t-1} + e_t$. This says that the price Y_t at time t is equal
to the previous price Y_{t-1} plus a random step e. If $Y_{100} = 405$ and $\rho = 1$, we
predict Y_{101} to be 405. Notice that with $\rho = 1$, the value of μ is irrelevant
and the series is a random walk. Our model can even represent the
independent deviations case if $\rho = 0$.

We now have in hand a model which, depending on the value of ρ,
can represent several plausible scenarios. The next step will be to
estimate ρ from the data.

We see that if $\rho = 1$, our prediction of Y_{t+1} is just Y_t. There would be
no point in looking at previous values of the series to compute a mean for
the series since the mean does not enter the prediction. In fact, if we were
predicting only with the information available up through time t, we

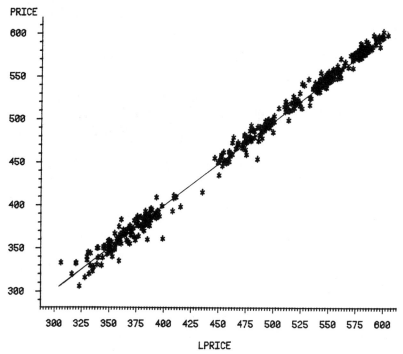

FIGURE 2   Price versus lag price.

would predict all future values to be the same as Y_t. This tells us that there is no useful predictive information to be had by comparing Y_t to the past values $Y_{t-1}, Y_{t-2}, \ldots$ and thus information as to whether Y_t is relatively high or low with respect to the historic mean of the Y's is irrelevant. Obviously, if this model turns out to fit the IBM stock price series well, we will not advocate the sell high-buy low strategy.

DATA

In the famous time-series book by Box and Jenkins (1976), we find 369 consecutive closing prices for IBM stock. We will use these to investigate the nature of stock prices and to see if our proposed buy low-sell high policy would be worthwhile. A plot of the data over time in Figure 1 shows a definite tendency for high values to be followed by high ones and low values by low ones. On the basis of this graph, it is evident that the value of ρ in our model $Y_t - \mu = \rho(Y_{t-1} - \mu) + e_t$ is not zero.

Our interest now lies in seeing if $\rho = 1$. A nonmathematical way to do

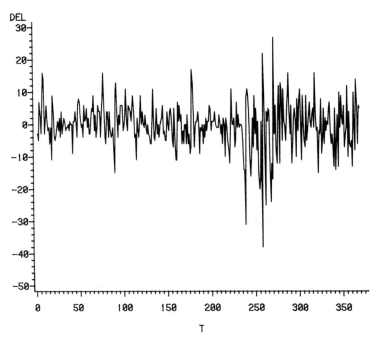

FIGURE 3 Differenced series.

this is to look at a plot of Y_t against Y_{t-1}. This means that we will plot 368 points with first (horizontal) coordinate Y_{t-1} and second (vertical) coordinate Y_t. Doing this for $t = 2, 3, \ldots, 369$ gives the plot shown in Figure 2.

If we imagine a line connecting the point (300,300) to the point (600,600), we see that this line runs pretty much through the middle of our plot. This line has slope 1. That means that any change in Y_{t-1} is accompanied by a roughly equal change in Y_t. Thus the coefficient ρ relating Y_t to Y_{t-1} must be approximately 1.

Let us try one more approach to the problem. We can divide the 368 values $Y_1, Y_2, \ldots, Y_{368}$ into the highest 92, the next highest 92, and so on. Thus we will have four groups delineated by the values of stock prices. Now, if we have a tendency to return to a mean level, we should expect low values Y_t to be followed by higher values Y_{t+1}. Also, we would expect high values Y_t to be followed by lower values Y_{t+1}.

Based on the foregoing reasoning, we will take our four groups and for each Y_t in the group, we will put a $+$ sign if Y_{t+1} is bigger than Y_t and a $-$ sign if Y_{t+1} is less than Y_t. Thus $-$ or $+$ shows whether Y_{t+1} is a decrease or increase from Y_t.

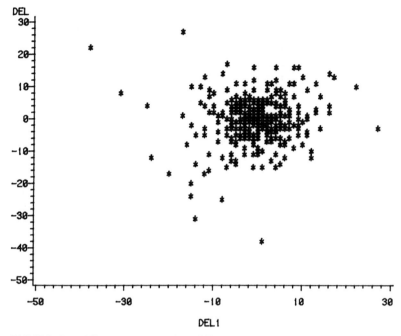

FIGURE 4   Differences versus lag differences.

How would we expect the + and − signs to be arranged if there is a tendency to return to the mean? When Y_t is above the mean, we would expect Y_{t+1} to be smaller than Y_t, so we would expect more − than + signs in the group with Y_t high. Similarly, if we had started at a low value of Y_t, we would expect Y_{t+1} to be higher than Y_t. On the other hand, if Y_{t+1} is equally likely to be greater or smaller than Y_t, we would expect roughly equal proportions of + and − signs in each of our four groups. The results for our data set are:

	−	+
Lowest Y_t's	40	52
Somewhat low	46	46
Somewhat high	42	50
Highest Y_t's	41	51

There seems to be little difference between the lowest Y_t group and the highest Y_t group. Again, we seem to have no outstanding evidence against the $\rho = 1$ assumption.

We note here that there are many more statistically sophisticated procedures for analyzing these data. Several of these procedures were used with basically the same result. At this point, we are entertaining the idea that $\rho = 1$, so our model is $Y_t = Y_{t-1} + e_t$, where e_t denotes a random deviation or "step" in our random walk.

CHECKING THE MODEL

If our model is really $Y_t = Y_{t-1} + e_t$, then $Y_t - Y_{t-1}$ equals e_t, so if we compute these differences $\Delta_t = Y_t - Y_{t-1}$ for $t = 2, 3, \ldots, 369$, we could plot these to see if they look "random." A plot of Δ_t versus t is given in Figure 3. We see no apparent patterns here (in contrast to the plot of Y_t itself). The daily changes in stock price seem to be between -15 and 15 for the most part and appear to fluctuate randomly about zero.

Further, we plot Δ_t against the previous values Δ_{t-1} to check for any remaining autocorrelation. It appears from the plot in Figure 4 that the changes Δ_t in the stock prices are, in fact, a random uncorrelated series. Figure 4 is analogous to Figure 2, with Δ_t now playing the role of Y_t. Note the contrast in the plots. All the evidence seems to show that the IBM prices form a random walk. Thus we have not come up with a money-making scheme for the stock market.

CONCLUSION

We have illustrated the analysis of a data set by formulating a model $Y_t - \mu = \rho(Y_{t-1} - \mu) + e_t$ which allows several scenarios of interest, namely $\rho = 0$, $0 < \rho < 1$, and $\rho = 1$. We have used graphical analysis to decide that the $\rho = 1$ scenario is consistent with the data at hand. When $\rho = 1$, the model reduces to $Y_t = Y_{t-1} + e_t$, which is called a random walk.

REFERENCE

Box, G. E. P., and G. M. Jenkins (1976). *Time Series Analysis Forecasting and Control.* Academic Press, New York.

29

The 100-Year Flood
and All That

Richard G. Heerdegen
Massey University
Palmerson North, New Zealand

INTRODUCTION

How often have you read that the stopbanks along a particular river have been designed to prevent the 100-year flood from overtopping them? Or that the extremely heavy rain that fell was estimated to have a return period of 1000 years? Or that the stormwater system of a city should cope with short, but intense downpours of rain 9 years out of 10?

These statements are a particular way of expressing estimates of uncommon occurrences in probability terms. The trouble with uncommon or rare events is that they do not conform to the normal probability distribution. As a consequence, a specialist field of study has developed to deal with rare events, generally referred to as extreme value analysis. Early attempts at predicting the probability of rare and unlikely events, such as the chance of being killed by a horse kick in the Prussian Army, led to the use of a distribution called Poisson to predict their probability. The major problem with certain types of rare events is not only determining the probability of their occurrence but also their magnitude. How large will the most extreme event be? What sort of probability can such an event really be assigned? One writer expressed it as a conundrum by stating that it is impossible that the improbable could never happen

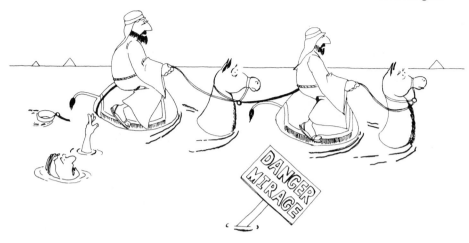

but easy to show that very improbable events are linked to such large periods of time that they will never be observed.

FLOODS IN THE MANAWATU RIVER

The Manawatu River is one of New Zealand's major rivers, with a drainage area at Palmerston North city of 3916 km². In its lower reaches it used to flood substantial areas, including parts of the city. Since the 1870's when the area was first settled, much work has been done to improve the efficiency of the channel and to build stopbanks to lessen the flood risk. Planning ordinances are also helping to minimize flood damage to property through wise land-use zoning.

Discharge records date back to 1929, which incidentally is one of the longest records in the country, and during the period 1929–1982 there were 148 occasions when the Manawatu River discharge exceeded 1000 $m^3 s^{-1}$ (see Figure 1). A discharge of this magnitude may be described as a flood because when the river reaches this level and is still rising, farmers are warned to move stock from the berm land between the river and the stopbanks. On average, a flood of this magnitude or greater occurs 2.7 times a year or about once in 19 weeks. Of course, not every month has an equal chance of receiving such a flood as Figure 2 shows.

As an example, of the 54 Octobers in the period of record, 10 had at least one flood. Therefore, the probability of at least one flood, $P(x)$, occurring in October is

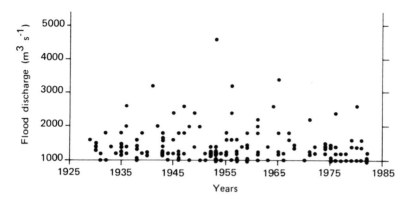

FIGURE 1   Magnitude of all floods $\geqslant 1000$ m^3 s^{-1} on the Manawatu River, Palmerston North, New Zealand, 1929–1982.

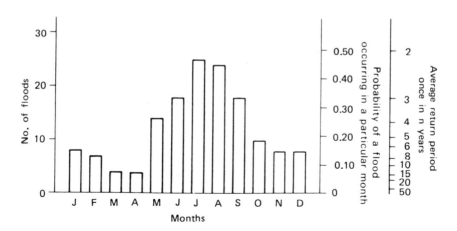

FIGURE 2   Monthly frequency of floods $\geqslant 1000$ m^3 s^{-1} on the Manawatu River, 1929–1982.

$$P(x) = \frac{\text{number of floods in October}}{\text{number of Octobers}}$$

$$= \frac{10}{54}$$

$$= 0.19$$

Looking at the problem another way, there were 10 floods in 54 Octobers, so on average the time between floods is 54/10 = 5.4 Octobers, which is 5.4 years. This interval of time is called the *return period* for floods in October. Thus on average, October can expect to receive a flood of at least 1000 m^3 s^{-1} once in about $5\frac{1}{2}$ years.

An analysis like this gives a reasonably clear picture of the expectation of receiving a flood in any one month. It shows that there is a pronounced seasonality in flooding, with a high probability of winter floods and the lowest probability in the late summer/early autumn. What it does not show is how high the floodwalls ought to be built to minimize the risk of flooding from the occasional very large flood.

But what those responsible for deciding on the degree of protection required from floodwalls and those who design them want to know is this: How large will the largest flood be in the next 100 years? Under typical urban conditions floodwalls are built to withstand a flood which on average might occur once in 100 years. This period is chosen because it reflects the degree of protection required for the assets at risk, in that a typical New Zealand timber-framed house is expected to last 100 years. Therefore, inherent in this question is the importance of the time span. Since duration in years is important, a selection from the data can be made to include only the largest flood for each year or *annual* floods for the period (see Figure 3). Note that the previous set of data included *all* floods above 1000 m^3 s^{-1}. If these annual floods are ranked and then plotted as a cumulative frequency curve (see Figure 4), a statement relating their occurrence to their size can be made [e.g., the largest flood (4546 m^3 s^{-1}) had a frequency of one in 54 years or 0.019 in 1 year or about twice per 100 years].

To make this into a probability statement, several assumptions are necessary. First, any probabilistic statement is based on records of *independent random* events. An analysis of the records shows that events separated by a year are independent, and in fact this is the reason for choosing annual floods rather than all floods. Second, implicit in a record of annual maximum floods is the belief that this "sample" from an infinite population is *representative* of that population and that no long-term changes in trend are present. Third, *inferences* made about the

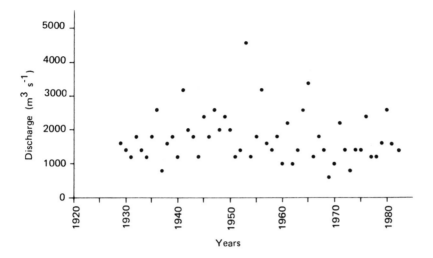

FIGURE 3   Magnitude of annual floods, Manawatu River, 1929–1982.

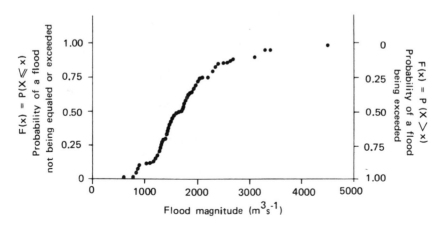

FIGURE 4   Cumulative probability distribution of the annual floods, Manawatu River, 1929–1982.

population from the sample can be said to follow the limitations of sampling theory.

SYNTHETIC FLOOD DATA

If then this "sample" of 50 events comes from an infinitely large population, then in a period of 1000 years there are 20 such samples and the chance of the largest flood in 1000 years occurring in one particular 50-year period is 1:20. It is a fact that in the entire 1000-year period, there will be only one event which is equaled or exceeded on average once in 1000 years. There will also be 10 floods exceeded on average once in 100 years, one of which will be the 1000-year event, and there will be 20 floods of 50-year average exceedence. In one 50-year period, then, there is a 1:2 chance that the largest event will be of at least 100-year average exceedence, and a 1:20 chance that *that event* is the largest of the 1000-year period.

For a large number of observations, an event of probability $P(x)$ will occur, on average, once in $1/P(x)$ occasions and will have a return period, T_r(years) of $1/P(x)$. Thus the flood that occurs once in 1000 years has a probability of occurence of 1/1000 or 0.001 and its return period is 1/0.001 or 1000 years. Similarly, the flood that is equaled or exceeded once in 50 years has a probability of occurrence of 1/50 or 0.02 and a return period of 1/0.02 or 50 years.

An interesting way to demonstrate this concept is to compile a synthetic record of 1000 years by drawing a random sample of 1000 events from a population that has the same mean and spread as the period recorded. The largest event in the 1000 synthetic "years" compiled by using these parameters is 6918 m^3 s^{-1} and the 20 largest events, representing floods which are exceeded, on average, once in 50 years, range from 3675 m^3 s^{-1} upward.

The synthetic 1000-year period was then divided sequentially into 10 100-year periods. Of those 10 centuries, four did not contain any of the 10 largest floods, three contained one, two contained two, and one had three (see Table 1). Although these results may initially seem a trifle odd, they do show that what happens in particular periods may differ considerably from what happens on average. Statistically, this variability is expected and is described by the Poisson distribution. Table 2 shows for a group of 10 centuries, the number of 100-year floods (i.e., the 10 largest floods in the 1000-year period) occurring within each century and demonstrates that the synthetic population really does conform to the Poisson distribution.

TABLE 1 Statistics from a Synthetic 1000-year Period of Annual Floods Divided into Successive Century Samples (m^3 s^{-1})

	Sample									
	1	2	3	4	5	6	7	8	9	10
Mean	1847	1680	1701	1957	1775	1881	1781	1757	1800	1785
Median	1744	1543	1548	1871	1630	1738	1682	1590	1534	1612
Maximum	4766	3889	3942	4184	3675	5275	4224	3956	4795	6918
Minimum	747	535	624	669	557	628	643	485	626	609
10 largest floods in 1000 years	4766	—	—	4184 4156	—	5275 4332 4214	4224	—	4795 4426	6918

TABLE 2   Poisson Table Comparing Expected Frequency of 100-Year Floods/
Century with Those Actually Occurring Within the 10 Sample Centuries

Number of events, K (i.e., 100-year floods/century)	Poisson probability, P(X = K) for mean = 1.0	Expected frequency (N = 10)	Actual frequency (i.e., number of centuries with K 100-year floods
0	0.37	3.7	4
1	0.37	3.7	3
2	0.18	1.8	2
3	0.06	0.6	1
4	0.02	0.2	0

The point of interest here is that presuming that the past 50 years reasonably represent a 1000-year period, then each 100-year period will, by merely reflecting random fluctuations, produce situations that are not identical. Even the largest flood (4546 $m^3\ s^{-1}$) which has occurred within the 54-year period of record, when compared with the entire synthetic 1000-year record, is exceeded by only four events, giving it a hypothetical recurrence value of 4/1000 or 0.0004 or a return period of 250 years.

This demonstrates the very dilemma that faces us. If the design requirements state that a reasonable estimate of the 100-year flood should be made, how large is it? A comparison of the current 54-year record with the synthetic data shows that the largest flood (i.e., a 54-year return period flood) might be thought of as closer to a 250-year return period event. The synthetic data show that its 50-year event (i.e., the twentieth largest event of the 1000 years), with a value of 3675 $m^3\ s^{-1}$, is only a little larger than the second largest event of the actual period since 1929, while the synthetic 100-year flood has a magnitude of 4155 $m^3\ s^{-1}$. However, a period of 1000 years of real data is not available to designers in order to derive their estimates. Rather, they rely on some statistical theory which will allow them to predict from the given data, however sparse those may be. If the theory is soundly based, the answers so produced will resemble those chosen from a very large population.

EXTREME VALUE ANALYSIS

Fortunately, extreme value analysis does just this. Using the available data, which consists of the largest annual floods in each year from the catchment in question, an estimate is made from the shape of the tail of

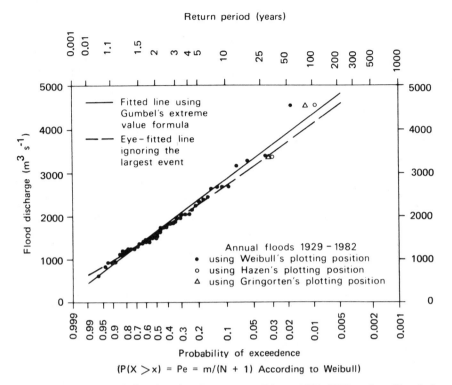

FIGURE 5   Annual flood series, Manawatu River, 1929–1982, using Gumbel graph paper.

the data's probability distribution, concerning the values that are not contained within it.

A series of independent observations, arranged in increasing order of magnitude from the largest to the smallest are termed order statistics, each of increasing rank. Generally speaking, order statistics around the median are normally distributed, but extreme values are not as evidenced by Figure 4. This means that the normal distribution, so widely used in other areas of statistics, is of no use in extreme value analysis. Consequently, a common empirical technique has been developed.

The annual flood series is ranked from largest to smallest and according to the plotting position used, which reflects the particular probability distribution employed, each event in the series is assigned a probability of exceedence (P_e) (or return period). The most universally used formula for calculating the return period plotting position (in years) of a particular flood peak (T_p) is that of Weibull:

$$T_p = \frac{N + 1}{m}$$

where N is the number of events in the series and m is the rank (largest to smallest) of each event in the series. The use of this formula is simple, widespread, and theoretically plausible in that zero and 100 percent probability do not occur.

Using the appropriate plotting position, the series of extreme events is then plotted on specially constructed graph paper, as in Figure 5, which is a graphical representation of the commonly used double-exponential probability distribution. If required, a line can be fitted to the points either by eye or according to a probability distribution, the parameters of which are derived from the data. The measured data will never conform to a straight line or smooth curve because of sampling variation. Since the fit of a probability function to such data is tantamount to smoothing it, the use of either method is justified. In most individual cases, a visually fitted line will allow for reasonable predictions. In the example used here, the value of the 100-year flood is about 4100 m^3 s^{-1} and the return period of the currently largest measured event, if it were to conform to the fitted line, would be about 200 years. (It is no coincidence that these figures correspond closely to those obtained from the synthetic data. The synthetic data were generated assuming a population for which the Weibull procedure is theoretically correct.)

DO THE EYES HAVE IT?

There are several points that need to be made here. The first relates to the plotting position, because to some extent it changes the shape of the plotted data and thus the form of any fitted line. Although Weibull's plotting position accords with our view that the return period of the largest event should be about the same as the period of record, it has been found to overestimate the magnitude of extreme events. Statistical analysis has shown that the largest event in a series generally has a long-term return period of 1.8 times the length of record, so that the largest event of the Manawatu River annual flood series would accordingly have a long-term return period of 94 years. As a consequence, various other plotting positions have been devised, each of which seeks to correct the bias observed in the Weibull formula, a bias which, as mentioned, has led to its overstating the frequency of occurrence of the larger events. The Hazen formula, for instance,

$$T_p = \frac{N}{m - 0.5}$$

has a plotting position for the largest event of twice the period of record, while Gringorten's formula,

$$T_p = \frac{N + 0.12}{m - 0.44}$$

which is used in the New Zealand study, compromisingly lies between the two. Thus the largest flood in the Manawatu River annual series could have a return period of 55 year (Weibull), 108 years (Hazen), or 82 years (Gringorten) (see Figure 5).

The second relates to the particular frequency distribution used if a mathematically fitted line to the data points is required. Many fitted distributions for flood frequency analysis have been tried over the years, and a considerable literature exists on the theory and use of such distributions in flood hydrology.

Numerous studies conducted in North America have used a three-parameter log-gamma distribution (referred to as log-Pearson type 3). The British study and its New Zealand counterpart both use the double-exponential distribution (also referred to as the Gumbel or general extreme value distibution), which for reasons that need not be explained here, becomes a two-parameter distribution that plots as a straight line on the Gumbel graph paper. The example in Figure 5 has the Gumbel function plotted on it. The change in the probability distribution from the curvature shown in Figure 4 to the more linear form in Figure 5 makes it much easier to assess the goodness of fit and the direction of any required extrapolation.

The third point is extrapolation itself. It is generally accepted that point estimates should not be made beyond the limits of data values without caution, in that the fitted function applies only to the data used. However, the whole reason for engaging in this method of flood magnitude prediction is to extrapolate beyond the data record, so a conflict between principle and practice arises. Since the tail of any distribution influences the derived parameters more than the central values, the trend of the line beyond the last data value is markedly affected by the values in the tail. It is generally compounded, too, by the relatively short data record available for most rivers. Furthermore, estimates of large flood discharges on most rivers are inevitably calculated empirically and not measured directly, thus producing possible errors of up to plus or minus 20%.

From the Gumbel probability distribution, estimates of the magnitude of an N-year return period flood can also be made. For the Manawatu River, the 100-year flood is calculated to have a magnitude of 4350 m^3 s^{-1} (see Figure 5), so the engineer now has at his or her disposal several estimates of how large a flood might be expected in a 100-year period. This does not preclude the use of other nonprobabilistic methods of determining flood size.

The major use of probabilistic estimates of flood size is made by engineers who are required to build structures according to some design estimate. Depending on the cost and necessity of providing greater protection from failure, engineering works will be constructed to withstand a particular-size event. Where the cost of failure is low, such as a stormwater system not being able to pass a sudden torrential downpour, the design standards are set to pass higher-frequency, lower-magnitude events, such as a 10-year or even a 2-year event. Highway bridges and floodwalls protecting urban areas are generally designed to withstand the 100-year flood, while concrete dams are built so that overtopping by a 500-year event does not take place. The concept of design standard, then, is one based on a probabilistic estimate, and is very widespread in engineering design.

THE RISK OF EXCEEDENCE

An interesting aspect of design to a probabilistically determined size, based on the Poisson distribution, is that the risk of exceedence by at least one event of the same return period as the design life, whatever it is, is 0.63. That is, that if a floodwall is built to withstand the estimated 100-year-return-period flood, there is a 63% chance that a flood of *that magnitude or greater* will occur in that 100 years. And in betting terms that is greater than a 50:50 chance. Thus in order to lessen the risk of such a flood overtopping the stopbank, it has to be built higher, and how much higher depends not only on financial and engineering limitations but on an acceptably defined risk.

Based on the known probability proposition, the risk, U, of at least *one T_r-year* flood occurring in a service period of r years can be determined by

$$U = 1 - \left(1 - \frac{1}{T_r}\right)^r$$

For example, the risk of at least one 100-year flood occurring in a 100-year period is

$$U = \left(1 - \frac{1}{100}\right)^{100}$$
$$= 0.634$$

TABLE 3   Return Period in Years of Floods (with a Probability of Exceedence Once in the Service Period) for Certain Service Periods and with an Accepted Risk of Exceedence

Risk of Exceedence U	Odds	Service Period r					
		2	5	10	20	50	100
0.75	3:4	1.7	4.0	6.7	14.9	35.6	72.7
0.63	3:5	2.0	5.0	10.0	20.0	50.0	100
0.50	1:2	3.4	7.7	14.9	29.4	72.6	145
0.25	1:4	7.5	17.9	35.3	70.0	174	348
0.20	1:5	9.0	22.9	45.3	90.1	225	449
0.10	1:10	19.5	48.1	95.4	190	475	950
0.01	1:100	198	498	996	1992	4975	9953

The odds of this flood occurring are thus better than 3:5, so to lower the risk of flooding, stopbanks would have to be built to withstand a larger T_r-year flood. How much larger depends on an acceptable level of risk. If the level of risk was lowered to a 1:4 chance ($U = 0.25$) of the stopbank being overtopped in 100 years, the magnitude of that flood can be calculated from the preceding equation rewritten as follows:

$$T_r = \frac{1}{[1 - (1 - U)]^{1/r}}$$
$$= \frac{1}{1 - (1 - 0.25)^{1/100}}$$
$$= 348 \text{ years}$$

On that basis, the stopbank would need to be built to withstand a flood with a return period of about 350 years in order to lessen the risk to 1:4 of overtopping at least once in a 100-year period. In the case of the Manawatu River, by reference to Figure 5 it can be seen that a 350-year return period flood will have a magnitude of approximately 5200 m³ s⁻¹, and it would be up to the design engineer to build the structure accordingly. Table 3 shows this particular detail together with other levels of risk and various service periods.

THE 100-YEAR FLOOD

There are two sides to the probability concepts embodied in floods. The first is that the notion of design floods of a given return period is based on sound probability theory. The flaw in most people's minds is that they believe that events of a particular magnitude *will* occur in a certain period, and that they will *only* occur once, and if that does not happen, the theory is wrong. Those with an understanding of the theory realize that, *on average*, and that has to be over a lengthy time span, certain-sized events will occur in a particular period of time. When, however, is never predictable. The second is that the concept of risk, with its implied probability, has to be clearly separated from the probability of occurrence of a flood of a particular magnitude.

But as for the notion of the 100-year flood, it is badly named. Many authors now use the term "probability of exceedence" $(1/T_r)$ rather than "return period," but one suspects that for most people this concept would have little meaning. The theory tells us that there is a better-than-even chance that at least one flood of that magnitude *or greater* will occur in a 100-year period, and that according to Poisson, 1 in 10 centuries will contain four floods of that magnitude or greater. I wonder what century we are living in? Or is it 100 years?

REFERENCES

Beable, M. E., and A. I. McKerchar (1982). *Regional flood estimation in New Zealand.* Water and Soil Technical Publication 20, Water and Soil Division, Ministry of Works and Development for the National Water and Soil Conservation Organisation, Wellington, New Zealand.

Chow, V. T. (1964). Statistical and probability analysis of hydrologic data: Part 1, Frequency analysis. In *Handbook of Applied Hydrology—A Compendium of Water Resources Technology* (V. T. Chow, ed.). McGraw-Hill, New York, pp. 8.1–8.42.

Cunnane, L. (1978). Unbiased plotting positions - a review. *J. Hydrol. 37*, 205–222.

Gumbel, E. J. (1958). *Statistics of Extremes.* Columbia University Press, New York.

Irish, J., and N. M. Ashkanasy (1977). Flood frequency analysis. In *Australian Rainfall and Runoff*, Chap. 9. The Institution of Engineers, Australia.

Langbein, W. B. (1960). Plotting positions in frequency analysis. In *Flood Frequency Analysis.* (T. Dalrymple, ed.). Manual of Hydrology: Part 3, Flood flow techniques. U.S. Geological Survey Water Supply Paper 1543-A.

Markowitz, E. M. (1971). The chance that a flood will be exceeded in a period of years, *Water Resour. Bull. 7* (1), 40–53.

National Environment Research Council (1975). *Flood Studies Report*, Vol. 1. National Environment Research Council, London.

Reich, B. M., and K. G. Renard (1981). Application of advances in flood frequency analysis, *Water Resour. Bull. 17* (1), 67–74.

Riggs, H. C. (1968). Frequency curves. In *Hydrologic Analysis and Interpretation* (Chap. A2, Book 4). Techniques of Water-Resources Investigations of the U.S. Geological Survey, Washington, D.C.

Tomlinson, A. I. (1980). *The frequency of high intensity rainfalls in New Zealand: Part I.* Water and Soil Technical Publication 19. Water and Soil Division, Ministry of Works and Development for the National Water and Soil Conservation Organisation, Wellington, New Zealand.

Ward, R. C. (1978). *Floods—A Geographical Perspective.* Macmillan, London.

U.S. Water Resources Council (1977). *Guidelines for Determining Flood Flow Frequency.* Bulletin 17A, Hydrology Committee, U.S. Water Resources Council, Washington, D.C.

30

War and Peace: Why Is Power Unstable?

Paul van Moeseke
Massey University
Palmerson North, New Zealand

INTRODUCTION: IS THERE A BALANCE OF POWER?

"You can do anything with bayonets," said Napoleon, "except sit on them." When the generals are through, the lawyers move in and draw up a treaty defining the new balance of power. Yet it appears that one cannot sit on treaties either: they tend to wear out fast.

Why is politics of both the international and the domestic varieties so unstable? Evidently because the so-called balance of power gets upset so often. The specter that is now haunting Europe, to paraphrase the opening line of Marx and Engels' *Communist Manifesto*, is the balance of terror, the nuclear edition of the balance of power. Intuitively, a power balance should be an arrangement stable against all possible *coalitions*. On the international scence such coalitions are called alliances; on the dometic scene, parties or pressure groups.

Recurrent instability raises the suspicion that in a typical power game, there simply may *be* no balance of power. In this chapter we offer a simple theoretical argument regarding why this should normally be so. The chapter is aimed at the general reader and assumes no prior knowledge of mathematics, statistics, or game theory. The only abstract requirement is the ability to add and compare a few numbers. The tool we use is the notion of a "distribution of power." Statistics is largely concerned with distributions, usually of physical, biological, social, or

economic characteristics. In the present chapter, however, we look at the distribution of power among individuals and coalitions. The distribution of power is the focus of politics even as the distribution of wealth (resources) is that of economics.

ALLIANCES AND PARTIES

The power and stability of alliances is of rather more than academic interest if for no other reason than that there is no point in pursuing a nonexistent balance of power. Modern European history, at least since the rise of the nation state, is a sadly repetitive search for this elusive balance. Periods of protracted war and interim treaties have led, every century or so, to a final convulsion followed by yet another grand design of universal balance, the most notable attempts being the Peace of Augsburg (1555, after the Reformation), the Treaty of Westphalia (1648, after the Thirty Years' War), the Treaty of Utrecht (1714, after the War of the Spanish Succession), the Conference of Vienna (1815, after the Napoleonic Wars), and the Treaty of Versailles (1918, after World War I), followed by the Locarno Pact (1925) and the Munich Agreement (1938) in the interwar period.

Every one of these grand designs came unstuck in the shifting kaleidoscope of coalitions whose instability is a constant source of miscalculation and war. To quote the *Encyclopaedia Britannica* on the preamble to the World Wars:

> To all intents and purposes the "balance of power" was eliminated when, in 1871, Bismarck united the Germanic peoples under the first Reich, and established Germany as a dominant power in Europe. The Treaty of Versailles with its system of guarantees and collective security was an attempt to restore the old theory of "balance of power" in modern dress.

Among the numerous shifts of alliances since 1917 one may note that the United States twice reversed its policy of neutrality in the European conflict, each time (April 6, 1917 and December 7, 1941) years after the outbreak of hostilities; that two of the three major Axis powers (Italy and Japan) had sided with the Allied powers in World War I, that Italy switched yet again in midwar (co-belligerent to the Allies from October 13, 1943), and that Russia and Germany were at war within less than 2 years (on June 22, 1941) of the conclusion of the Soviet-German Nonaggression Pact (August 23, 1939). Again, in today's awkward *ménage à trois* of China, the United States, and the USSR, the potential alignment, and realignment, of China with either superpower constantly threatens to upset the balance of the postbellum world.

On the domestic scene, of course, administrations—whether coalition governments in West Europe's parliamentary democracies, central party committees in the socialist camp, or juntas in the Third World—can be overturned, or replaced, by shifts in the coalitions of parties and pressure groups that do, or do not, support them. Nor are the two-party regimes of the Anglo-Saxon type exempt since existing policies have to be readjusted, abandoned, or reversed and cabinets reshuffled at every new lineup of factions both within the ruling party and across parties. A striking instance is the Liberal–Social Democrat alliance in Britain.

THE POWER GAME

We adopt, or adapt, a few elementary notions from game theory to describe the power game. The most important one is that of the *outcome* of a game, which is simply the final *distribution* of power. We assume there are n participants, or *players*, numbered $1, 2, \ldots, n$, and that each wants to end up with as large a slice of the cake as possible.

The entire cake is equal to 1 unit and the slices are represented by fractions, for example,

$$0.40, 0.30, 0.30 \qquad (1)$$

meaning that player 1 ends up with 40% of the cake, players 2 and 3 with 30% each. The individual slices or fractions are generally called *payoffs*. Clearly, these numbers define an outcome of the power game (no matter how arrived at). Statistically, the outcome is a distribution of payoffs (in this case, slices of a cake): they add up to 1 and none of them are negative, of course.

More generally, the outcome of a power game with n players is a distribution

$$x_1, x_2, \ldots, x_n \tag{2}$$

of (nonnegative) numbers adding up to 1:

$$x_1 + x_2 + \cdots + x_n = 1 \tag{3}$$

where x_1 is the payoff to player 1, x_2 the payoff to player 2, ..., or, in general, x_i is the payoff to player i. Equality (3) is conveniently abbreviated to

$$(\text{sum } x_i) = 1$$

The cake model covers a multitude of sins, but to fix the ideas the reader may refer to the following highly simplified situations. On the *national* scene one may think of the cake as the cabinet, with a given number of cabinet posts or ministers, and the players as the political parties represented in parliament. If there are 20 ministers and three parties, the above outcome (1) is a coalition goverment where $(0.40 \times 20 =)$ 8 minsters belong to the first party and $(0.30 \times 20 =)$ 6 ministers to each of the other two parties. On the *international* scene the cake simile might symbolize the map of Europe; the players are the nations involved, the outcome is the distribution of territory.

It pays to keep the examples simple. In fact, they can be made more realistic without affecting the reasoning below: cabinet posts, for instance, can be weighted according to relative importance or departmental budgets; and territory can be replaced by population or national income, or weighted by strategic location. For the argument that follows it suffices that the total cake is given, whether it is homogeneous or not, and that neither its size nor composition are affected by the game. (In the language of game theory we are dealing with *constant-sum games*.)

Power, hence problems of the political type, play an equally important role in the economy, where oligopolistic competition and the struggle for market shares, as well as direct restraint of competition, result in frequent mergers and takeovers.

CONFLICT AND COALITION

Take an arbitrary distribution (2). When will it represent a balance of power or equilibrium, which we could call a "solution" to the power game?

Evidently not if any single player is powerful enough to upset it. For instance, a party with an overall majority in parliament need not ratify any government unless that party holds *all* cabinet posts. But even if no one player can upset the distribution, equilibrium is not guaranteed: even if no party has an absolute majority any pair of parties holding more than 50% of the seats between them can keep all other parties out of the cabinet. Generally, the same holds for any coalition of parties with an absolute majority.

We are therefore led to the following definition of the *balance of power*: it is a distribution (outcome) that cannot be upset by *any* coalition of players. This implies, in particular, that it cannot be upset by any single player either, since he or she can be thought of as a "one-person coalition."

In the above three-party parliamentary example the list of all possible coaltitions is

$$[1], [2], [3], [1,2], [1,3], [2,3], [1,2,3] \tag{4}$$

where, for example, the notation [1,3] denotes the coalition of players (parties) 1 and 3. In particular, the last group is the "total" coalition, which occurs, for instance, in a so-called government of national unity such as (1). (The reader may want to check that if there are n players, there are $2^n - 1$ possible coalitions.) Coalitions other than one-person and total coalitions (i.e., coalitions with at least 2 and fewer than n members) can conveniently be called *proper* coalitions.

When can a coalition upset a given distribution? Intuitively speaking, if it can secure for its members a larger total amount than the sum of their payoffs, because then all its members could be made better off. So what is the *value* of a coalition to its members? We define it as the share of the cake it can insist on in any event, regardless of the actions of all other players.

If, in the parliamentary example, the first two parties have an overall majority, they need not accept distribution (1) since they can secure *all* cabinet posts (i.e. the whole cake) for themselves. The *value* of the coalition [1,2] therefore equals 1; we write

$$value[1,2] = 1$$

and note that distribution (1) will not stand simply because $0.40 + 0.30 < 1$ or

$$x_1 + x_2 < \text{value}[1,2] \tag{5}$$

More generally, the value of any coalition C is denoted value $[C]$. The coalition of all players *outside* C is [not C] and its value is denoted as value[not C]. Since the worst that can happen to any coalition is no cake at all, clearly, for all C, value$[C] \geqslant 0$ and since no more than the whole cake can be claimed:

$$\text{value}[C] + \text{value}[\text{not } C] = 1 \quad \text{(for all } C) \tag{6}$$

CUTTING THE CAKE

The reader is no doubt familiar with the classical two-person cake problem where player 1 cuts and player 2 chooses: Mr. 1 will cut the cake exactly in half—otherwise Mr. 2 will naturally take the bigger slice. (In particular, if Mr. 1 does *not* cut the cake, that is, if one of the slices is 0, then Mr. 2 will take the whole cake!) The distribution $x_1 = \frac{1}{2}$, $x_2 = \frac{1}{2}$ is clearly stable and neither player will upset it.

Now let us disrupt the culinary peace by admitting a third player: Mr. 1 still does the cutting, if any; Mr. 2 chooses first, followed by Mr. 3. Mr. 1 again gets what is left: he can, however, always make sure of getting $\frac{1}{3}$ simply by cutting equal slices so that

$$\text{value}[1] = \tfrac{1}{3} \tag{7}$$

(If the cuts were unequal, he would be left with the smallest, certainly less than $\frac{1}{3}$.) Similarly, Mr. 2, who has first choice, will also get at least $\frac{1}{3}$:

$$\text{value}[2] = \tfrac{1}{3} \tag{8}$$

(and more whenever the cuts are unequal). Finally, Mr. 3 cannot, *by himself*, make sure of getting anything at all:

$$\text{value}[3] = 0 \tag{9}$$

for if the other two collude, Mr. 1 will *not* cut so that Mr. 2 can take all (and split the spoils with Mr. 1 later). This is the same thing as saying

$$\text{value}[1,2] = 1$$

which follows also from equations (6) and (9). Similarly, applying equation (6) to (7) and (8) yields

$$\text{value}[2,3] = \tfrac{2}{3}$$
$$\text{value}[1,3] = \tfrac{2}{3}$$

DISTRIBUTING THE CAKE

What is probably the most famous, and certainly the most fatuous, utterance on pastry politics was Queen Marie-Antoinette's comment, when told that the people of Paris had no bread: "Let them eat cake!" Although history does not record exactly what kind of distribution Her Majesty had in mind, let us propose an arbitrary distribution to start with, for instance the egalitarian one:

$$x_1 = x_2 = x_3 = \tfrac{1}{3}$$

Will it stick? Evidently not, since value[1,2] = 1, so that coalition [1,2] can grab the whole cake as against the $\tfrac{1}{3} + \tfrac{1}{3} = \tfrac{2}{3}$ share presently allotted to its members. In other words, as in the parliamentary example (5) above, it will not stick because

$$\tfrac{2}{3} = x_1 + x_2 < \text{value}[1,2] = 1 \tag{10}$$

Since players 1 and 2 could get the whole cake by collusion, it would be natural to think they can go ahead, say by splitting 50:50.

This would yield the new distribution

$$x_1 = \tfrac{1}{2} \quad x_2 = \tfrac{1}{2} \quad x_3 = 0$$

making both coalition partners better off. But that will not hold either: Mr 3, can bribe either partner, for example, Mr. 1 (and expel the other partner) because now

$$\tfrac{1}{2} = x_1 + x_3 < \text{value}[1,3] = \tfrac{2}{3} \tag{11}$$

and it is sufficient for Mr. 3 to propose that they split the difference $\tfrac{2}{3} - \tfrac{1}{2} = \tfrac{1}{6}$ in half (say) to arrive at

$$x_1 = \tfrac{7}{12}, x_2 = \tfrac{1}{3}, x_3 = \tfrac{1}{12}$$

(This means that Mr. 1 cuts equal slices on the understanding that he gets most of Mr. 3's slice afterward.) The reader can easily verify that this distribution can in turn be upset by two coalitions: [1,2] and [2,3]. The latter has most to gain (i.e., $\tfrac{2}{3} - \tfrac{1}{3} - \tfrac{1}{12} = \tfrac{1}{4}$): splitting the difference leads to

$$x_1 = \tfrac{1}{3} \quad x_2 = \tfrac{11}{24} \quad x_3 = \tfrac{5}{24}$$

and one can go on.

Will this process ever stop? More generally, is there any arbitrary distribution whatever that cannot be upset by *any* coalition? The short answer is no, so that, in general, disruption will go on forever regardless of initial distribution or later division agreements (50:50 in the example).

The proof and detail of this rather shattering result is stated in the next section, which can be skipped by the reader interested only in the conclusions.

NECESSARY INSTABILITY

When can a coalition block (upset, disrupt) a distribution? Intuitively speaking, if it can secure for its members a greater total amount than the sum of their proposed payoffs as in equations (10) and (11) because then all its members can be made better off.

Conversely, a distribution is stable against all coalitions if instead of equations (10) and (11) we have

$$x_1 + x_2 \geqslant \text{value}[1,2]$$
$$x_1 + x_3 \geqslant \text{value}[1,3]$$

and so on, for all possible coalitions. We can summarize by stating that a distribution is stable if, for all coalitions,

$$(\text{sum } x_i \text{ in } C) \geqslant \text{value}[C] \tag{12}$$

where the left is short for "sum of the x_i for all players i in C."

Recall that the value[C] of a coalition (which is really its power to block) is measured by a nonnegative number such that

$$\text{value}[C] + \text{value}[\text{not } C] = 1 \qquad \text{(for all } C) \tag{13}$$

We assume that, in realistic power games, all coalitions are at least as strong as, and some strictly stronger than, the sum of their members (otherwise coalitions would be pointless):

$$\text{value}[C] \geqslant (\text{sum value}[i] \text{ in } C) \qquad \text{(for all } C) \tag{14}$$

$$\text{value}[C] > (\text{sum value } [i] \text{ in } C) \qquad \text{(for at least one } C) \tag{15}$$

where the right is short for "sum of the value[i] for all players i in C." We can now prove that equations (13), (14), and (15) imply that there is *no* stable distribution (outcome): for if one assumes there is, condition (15) can be shown to be violated.

Proof

Assume that there exists a distribution $x_1, x_2, \ldots, x_n$ stable against all coalitions so that by equation (12)

$$(\text{sum } x_i \text{ in } [\text{not } C]) \geqslant \text{value}[\text{not } C] \quad (\text{for all } C) \qquad (16)$$

since the alliance of all players *not* in C is also a coalition. Addition of equations (12) and (16) yields

$$(\text{sum } x_i) \geqslant \text{value}[C] + \text{value}[\text{not } C].$$

But this holds with *equality* since both sides equal 1. Hence equations (12) and (16) also hold with equality since their sum does, so that

$$(\text{sum } x_i \text{ in } C) = \text{value}[C] \quad (\text{for all } C) \qquad (17)$$

In particular, for any one-man coalition [i],

$$x_i = \text{value}[i] \qquad (18)$$

and, by combining equations (17) and (18),

$$\text{value}[C] = (\text{sum value}[i] \text{ in } C) \quad (\text{for all } C)$$

which contradicts (15). Hence our assumption of the existence of a stable distribution was false.

CONCLUSIONS

1. The *major* conclusion is a disturbing one: there are *no* stable alliances in the power game, an inference that seems to be borne out by the perpetual shifting of coalitions and the attendant political instability adverted to in the introduction.

To revert from cakes to parliamentary business, consider the relatively frequent case where neither of the two major parties has an absolute majority in parliament with a minor third party holding the balance. Clearly,

$$\text{value}[1] = \text{value}[2] = \text{value}[3] = 0$$
$$\text{value}[1,2] = \text{value}[1,3] = \text{value}[2,3] = 1$$

where the first line means that no single party can form a government and the second that any two parties can. It is easy to see that any distribution whatever of cabinet posts between the ruling parties can be disrupted by the third party, as the West German Liberals demonstrated by their spectacular switch from Social Democrats to Christian Democrats in 1982.

The absence of a power equilibrium among three or more parties or partners has sometimes been compared with the nonexistence of a general solution to the three-body problem in physics (i.e., the motion of three bodies moving under no influence but their mutual gravitation).

One may think of the orbits of triple-star systems or of Earth's orbit around the sun, perturbed by Jupiter if that planet's mass were much larger or its orbit much closer to that of Earth.

2. The *minor* conclusion is more cheerful: stability is possible if equation (15) does *not* hold. Since coalitions are then pointless, such games are called *inessential*. The two-person cake model and, more generally, a broad class of two-party games are of this type. So is the political problem where one party has an overall majority in parliament, and, more generally, the class of n-party games where party i is dominant (value$[i]$ = 1).

Both cases are illustrated by the de facto partition of Europe into East and West blocs, which has indeed assured a significant measure of stability for nearly 40 years, as has the preeminence of either superpower within its own bloc, however undesirable the partition may be on other grounds.

The reader interested in game theory may note that the proof above is an adaptation of a result in the theory of constant-sum cooperative games to the effect that the *core* of the game is empty (see Luce and Raiffa, 1957). The core, defined by Gillies as quoted there, is the set of distributions stable in the sense of this article. Technical references and recent result of coalition theory in economics are found in an article by the author (1979), where it is shown that the purely competitive economy is stable because it is an inessential game. Although the theory of games was originally developed by von Neumann and Morgenstern (1944) in an economic framework, economic applications are inherently more complex than pure power play since economics by definition *also* involves outcomes in terms of material efficiency.

ACKNOWLEDGMENTS

This chapter was written during my tenure of the Professorial Fellowship in Economic Policy of the Reserve Bank of Australia, whose generous support is gratefully acknowledged.

I am greatly indebted to Professor William Parker of Yale University for commenting in detail on an earlier draft and pointing out some errors and ambiguities in the numerical examples. I further thank Dr. Peter Read of Massey University for a critical reading of a later version.

REFERENCES

Luce, R., and H. Raiffa (1957). *Games and Decisions*. Wiley, New York.

Moeseke, P. van (1979). Value cores for finite agents. *Econ. Rec.* 55, 76–81.

von Neumann, J. and O. Morgenstern (1944). *Theory of Games and Economic Behavior*. Princeton University Press, Princeton, N.J.

Index

About the Editors

Richard J. Brook is Reader in Statistics at Massey University in Palmerston North, New Zealand. Involved in statistical consulting within and outside of the university, he has published articles in international statistical, education, psychological, and management journals, and is the coauthor, with Gregory C. Arnold, of *Applied Regression Analysis and Experimental Design* (Marcel Dekker, Inc.) He is a member of the American and New Zealand Statistical Associations and the Biometrics Society, and a Fellow of the Royal Statistical Society. Dr. Brook received the B.Sc. degree (1956) from Adelaide University in Australia, M.A. degree (1970) from the University of Oklahoma, and Ph.D. degree (1972) from North Carolina State University.

Gregory C. Arnold is Senior Lecturer in the Department of Mathematics and Statistics at Massey University in New Zealand, where he helped establish an internal statistics program as well as a full undergraduate degree correspondence course. He is also an experimental design consultant for the university and nearby research institutions. He is a member of the American and New Zealand Statistical Associations and Biometrics Society. Mr. Arnold holds the B.Sc. (1966) from Victoria University in Wellington, New Zealand.

Thomas H. Hassard is Associate Professor of Biostatistics at the University of Manitoba in Winnipeg, Canada, where he teaches in the Medical School and is Director of its Biostatistical Consulting Unit. The author of articles on the application of statistics to medicine and other fields, he is an editorial board member of several medical journals. He is a member of the British Computer Society and Biometrics Society, and a Fellow of the Royal Statistical Society and Institute of Statisticians. Dr. Hassard received the B.Sc. (1970), M.Sc.(1972), and Ph.D.(1978) degrees from Queens University in Belfast, Northern Ireland.

Robert M. Pringle is a consultant statistician for the Ministry of Agriculture and Fisheries in Palmerston North, New Zealand. A member of the International Statistical Institute, he has published a text on theoretical statistics and articles on theoretical statistics, applied statistics, and related topics. Dr. Pringle received the B.Sc. (1965) and Ph.D. (1969) degrees from the University of Natal in South Africa.